THE CAMBRIDGE COMPANION TO
THE PHILOSOPHY OF BIOLOGY

The philosophy of biology is one of the most exciting new areas in the field of philosophy and one that is attracting much attention from working scientists. This *Companion*, edited by two of the founders of the field, includes newly commissioned essays by senior scholars and by up-and-coming younger scholars who collectively examine the main areas of the subject – the nature of evolutionary theory, classification, teleology and function, ecology, and the problematic relationship between biology and religion, among other topics. Up-to-date and comprehensive in its coverage, this unique volume will be of interest not only to professional philosophers but also to students in the humanities and researchers in the life sciences and related areas of inquiry.

David L. Hull is an emeritus professor of philosophy at Northwestern University. The author of numerous books and articles on topics in systematics, evolutionary theory, philosophy of biology, and naturalized epistemology, he is a recipient of a Guggenheim Foundation fellowship and is a Fellow of the American Academy of Arts and Sciences.

Michael Ruse is professor of philosophy at Florida State University. He is the author of many books on evolutionary biology, including *Can a Darwinian Be a Christian?* and *Darwinism and Its Discontents*, both published by Cambridge University Press. A Fellow of the Royal Society of Canada and the American Association for the Advancement of Science, he has appeared on television and radio, and he contributes regularly to popular media such as the *New York Times*, the *Washington Post*, and *Playboy* magazine.

(*Continued after Index*)

To the memory of Ernst Mayr, 1904–2005

The Cambridge Companion to

THE PHILOSOPHY OF BIOLOGY

Edited by

David L. Hull
Northwestern University

Michael Ruse
Florida State University

CAMBRIDGE
UNIVERSITY PRESS

CAMBRIDGE
UNIVERSITY PRESS

32 Avenue of the Americas, New York NY 10013-2473, USA

Cambridge University Press is part of the University of Cambridge.

It furthers the University's mission by disseminating knowledge in the pursuit of education, learning and research at the highest international levels of excellence.

www.cambridge.org
Information on this title: www.cambridge.org/9780521616713

© Cambridge University Press 2007

First published 2007
Reprinted 2008, 2009

A catalogue record for this publication is available from the British Library

Library of Congress Cataloguing in Publication data

The Cambridge companion to the philosophy of biology / edited by David Hull and Michael Ruse.
  p. cm.
Includes bibliographical references and index.
ISBN-13: 978-0-521-85128-2 (hardback)
ISBN-13: 978-0-521-61671-3 (pbk.)
1. Biology – Philosophy.   I. Hull, David L.   II. Ruse, Michael.   III. Title.
QH331. C285   2007
570.1 – dc22   2006025898

ISBN 978-0-521-85128-2 Hardback
ISBN 978-0-521-61671-3 Paperback

CONTENTS

CONTRIBUTORS

ANDRÉ ARIEW is an associate professor of philosophy at the University of Missouri–Columbia. He has written on various topics in the philosophy of biology including teleology, innateness, fitness, and the structure of natural selection explanations.

FRANCISCO J. AYALA is University Professor and Donald Bren Professor of Biological Sciences at the University of California, Irvine. On 12 June 2002, President George W. Bush awarded him the National Medal of Science at the White House. From 1994 to 2001, Ayala was a member of the U.S. President's Committee of Advisors on Science and Technology. He has been president and chairman of the board of the American Association for the Advancement of Science (1993–96) and president of Sigma Xi, the Scientific Research Society of the United States (2004–05).

ROBERT N. BRANDON is professor of philosophy at Duke University. He is the author of *Adaptation and Environment* and *Concepts and Methods in Evolutionary Biology* (Cambridge Studies in Philosophy and Biology).

DAVID J. BULLER is Presidential Research Professor in the Department of Philosophy at Northern Illinois University. He is the author of *Adapting Minds: Evolutionary Psychology and the Persistent Quest for Human Nature* (MIT Press/Bradford, 2005) and the editor of *Function, Selection, and Design* (SUNY Press, 1999).

LINDLEY DARDEN is professor of philosophy in the Committee for Philosophy and the Sciences and in the Program in Behavior, Evolution, Ecology, and Systematics at the University of Maryland,

College Park. She is the author of *Reasoning in Biological Discoveries: Mechanisms, Interfield Relations, and Anomaly Resolution* (Cambridge University Press, 2006) and *Theory Change in Science: Strategies from Mendelian Genetics* (Oxford University Press, 1991). She served as president of the International Society for History, Philosophy, and Social Studies of Biology in 2001–03. She and Carl F. Craver coedited the June 2005 "Mechanisms in Biology" issue of *Studies in History and Philosophy of Biological and Biomedical Sciences*.

ZACHARY ERNST is a member of the Department of Philosophy at the University of Missouri–Columbia. His research is on evolutionary game theory, the application of game theory to evolutionary biology, formal logic, and automated theorem proving.

PETER GODFREY-SMITH is professor of philosophy at Harvard University. He has degrees from Sydney University and the University of California, San Diego, and works mainly in the philosophy of biology and philosophy of mind. He is the author of *Complexity and the Function of Mind in Nature* (1996) and *Theory and Reality: An Introduction to the Philosophy of Science* (2003).

PAUL E. GRIFFITHS is ARC Federation Fellow and Professor of Philosophy at the University of Queensland. He is the author, with Kim Sterelny, of the textbook *Sex and Death: An Introduction to the Philosophy of Biology* (Chicago, 1999). From 2002 to 2005 he and Karola Stotz were principal investigators of the National Science Foundation–funded project "Representing Genes" at the University of Pittsburgh. This interdisciplinary project used Web-based survey methods to examine differing understandings of the gene in contemporary bioscience.

VALERIE GRAY HARDCASTLE is professor of science and technology in society and associate dean of the College of Liberal Arts and Human Sciences at Virginia Tech. Her area of research interest lies at the intersection of psychology, psychiatry, neuroscience, and philosophy. She has published numerous articles and books on the relation between neuroscientific data and psychological theories, most recently *The Myth of Pain* (MIT Press) and *Understanding Brain Activity* (forthcoming).

CHRISTOPHER HORVATH is associate professor of philosophy and biological sciences at Illinois State University in Normal. He holds a Ph.D. in philosophy from Duke University. Professor Horvath has published papers on phylogenetic systematics, evolutionary psychology, homosexuality, the evolution of human sexuality, and gender studies.

MAUREEN KEARNEY is associate curator in the Department of Zoology and head of the Division of Amphibians and Reptiles at the Field Museum of Natural History. She is also a member of the Committee on Evolutionary Biology and lecturer in the Biological Sciences Collegiate Division at the University of Chicago. Her research focuses on the evolution, comparative anatomy, and development of reptiles, using a phylogenetic perspective. She is also interested in the theory and methods of phylogenetic analysis.

MANFRED D. LAUBICHLER is a theoretical biologist and historian of biology. He works on conceptual problems of evolutionary developmental biology, the development and evolution of social insects, and the history of theoretical and developmental biology. He is assistant professor of theoretical biology and history of biology in the School of Life Sciences at Arizona State University and a member of the Centers for Biology and Society and Social Dynamics and Complexity. He is associate editor of *Biological Theory* and the *Journal of Experimental Zoology, Part B: Molecular and Developmental Evolution*. Together with Jane Maienschein he directs the Embryo Project, an international network for the study of the history of developmental biology. He is coeditor of *From Embryology to Evo Devo* (with Jane Maienschein, forthcoming from MIT Press), *Modeling Biology* (with Gerd Müller, forthcoming from MIT Press), and *Hochsitz des Wissens: Das Allgemeine als wissenschaftlicher Wert* (with Hans-Jörg Rheinberger and Peter Hammerstein, 2006, Diaphanes Verlag).

TIM LEWENS is a lecturer in the Department of History and Philosophy of Science at the University of Cambridge, where he is also a Fellow of Clare College. He is the author of two books – *Organisms and Artifacts: Design in Nature and Elsewhere* (MIT Press) and *Darwin* (Routledge).

ELISABETH A. LLOYD is Arnold and Maxine Tanis Chair of History and Philosophy of Science and Professor of Biology at Indiana University, Bloomington. She is the author of *The Case of the Female Orgasm: Bias in the Science of Evolution* (Harvard University Press, 2005) and *The Structure and Confirmation of Evolutionary Theory* (Greenwood Press, 1988; Princeton University Press, 1994), as well as coeditor of *Keywords in Evolutionary Biology* (Harvard University Press, 1992) with Evelyn Fox Keller.

JANE MAIENSCHEIN is Regents' Professor and Parents Association Professor at Arizona State University, where she is director of the Center for Biology and Society. President-elect of the History of Science Society, she specializes in the history and philosophy of biology and the way that biology, bioethics, and biopolicy play out in society. Her most recent book is *Whose View of Life? Embryos, Cloning, and Stem Cells* (Harvard University Press).

GREGORY M. MIKKELSON has published works of poetry, philosophy, and science. These, along with many of his other activities and ambitions, could fairly be described as "naturalistic" in several senses of that word. His main goal in life is to find harmony between the wild and the tame.

ROBERTA L. MILLSTEIN specializes in the philosophy of science and the history and philosophy of biology. Her research has focused on conceptual and epistemological issues within evolutionary theory and has addressed questions such as, Is evolution indeterministic? What is the most appropriate interpretation of probability for evolutionary theory? What role, if any, does causality play in evolutionary theory? Is natural selection a mechanism in any of the senses recently propounded by philosophers of science? Can natural selection be distinguished from random drift? Millstein is currently an associate professor in the Department of Philosophy at California State University, East Bay (formerly Hayward).

ROBERT T. PENNOCK is professor of history and philosophy of science at Michigan State University, where he is on the faculty of the Lyman Briggs School of Science, the Philosophy Department, the Department of Computer Science, and the Ecology, Evolutionary Biology and Behavior graduate program. His research interests are in philosophy of biology and in the relationship of epistemic and ethical values in science. His book *Tower of Babel: The Evidence against the New Creationism* has been reviewed in more than fifty publications; the *New York Review of Books* called it "the best book on creationism in all its guises." Dr. Pennock also does scientific research on experimental evolution and evolutionary computation, some of which was featured in a cover story in *Discover* magazine. Pennock speaks regularly around the United States on issues of science and values and was named a national Distinguished Lecturer by Sigma Xi, the Scientific Research Society.

GRANT RAMSEY is a doctoral candidate in the Department of Philosophy at Duke University. His primary interests include the philosophy of evolutionary theory and the relationship between culture and biology. Among his publications are articles on the concept of biological fitness, the evolution of culture, animal innovation, and plant ecology.

ROBERT J. RICHARDS is the Morris Fishbein Professor of the History of Science at the University of Chicago, where he is professor in the departments of Philosophy, History, and Psychology and director of the Fishbein Center for the History of Science and Medicine. He is the author of *Darwin and the Emergence of Evolutionary Theories of Mind and Behavior* (University of Chicago Press, 1987), *The Meaning of Evolution: The Morphological Construction and Ideological Reconstruction of Darwin's Theory* (University of Chicago Press, 1992), and *The Romantic Conception of Life: Science and Philosophy in the Age of Goethe* (2002). He has just completed *The Tragic Sense of Life: Ernest Haeckel and the Battle over Evolutionary Thought*.

JASON SCOTT ROBERT is assistant professor of life sciences in the School of Life Sciences at Arizona State University (ASU). He teaches in the Bioethics, Policy, and Law Program within the Center for Biology and Society and is also affiliated with the Consortium for Science, Policy, and Outcomes. Prior to joining the faculty at ASU, Robert was assistant professor and Canadian Institutes of Health Research New Investigator in the Department of Philosophy at Dalhousie University. He has published many articles in the philosophy of biology and bioethics, and his first book is

Embryology, Epigenesis, and Evolution: Taking Development Seriously (Cambridge University Press, 2004).

ALEXANDER ROSENBERG joined the Duke faculty in 2000. Previously he was professor of philosophy at Syracuse University and the University of California, Riverside, and Director of the Honors Program at the University of Georgia. He has been a visiting professor and Fellow at the University of Minnesota, as well as the University of California, Santa Cruz, and Oxford University. He has held fellowships from the National Science Foundation, the American Council of Learned Societies, the National Humanities Center, and the John Simon Guggenheim Foundation. In 1993 Rosenberg received the Lakatos Award in the philosophy of science. He was the Phi Beta Kappa–Romanell Lecturer for 2006–07. Rosenberg is the author of ten books, the latest of which is *Darwinian Reductionism or How to Stop Worrying and Love Molecular Biology* (University of Chicago Press), and approximately 170 papers in the philosophy of biology; the philosophy of cognitive, behavioral, and social science (especially economics); and causation. Rosenberg is also codirector of Duke's Center for the Philosophy of Biology.

SAHOTRA SARKAR teaches in the Section of Integrative Biology and the Department of Philosophy at the University of Texas at Austin. He is the author of *Biodiversity and Environmental Philosophy: An Introduction* (Cambridge Studies in Philosophy and Biology) and *Genetics and Reductionism* (Cambridge Studies in Philosophy and Biology).

ROBERT A. SKIPPER JR. works in the history and philosophy of biology and science more generally. In the history and philosophy of biology, Skipper works on theory assessment in population genetics and conceptual problems at the foundations of evolutionary genetics more generally. Within philosophy of science, he uses evolutionary biology to drive his research on scientific explanation, confirmation, and theory change. He received a Ph.D. in history and philosophy of science from the University of Maryland, College Park, and is presently assistant professor of philosophy at the University of Cincinnati.

KIM STERELNY is professor of philosophy at Victoria University in New Zealand and at the Australian National University in Canberra. He is the author of many books on the philosophy of science, including *Sex and Death,* cowritten with Paul Griffiths. He is a winner of the Lakatos Prize for the best book of the year in the philosophy of science (*Thought in a Hostile World*: *The Evolution of Human Cognition*, published in 2003).

KAROLA STOTZ was trained in human biology and social sciences at the University of Mainz and received her Ph.D. in philosophy from the University of Ghent. She is currently a postdoctoral Fellow in the Cognitive Science Program at Indiana University, Bloomington. From 2002 to 2005 she and Paul Griffiths were principal investigators of the National Science Foundation–funded project "Representing Genes" at the University of Pittsburgh. This interdisciplinary project used Web-based survey methods to examine differing understandings of the gene in contemporary bioscience.

PREFACE

The philosophy of biology is one of the most vigorous and exciting areas in modern philosophy. There are many active researchers and their students, there is a plethora of new ideas and suggestions, there are good-quality, dedicated outlets for the work – notably the journal *Biology and Philosophy* – and there are organizations – notably the International Society for the History, Philosophy, and Social Studies of Biology – that embrace and encourage the workers in the field. This heady and forward-looking community is of recent vintage. Although in the history of philosophy some of the very greatest thinkers – Aristotle in the ancient world and Immanuel Kant in the modern world – had things of great importance to say about the life sciences, for much of the past two hundred years biology was ill served and little regarded by philosophers. This will seem strange. After all, in the nineteenth century came the great evolutionary theory of Charles Darwin, expounded in his *On the Origin of Species by Means of Natural Selection* in 1859, and in the twentieth century there was the coming of molecular biology, as represented by the double helix discovered in 1953 by the American James Watson and the Englishman Francis Crick. Generally, however, it was physics that caught the attention of philosophers. Biology was often neglected or worse, being put to use by people with deeply antiscientific metaphysical agendas, especially the vitalists at the beginning of the twentieth century.

Conditions started to change about forty years ago, in the 1960s, at first slowly and then with increasing speed as the topic attracted attention. A number of young philosophers of science took note of the exciting developments in the biological sciences, not just molecular advances but also the ways in which traditional areas (notably evolutionary biology) were now developing new ideas and attracting first-class minds. These philosophers sensed that there was an important part of science that was neglected, and they moved to understand and work on the conceptual problems that are always raised by empirical studies. At the same time, a number of biologists started to turn seriously to philosophy for help in articulating issues in their science, realizing that good empirical work demands sound philosophical bases to the theories and models that guide research. Spurring and stimulating each other, the philosophers and biologists worked on such issues as the nature of evolutionary explanation and the role played by Darwin's mechanism of natural selection; the extent to which biology is an autonomous science and whether issues like teleology and historicity mark the life sciences as something irreducibly different from the physical sciences, or whether ultimately these are matters that drop away in a mature science; problems of classification, both about the basic units of division (notably species) and about the proper way to conceptualize life's history (the coming of phylogenetic systematics, or cladism, was a major issue here); the relationship between the older Mendelian genetics and the newer molecular genetics, and whether this was a case of theory absorption (reduction) or of theory change (replacement); as well as related topics. Perhaps reflecting the interests of the early researchers, perhaps reflecting the fact that it is simply of great philosophical interest in its own right, evolutionary theory tended to dominate discussions, and indeed you will sense that this tendency persists to this day.

In the past four decades, biology itself has continued to advance in many exciting ways. The already-mentioned revolution in classification, systematics, brought on fiery debates about the nature and intentions of workers in the field, and ever-increasing sophistication as molecular techniques became readily available, backed by the increasing power of computer programs. There was the development of the evolutionary approach to social behavior, sociobiology, with highly controversial attempts to expand the science from other animals to us humans. There were debates about the history of life as revealed through the fossil record, and whether pure Darwinism is adequate as an explanation, or whether new approaches (particularly the theory of jerky change, punctuated equilibrium) demand new theoretical approaches. Most recently, there has been the revival of embryology and consideration of the ways in which this can be improved and extended through molecular findings. The field of evolutionary development, or "evo-devo," has attracted some of the best biological minds of the generation.

Philosophy has responded to these developments in the biological sciences. They offer new challenges to those who are interested in deep conceptual issues of science, and at the same time they offer new insights into some of the perennial questions of philosophy itself, about knowledge and truth and about action and morality. It is this response that we as editors have tried to capture in this *Companion*. We have tried to give the reader a sense of the exciting work that today characterizes the discipline or subdiscipline of the philosophy of biology. We have not tried simply to give a survey or a textbook introduction. There are already good works of this ilk, some written by contributors to this *Companion*. Rather we have tried to give a sense of the issues that engage today's philosophers of biology and an understanding of how these issues are tackled. We have asked our contributors to write in a way and at a level that a nonexpert would find interesting and understandable, but at the same time we have emphasized that this should not be at the expense of trying to tackle complex problems and showing by example of work in action rather than simply through overall surveys.

We did not set out to impose formal divisions on the subject and we would like to think that all of the contributions to the *Companion* are freestanding, in that one could read any one in isolation from the others. However, the reader will sense that there is a kind of informal flow to the issues, with one topic leading naturally to another, and we have tried to reflect this in the order in which the contributions are presented. We start with Darwin's theory of evolution through natural selection, or rather with its modern-day successor. For Darwin, as for today's evolutionists, the important point about selection is that it explains not just change but change of a particular kind, namely, in the direction of adaptive advantage – it explains the eye and the hand and all of the other organic features that were at the heart of the traditional argument from design for the existence of God. It is this topic of adaptation that is the focus of Tim Lewens's contribution, as he teases apart the meaning of the term in modern science and tries to assess its significance as something in need of explanation.

From adaptation, with the contribution by Roberta L. Millstein and Robert A. Skipper Jr., we move to the structure and nature of modern evolutionary thinking. After Darwin, the most important advance in such thinking accompanied the discovery and development of genetics, the theory of heredity. This transformed our ideas about the past, especially after the so-called population geneticists embedded natural selection in a theoretical framework based on Mendelian genes interacting in groups. Millstein and Skipper discuss how this happened and then go on to look at some consequences, especially whether the importance of selection is now diminished by other factors, notably Sewall Wright's notion of genetic drift, essentially a matter of random change caused by contingent factors overwhelming the systematic effects of differential reproduction. From here we move naturally and smoothly to Elisabeth A. Lloyd's essay, which takes up a topic that plagued Darwin himself and which has raised much discussion in recent years, about the level or levels at which natural selection may be said to act, in particular whether it is always something between individuals, perhaps even between "selfish genes," or whether it can and does act significantly at higher levels between groups. Then, completing this part of the *Companion*, we have an analysis by Robert N. Brandon and Grant Ramsey of some of the most interesting issues that emerge from modern evolutionary thinking, namely, those that concern the statistical nature of the theory and how it works with and tries to explain the actions and effects not of individuals working alone but of individuals in groups and of the cumulative results.

Next comes a batch of articles dealing with some of the issues in the philosophy of biology that relate to traditional questions in the philosophy of science. Much time has been spent on questions about whether new theories are absorbed into old theories (reduction) or whether they simply push them aside and replace them. The essay by Paul E. Griffiths and Karola Stotz opens the way for discussion of this topic in biology with respect to the nature of genetics and how concepts of the gene have changed through the years, especially in the light of the coming of molecular techniques and theories. Complementing this piece is the contribution by Peter Godfrey-Smith, taking up the topic of biological information and of how and in what sense something like the gene (or in its modern guise, the DNA molecule) can be said to carry such information. After this, Alexander Rosenberg gives us a more general discussion of reduction in biology and the ways in which one can and should expect continuous change between theories of different times and levels. From here we move to Lindley Darden, who looks at explanations in biology and at how models are an important factor in trying to explicate mechanisms. André Ariew then brings this kind of discussion to a conclusion by surveying and discussing a perennial question for the philosopher of biology: whether in some sense evolutionary understanding is teleological or forward-looking in a way alien to the physical sciences, and whether in some sense this represents a nonreducible factor that enters into all explanations in the life sciences.

Thanks to people like the late Stephen Jay Gould, there has been much written recently by biologists about the relationships between microprocesses of change and the overall macronature of life's history. Can the latter be explained in terms of the former? This is the topic of Kim Sterelny's contribution, which focuses on a notion that he calls "minimalism," namely, the thesis that small-scale changes can explain all. The ways in which this topic ties in with some of the earlier contributions about reduction hardly need stressing. The systematist Maureen Kearney then discusses the perennial problem of biological classification and the extent to which it can and should reflect life's history. Her contribution reflects some of the (earlier-mentioned) major debates that have occurred in this area, particularly about the nature of biological species and whether they are to be considered as individuals or as groups, and then about the techniques and significance of the apparently all-conquering taxonomic approach of the cladists with their interesting (and controversial) assumptions about such notions as simplicity or (as they call it) parsimony.

Biologically speaking, humans may or may not be the most important of all organisms. They are certainly the organisms of most interest to us humans! The geneticist Francisco J. Ayala gives important background to the problems facing researchers into the evolution of humans. In the light of the completed mapping of the human genome, he stresses that we still face issues about how we sense and feel, how the mind emerges and its connection to the material body, and how and why humans emerged from the ape line. David J. Buller then rows into controversial waters surrounding the new science of human sociobiology or (as it is now often called) evolutionary psychology. Is human nature to be explained as a function of the genes as sifted by natural selection, or is this approach altogether too simplistic? In her discussion of neurobiology, Valerie Gray Hardcastle takes up in more detail some of the issues raised by Ayala. She stresses how in attempts to understand the functioning of the brain and its relationship to mind, methodological and metaphysical questions arise, and often it is not easy to tease out the questions that should be asked and the relevance of empirical findings that result from research.

Next, going from the general to the more particular, Christopher Horvath turns to the much-discussed topic of human sexual orientation. He stresses the way in which when we get to controversial aspects of human nature it is not easy to disentangle strongly held social and moral beliefs from more objective scientific findings and theories. Horvath's contribution illustrates the way in which modern philosophy of science, philosophy of biology in particular, has moved from participating in purely theoretical discussions to addressing issues of immediate societal interest and concern – and so to moral theory and behavior. For almost all of the twentieth century, thanks particularly to the devastating critique in G. E. Moore's *Principia Ethica* (1903), attempts to relate biology to ethics were regarded as the philosophical equivalent of a bad smell – not just wrong, but in some sense unclean. Matters have changed dramatically in recent years, thanks to the development of such areas as sociobiology but also thanks to formal work by economists and philosophers and others. Very significant has been work on the topic of game theory and how it applies to human evolution. In his contribution, Zachary Ernst introduces us to this topic.

For much of the nineteenth century and well into the twentieth century, embryology was an important part of the biological scene. Then with the coming of genetics and the move to molecular topics, it rather dropped from sight, being regarded as somewhat of a descriptive topic of little theoretical interest. In the past two or three decades, as mentioned, conditions have changed dramatically as molecular biologists and evolutionists have joined forces to look at development. Today evolutionary development, or evo-devo, is one

of the hottest areas of biological research. The historian and philosopher of science Jane Maienschein opens our cluster of articles on
this topic, looking at the changing meaning of the notion of an
embryo, how it evolved from rather crude notions in the eighteenth
century to a variety of sophisticated concepts today. She stresses in a
fashion akin to that of Horvath that, in discussions of this nature, it
is often difficult to distinguish claims of fact from deeply held moral,
social, and religious beliefs. Manfred D. Laubichler follows with a
more detailed discussion of evo-devo and stresses how it opens up
questions of great philosophical interest. He himself focuses on the
problem of innovation and of how new features get produced
and introduced into populations. Is this a challenge to traditional
Darwinian explanations that emphasize the all-sufficiency of natural
selection? Jason Scott Robert concludes this part of the *Companion*
by looking at how molecular biology has brought new insights. He
talks of "systems biology," the area of study that looks at how the
genes get translated and used to build the functioning organism,
referring explicitly to ethical issues that are raised and showing that
at this point philosophy of biology touches on and in respects blends
with the sorts of concerns that occupy bioethicists.

In the early years of contemporary philosophy of biology, ecology
was curiously and unfortunately overlooked. Although envir-
onmentalists were raising important issues, most of which called out
for detailed analytic scrutiny, philosophers of biology were unre-
sponsive. Things have now changed, and there is increasing interest in
the issues raised by the interactions of organisms on a daily (rather
than historical and evolutionary) basis. Gregory M. Mikkelson gives
us a background to ecology and the philosophical issues that it raises.
He shows how there are questions about the notion of hierarchy, from
individuals up to groups and then to whole systems; about the relationship of ecology to evolutionary questions; about the very notion of
law-governed explanations in ecology (is it more of a subject dealing
with the unique and the contingent?); and about the extent to which
ecologists should be naturalists, observing what is happening in the
wild, and the extent to which they should be experimenters, manipulating situations and trying to predict outcomes. Sahotra Sarkar
follows by looking at concepts of ecological diversity, and how in their
new guise of "biodiversity" many fascinating philosophical questions
about understanding and measurement get raised.

Coming to the end of the *Companion*, we have next Robert T. Pennock, who takes up the question of biology and religion. Pennock has been much involved in the fight against creationists, particularly in their new incarnation as so-called intelligent design theorists. His contribution reflects this activity, but he aims to put his discussion in the broader context of biology and religion generally, and thus he is led into issues such as the meaning of morality in a post-Darwinian world, as well as the autonomy of religion itself. Is it just an adaptation like any other and does this have implications for its truth value? Pennock raises the issue of biology, religion, and sexual orientation, and it is interesting to compare his thinking with the earlier contribution of Christopher Horvath on the same topic. Finally, through an examination of the thinking and use of language of the nineteenth-century biologist Ernst Haeckel, the historian and philosopher of science Robert J. Richards takes on questions about understanding science by the study of the history of the subject. We think this a particularly appropriate contribution with which to end the *Companion*. The relationship between the history of science and the philosophy of science has not always been as intimate and fruitful as one might desire, but the philosophy of biology has been exceptional in the ways in which it has (as many of our contributions show) drawn on the history of biology for insights about the conceptual nature of present-day biology. Richards's contribution shows the value of this practice and why it should be cherished and encouraged as a mode of inquiry by the next generation of philosophers of biology.

We hope that you enjoy the collection. Good philosophy, like good science, is never finished. If you are lucky, you have more questions at the end of the day than at the beginning. We invite you to engage with our contributors and to add to the exciting advances in the philosophy of biology. We invite you also to take a moment and think about the man to whose memory this volume is dedicated. Ernst Mayr, who died in 2005 in his hundred-and-first year, was a German-born taxonomist who emigrated to America and became one of the most important and influential evolutionists of the twentieth century. He was always interested in philosophy and was the leader among those mentioned earlier who worked from the side of science to develop the newly invigorated field of the philosophy of biology. This volume in your hands started in 2001 as the first conference at Florida State University sponsored by the legacy of William H. and Lucyle T. Werkmeister.

Mayr, very old, still spent his winters in Florida and was invited to the conference. He accepted, but the organizers were warned that he would stay only a short while, that much attention must be paid to the needs of old age, and that he would be too fatigued to attend any social events. He arrived and within minutes was on his feet making points, leveling objections, and stressing the correct (that is, Mayrian) way of seeing things. On the Saturday, he started at nine o'clock in the morning and insisted on joining the participants in an afternoon trip to Wakulla Springs, a local beauty spot of unspoiled river with alligators, fish, and many birds (the site of the original Tarzan movies as well as the classic science fiction movie *The Creature from the Black Lagoon*). Mayr stood in the prow of the boat, identifying birds and explaining the differences since his last trip to the spot in 1931. At ten o'clock that night, the exhausted organizers finally had to shuffle Mayr out of the door and insist that he go home to bed! Ernst Mayr was a great scientist, but more than this, he was a mensch. For the editors especially, who frequently were the focus of Mayr's scolding, it is a real privilege to acknowledge our debt and our love.

More immediately, the editors thank Jason Zinser, who was our assistant on this project, and Alan Casselman, who worked on the bibliography. Sadly, because of his too-early death, we cannot thank Terry Moore at the Press, who responded with such enthusiasm to the idea of this volume, but we do remember him with gratitude. Beatrice Rehl, who took over the project, has supported and helped us in an exemplary way. We are grateful to our production editor, Janis Bolster, and our indexer, Lin Maria Riotto. Finally, the editors want to thank each other. We have been fellow philosophers now for forty years and good friends from the beginning. Never a cross word has been exchanged between us. Deliberately, we did not ourselves contribute to the volume, wanting rather to commission and promote the work of others. It is the field or discipline of the philosophy of biology that is our main creation, and inasmuch as this volume shows that we have succeeded, it is because each of us knows how much we owe to the other.

1   Adaptation

Eyes have long fascinated those who study the natural world. Cleanthes – the natural theologian protagonist of Hume's *Dialogues Concerning Natural Religion* – invites his interlocutor to 'consider, anatomize the eye: Survey its structure and contrivance; and tell me, from your own feeling, if the idea of a contriver does not immediately flow in upon you with a force like that of sensation' (1990, 65). Darwin, too, counted the eye among what he called 'organs of extreme perfection'. Placing himself squarely within the tradition that runs from natural theology, through Darwin, to a certain style of modern biology, Maynard Smith writes that 'the main task of any theory of evolution is to explain adaptive complexity, that is, to explain the same set of facts that Paley used as evidence of a creator' (1969, 82). More recently still, Dawkins (1986) is impressed, also, with a force like that of sensation, by how well suited – how well adapted, that is – the eye is to its purpose. Like Paley, he thinks eyes are better pieces of work than watches, although unlike Paley he regards their artificer as blind.

An essay on adaptation could fill volumes. One might begin by asking how adaptation is to be explained. Immediately we would need to answer the prior question of what the proper definition of adaptation is, and we would also have to get clear on the nature of the diverse candidate processes – natural selection, self-organisation, macromutation, development, divine design – sometimes tabled as potential explanations. We might go on to ask in what senses adaptations are purposive, and whether they all share some single ultimate purpose, such as the proliferation of an organism's genes. Once the nature of adaptation is pinned down, we could move on to consider the questions of whether adaptation is ubiquitous or rare,

1

and whether there might be important nonadaptive phenomena in the biological world that an exclusive concern with adaptation might lead us to overlook. In short, a thorough study of adaptation would need to address most of the topics covered in this *Companion* – teleology, the units of selection, development, and others. Here, then, I will restrict myself to brief discussions of four questions. How should we *define* adaptation, how should we *explain* adaptation, how can we *discover* adaptation, and how *important* is adaptation?

1.   DEFINING ADAPTATION

In the analytical table of contents of his landmark work *Adaptation and Natural Selection*, George Williams claims that 'evolutionary adaptation is a special and onerous concept that should not be used unnecessarily, and an effect should not be called a function unless it is clearly produced by design and not by chance ... Natural selection is the only acceptable explanation for the genesis and maintenance of adaptation' (1966, vii). I want to take some time in the first two sections of this essay to pick these comments apart.

What, precisely, is the special and onerous concept of adaptation? As a preliminary, we should take Elliott Sober's (1984, 196) advice and distinguish products from processes. Consider an example: marriages produce marriages. This sounds peculiar, until we remember that 'marriage' can refer either to the process of getting hitched or to the blissful union that is the product of that process. Similarly, 'adaptation' can refer either to the process by which organisms become well suited to their environments, or it can refer to the organic traits that are the end results of this process. Unless I stipulate otherwise, I will be talking about adaptation as a product in this essay.

Broadly speaking, there are three quite different styles of definition of the adaptation concept. First, we could give a rough indication of what adaptation means by pointing to some of its instances – things like the eye, or the wing. Such definition by example, certainly when the examples are few, tells us little about how we should apply the concept. At this point, a second style of definition may appear. Adaptation is a concept used in modern biology, yet modern biologists sometimes define the term in an informal way that echoes natural theology's conception of organisms as designed objects. Williams gives just such a definition in the quotation we just saw: 'An effect should not be called a function unless it is clearly produced by design and not by chance'. This distinction between what an object's *effects* are and what its *functions* are makes clear sense when we are talking about tools designed by agents. A screwdriver may be good at levering lids from paint tins, but that is not what the screwdriver is *for* – that is not its function – because the screwdriver was not designed to lift lids from paint tins. Williams's definition expresses his view that adaptations are traits that are *for* something. For Williams, therefore, the question of whether some trait is an adaptation should depend on its design history. But Williams is no creationist: the design history in question is the evolutionary history of the trait.

Williams's comment explains why many biologists draw a distinction between adaptive traits and adaptations. Adaptive traits augment fitness in some way or another – we might also use Mayr's (1986) term and say that they have the property of *adaptedness*. The adaptedness of a trait is not sufficient for the trait to be an adaptation, because the trait, like the screwdriver, may not have the right kind of history.

Richard Dawkins also defines adaptation in terms of good design, and he defines design, in turn, as that which gives only the appearance of intelligence: 'We may say that a living body or organ is well designed if it has attributes that an intelligent and knowledgeable engineer might have built in order to achieve some sensible purpose, such as flying, swimming, seeing, eating, reproducing, or more generally promoting the survival and reproduction of the organism's genes' (1986, 21). For Dawkins, as for Maynard Smith, the way to define adaptation is in terms of what a natural theologian might have counted, mistakenly, as evidence of intelligence.

It is hard to square Williams's claim that adaptation is a special and onerous concept for modern evolutionary biology with all these covert uses of what appear to be natural theological notions in the definition of that concept. If adaptation could be defined only as something that the superstitious would take as evidence for a designer, then the best thing for modern biology to do would be to eliminate the adaptation concept altogether on the grounds that it is part of a natural theological worldview we no longer share. Because few, if any, biologists or philosophers could tolerate the elimination of the adaptation concept from biology, a move to a definition in the third, more formal, style is widely preferred.

2.   HISTORICAL DEFINITIONS OF ADAPTATION

Formal definitions of adaptation tend to divide into historical and nonhistorical varieties. A formal definition that is endorsed by many philosophers (although not by so many biologists) is Sober's (1984, 208):

A is an adaptation for task T in population P if and only if A became prevalent in P because there was selection for A, where the selective advantage of A was due to the fact that A helped perform task T.

One of the reasons why a definition like this is attractive is that it promises to tidy up Williams's claim that adaptations are the result of design rather than chance. What is required, if this claim is to be made respectable, is some evolutionary process that can play the role of design. Sober achieves this by defining adaptation as the product of a natural selection process, a process that can be distinguished from the mere chance appearance in a population of the trait in question.

Sober's definition leads to some awkward results, especially if assessed by its success in grounding the notion that adaptations are produced by design. First, a trait can be an adaptation for some task even when the first occurrence of the trait is an entirely fortuitous affair that has nothing to do with selection. This is a consequence of the definition of 'selection for' a property. Suppose a pair of wings arises, fully formed and fully functional, in a population of flightless foxes. These wings help their prodigiously lucky bearer to catch chickens more effectively than other foxes, and as a result the flying fox is far fitter than its fellows. Baby foxes inherit the wings of their parents, and wings soon become prevalent in the population. In this (intentionally absurd) scenario, there is selection for flying in virtue of the fact that wings increase their frequency in the population because they allow flying. Hence wings are adaptations for flying by Sober's definition, even though the metaphors of selection designing, building, or shaping the trait are hard to apply. This is hardly a fatal objection to Sober: modern biology can get by perfectly well with an adaptation concept that jars some of our intuitions about when it is appropriate to speak of 'design' or 'shaping'. Even so, we will see that for some biologists, adaptations are understood as traits that have been (in some sense) shaped, built, or modified by selection, not merely traits whose frequency has increased because of selection.

Sober's definition helps us to make some sense of Williams's claim that adaptations are not products of chance, but in doing so it causes problems for Williams's follow-up assertion that selection is the only permissible explanation of adaptation (a claim that Richard Dawkins [1996] also makes). It makes that second claim true, but vacuously so. It is hard to portray Darwin's intellectual break-through as the realisation that adaptation is best explained by nat-ural selection, if adaptation is simply defined as a product of a selection process. Fisher (1985, 120) makes the point forcefully: 'Defining the state of adaptation in terms of its contribution to current fitness, rather than origin by natural selection, is essential if natural selection is to be considered an explanation of adaptation.'

Fisher's argument can be resisted. We can keep hold of Sober's definition of adaptation while rephrasing our understanding of Darwin's breakthrough in more particularist terms: Darwin realised that natural selection could explain the organisation of eyes, wings, instinctive behaviours, and many other specific traits. None of these claims is vacuously true, even if the general claim that natural selection is the only explanation of adaptation is. We might also consider replacing Williams's general assertion with the rephrased claim that selection is the only permissible explanation of adapt-edness, where adaptedness is defined nonhistorically in terms of a contribution to fitness. Whether this revised claim is true would require further assessment, but it seems clear that it is not trivially true.

There are other problems that have driven some biologists (e.g., Reeve and Sherman, 1993) to prefer nonhistorical definitions. Consider a trait that becomes prevalent in a population by chance, but that is subsequently maintained at a high frequency in the popula-tion because of its superior fitness compared with alternatives. Sober's definition denies that the trait is an adaptation, for selection has not made it prevalent, even though maintaining selection does subsequently explain why it remains prevalent. Conversely, Sober might have to accord the status of adaptation to traits that have spread through a population in virtue of some effect, but that have not had that effect for several generations. We might have to say that the human appendix is an adaptation for digestion. Yet these kinds of traits are more usually thought of as vestiges, not adaptations.

These problems are not fatal to Sober either – the obvious solution is to keep a historical definition, but one that looks only to quite recent selection history, including selection that maintains the frequency of a trait in virtue of one of its effects. Sober himself considers some analogous moves to weaken the original definition while retaining its historical element (1984, 198). Although a revised definition of this form upholds a conceptual distinction between being an adaptation for E and promoting fitness by E-ing, very few actual traits, so long as they are inherited, will fall into the latter category without also falling into the former. The revised historical definition helps to ground a function/effect distinction that non-historical accounts will have trouble maintaining, but the satisfaction of this desideratum may seem like a philosophical indulgence when viewed from the perspective of biological inquiry, especially once we see how rarely the conceptual distinction will make any practical difference. The biologists Endler and McClellan prefer to use adaptation to indicate current contribution to fitness on just these grounds:

It is important to distinguish between traits that were always selected for one function ('adaptations') from those which were originally selected for another function and by chance can be used in a new way ('exaptations' for the new function). We use adaptation in both senses because as soon as a new function for a trait occurs, natural selection will affect that trait in a new way and change the allele frequencies that generate that trait. (1988, 409)

This comment is likely to mislead, because the historical definition of adaptation preferred by many philosophers is not the same as that of Gould and Vrba (1982), whose distinction between adaptation and exaptation Endler and McClellan are referring to here. Gould and Vrba's definition of adaptation, like Williams's, appeals not just to selection for some property, but to a stronger notion of shaping, or structural modification, consonant with the everyday concept of design. An adaptation, for Gould and Vrba, 'was built by

natural selection for the function it now performs' (53). 'Exaptations', on the other hand, have not been shaped by selection for the
tasks they now perform. A definition of adaptation in terms of recent
(maintaining) selection will make almost all exaptations for some
function adaptations for that same function. Some philosophers
have questioned the coherence of the adaptation/exaptation distinction (e.g., Dennett 1995, 281); however, providing we can make
sense of the contrast between being shaped for a function and being
selected for a function, and providing our definition of adaptation
appeals to shaping, this scepticism is premature.

3.   NONHISTORICAL DEFINITIONS OF ADAPTATION

Reeve and Sherman have articulated the most thorough defence of
a nonhistorical definition of adaptation:

An adaptation is a phenotypic variant that results in the highest fitness
among a specified set of variants in a given environment. (1993, 9)

Why insist that an adaptation must be the fittest of a set of variants? Which variants go into that set? To answer these questions,
we need to look at the primary goal of Reeve and Sherman's definition, which is to develop a concept suitable for answering questions about what they call 'phenotype existence'. They distinguish
these kinds of questions from those about 'evolutionary history'.
Students of phenotype existence ask 'why certain traits predominate
over conceivable others in nature, irrespective of the precise historical pathways leading to their predominance, and then infer
evolutionary causation based on current utility'. Practitioners of
evolutionary history, on the other hand, 'seek to infer the origins and
phylogenetic trajectories of phenotypic attributes, and how their
current utility relates to the presumed functions in their bearers'
ancestors' (2).

There are two slightly different rationales for appealing to a range
of conceivable variants in defining adaptation. The first has to do
with establishing the selective history of the trait (i.e., 'evolutionary
causation'). The second has to do with establishing the trait's causal
contribution to survival and reproduction (i.e., its 'current utility').
Beginning with the first rationale, Reeve and Sherman want the
claim that a trait is an adaptation to be evidence for, rather than (as it

is for Sober) synonymous with, any further claim about evolutionary causation. If a trait is fitter than all the hypothetical alternative traits we are considering, then the chances are it also outcompeted the actual alternative traits in the population. It is therefore likely that selection explains its presence. That is why we should exercise restraint in the hypothetical alternatives we include in the con-sidered set – they need not include all and only actual competitors, but they should reflect likely competitors: 'A suitable choice requires only that the set contains phenotypes that might plausibly arise' (Reeve and Sherman 1993, 10). Reeve and Sherman's insis-tence that a trait be recognised as an adaptation only if it is the *fittest* of the phenotype set is not obligatory given the goals of this kind of evolutionary research: a trait can make a significant contribution to fitness – including the greatest contribution among actual variants present and past – even when some other plausible trait might have been better still.

The second reason for appealing to a range of conceivable variants in defining adaptation has considerable metaphysical interest, especially as a case study in the problems of causation. On the nonhistorical approach, to ask whether human eyes are adaptations is to ask whether they make a causal contribution to fitness, and if so, what that contribution is. It might seem that there is no need to specify a set of alternative possible eyes in order to answer this question; we need only consider the question 'What would we be like without eyes?' The problem is that this question has many plausible answers. We might say that vision is so important that if we had no eyes, we would have some other kind of sensory apparatus instead. If we say this, we will say that eyes are not adaptations for providing sensory information, for we would do just as well in that respect without eyes. Alternatively, we might say that if we had no eyes, we would be dead, as a result of infection in our empty eye sockets. If we say this, we will say that eyes are adaptations for preventing infection reaching the eye sockets. Both answers seem silly, but such silliness seems to result from asking, without con-straint, 'What would we be like without eyes?'

These problems about how to say what the causal contribution of some part is to a whole are not specific to biology. Consider my laser printer. Our inclination is to say that the ink cartridge contributes to the workings of the whole by dispensing ink. But what allows us to say this? After all, it is not true that the only thing that would be different if my printer had no cartridge is that it would dispense no ink. Paper would not pass through the printer, either. If we understand counterfactual conditionals in the manner of David Lewis (1973a), then counterfactuals are made true by states of affairs at the nearest 'possible worlds' where the counterfactuals' antecedents are true. Roughly speaking, a possible world is a way things might have been. The statement 'Were Beckham to have got the penalty, England would have won Euro 2004' is true just in case those nearest worlds (i.e., the worlds most similar to way things actually are) where Beckham gets the penalty are also worlds where England wins Euro 2004. Now the nearest world at which my printer has no cartridge is, presumably, one where I have removed the cartridge to shake it, or some such. At this world, the printer will not function at all. Are we to say, then, that the function of the ink cartridge is to enable paper to pass through the printer?

Comparing actual eyes with a clearly specified set of alternative traits seems to be a good way of circumventing these problems for causal analysis. One might wonder, though, exactly what the role is of specifying alternatives in the determination of a trait's causal contribution. This method is somewhat at odds with contemporary counterfactual views of causation (e.g., Lewis 1973b). According to these theories, causation is indeed bound up with ways the world might have been, but we determine, say, the causal impact of a brick's flight through a window not by specifying alternative flight paths, but by specifying which actual event, or which actual fact, we are interested in understanding causally, and asking what would have been the case had that event not occurred, or had that fact not obtained. On this view, alternative flight paths follow from a specification of the fact or event of interest to us; the specification of alternatives is not a preliminary to causal analysis of some fact or event.

Generalising this method to the analysis of systems, the first step to determining a causal contribution of a part to a system is not the specification of alternative possibilities for what the part might have been like; rather, it is to specify what aspect of the part we are interested in. The effect of this is to move us away from asking blunt questions about the causal contributions of parts (organs, for example, in biology), and towards asking questions about the contributions of traits understood as finely specified facts about systemic organisation. In the case of the printer, we can ask, for example, what the effect of the cartridge's having ink at such and such density might be. Immediately we dismiss the world where the printer has no ink cartridge as irrelevant to answering our causal question; the nearest world where the ink density is different is not one where there is no ink cartridge at all. So one of the roles for specifying a range of alternative traits is to draw out, through comparison, the aspect of the trait under consideration that we are interested in. Note that we need not suppose that any actual printer cartridge has existed with ink at a different density for an appeal to such counterfactual circumstances to have legitimacy in causal reasoning. That is why, to the extent that Reeve and Sherman's phenotype set is supposed to focus attention on specific aspects of actual traits by positing ways the world might have been if the trait had been different in those respects, they are quite right not to insist that membership of the set be restricted to actual traits.

4.   EXPLAINING ADAPTATION

Does natural selection explain adaptation? We have already seen how this question runs into the definitional problems of the earlier sections. If adaptation should be defined as a product of selection, then the claim that selection explains adaptation is secured by definition alone. Let us ask, instead, whether selection explains adaptedness – understood as contribution to fitness – while withholding judgement on the question of whether we should opt for a nonhistorical definition of adaptation itself.

Our first job is to distinguish the question of whether selection explains the spread of traits from the question of whether selection explains the origin of traits. In the first section I gave the example of a wing that arises fully formed through macromutation, and that then spreads through a population. In this scenario, selection explains why the trait increases its frequency, and it also explains the increase in adaptedness of the population, but it does not explain the appearance of the first wing token. If selection never explains the origin of trait types, then Darwin's innovation is not as great as it seems. People like Paley were puzzled at how such things as eyes could come to exist at all; the response that once one eye exists, eyes

will tend to become prevalent is not enough to satisfy the demand
for an explanation of the first appearance of such an extraordinary
thing. And note that if one replies that eyes are really rather likely to
arise through macromutation after all, or that one should simply
look to chance to explain the first instance of the trait, then there is a
sense in which one has not truly offered a selective explanation for
the existence of eyes; rather, one has appealed to something like
a general law of generation of organic form, or to blind chance.

Selection can explain the origin of novel trait types by causing
tokens of other traits to spread (Endler 1986). Selection is creative.
Consider the following case, borrowed from Neander (1995). Imagine
that there are three 'genetic plans', P1, P2, and P3. P3 gives us a fully
functional eye. P2 yields a slightly inferior eye, and P1 is even worse.
A population exists in which all members have P1, bar a few with
P2. As P2 increases in frequency in the population, then the chances
of a variant arising with P3 may increase too. In case this is not clear,
think of an example; if I can persuade more people to buy lottery
tickets, then I increase the chances that someone will win. Similarly, if the number of genetic loci where a favourable mutation
might occur increases as a result of selection, then selection can
explain the first appearance of a favourable mutation of that type.

Sticking with the lottery example, although increasing the
number of tickets explains why someone wins, it does not explain
why Emma wins, rather than Nicole, who has also bought a ticket;
increasing the number of tickets makes Emma no more likely to win
than Nicole. Similarly if selection, by increasing the number of
organisms with P2, explains why P3 eventually arises, it does not
follow from this that selection explains why Sam first acquires the
P3 mutation, rather than Suzy, who is also descended from a P2
individual. The claim that selection is creative does not straight-
forwardly resolve the related question (addressed by Sober 1995 and
Neander 1995, among others) of whether selection explains why
individual organisms have the traits they do.

Now that we have seen how selection explains adaptedness, we
can also see that selection does not explain adaptedness alone
(Lewens 2004). There are a number of assumptions hidden in our
earlier scenario about the emergence of eyes. If increasing the
number of P2 variants is to increase the chance of a P3 variant
arising, it must be the case that the P3 plan is more likely to arise

from the P2 plan than from the P1 plan, for suppose that while P2 is fitter than P1, and P3 is fitter than P2, P3 is mutationally closer to P1 than to P2. If that is the case, then selection will make the emergence of the P3 plan less likely as it increases the frequency of the P2 plan. What is required for selection to have its cumulative character, whereby it leads to progressive adaptation, is for fitness orderings of phenotypes to map onto the mutational 'distances' between the plans that code for those phenotypes. That, in turn, is a requirement on the organisation of individual organisms; indeed, the familiar point that selection can cause populations to get stuck on the 'local optima' of fitness landscapes is another way of making the point that organisms may be constructed in such a way that mutations for highly functional versions of some organ may sometimes be more likely to arise from forms of low function than from forms of intermediate function.

Selection does not explain adaptedness alone, because selection has the character that makes it cumulative only when it acts on systems with the right kind of organisation. This is not the place to say what those organisational conditions might be, but it is the place to point out that this makes organic organisation an element of the explanation of adaptedness itself (Lewens 2004, Walsh 2000). We also have reason to believe that 'drift' or 'sampling error', whereby populations (especially small populations) change in ways that do not reflect the fitnesses of the traits in the population, can help to explain adaptedness. Drift can prevent populations from getting stuck on local optima, and in that sense the existence of sampling error, whereby lower-fitness forms sometimes replace higher-fitness forms, will occasionally increase the chances that the population will arrive at an adaptive peak that is higher still. So drift, also, can explain adaptedness.

5.   DISCOVERING ADAPTATION

Let us move now from the metaphysics of adaptation to its epistemology. How can we determine whether a trait is an adaptation? One popular approach to this question once again looks back to natural theology for inspiration (Lauder 1996 gives a useful survey of these moves). It used to be thought that the good design of an organism indicated intelligent design; if good design is instead best explained by natural selection, then good design is evidence for the action not of the hand of the divine artificer, but the hand of selection. Thus Pinker and Bloom (1990, 707) write that 'evolutionary theory offers clear criteria for when a trait should be attributed to natural selection: complex design for some function, and the absence of alternative processes capable of explaining such complexity'.

Is this a good inference? Partly it depends, once again, on the strength we read into the claim that a trait should be 'attributed to natural selection'. We have already seen that if a trait promotes fitness in some population through effect E, it is highly likely that there has been selection for E. This requires only that there has been some heritable variation in past generations with respect to the performance of E. Yet some intend by adaptation not merely 'trait that has been selected for its function' but the far stronger 'trait that has been built for its function'. Hence Gould and Lewontin's worry that the usefulness of a trait in some respect does not entail, nor even make probable, the claim that the trait is an adaptation for that effect: 'male tyrannosaurs may have used their diminutive front legs to titillate female partners, but this will not explain *why* they got so small' (1979, 581).

We can also explain why Lewontin (1984) is concerned about the potentially misleading label 'adaptation'. If we stress parallels between natural selection and natural theology we are likely to think of selection as akin to a 'Blind Watchmaker' – selection as a craftsman shapes organic form to meet environmental problems. But adaptedness is not always produced by selection acting in this 'shaping' mode. One alternative is that members of a population slowly migrate until they find a habitat in which some preexisting, unchanged, trait enhances fitness. This is a natural selection explanation of sorts, for the relational trait of 'being in a better habitat' increases its frequency as the members of this population change their environments. But it is nonetheless misleading to say that the fact of adaptedness indicates selective design in this scenario, and we would be wrong to infer selective shaping, although not selection for, from the functionality of the trait.

There is a second, and more contentious, role for appeals to good engineering design in establishing the existence of adaptation. In the first mode, we know that the trait promotes fitness, and we infer that selection explains this state of affairs. In the second mode, we infer from the structure of the trait both a likely fitness contribution and selection for that contribution. This is the way that Dennett characterises the inference from the fossil remains of *Archaeopteryx* to the joint claims that *Archaeopteryx* was able to fly, and that its wings were selected for flight: 'An analysis of the claw curvature, supplemented by aerodynamic analysis of wing-structure, makes it quite plain that the creature was well-designed for flight' (1995, 233).

It is at this point that the objection that adaptationist hypotheses are 'just-so stories' (Gould and Lewontin 1979) is most likely to rear its head. The problem of just-so stories is a problem of under-determination – it is simply too easy to 'make evolutionary sense' of a trait by showing how it might have contributed to fitness. Con-sider the crest of the *Corythosaurus casuaris*, an example beautifully discussed by Turner (2000). *Corythosaurus* was a 'duckbilled dino-saur' (a hadrosaur), whose fossil remains suggest that its skull bore a curious crest, shaped like a Corinthian soldier's helmet. These crests are hollow, and air inhaled through the nostrils would have passed through the cavities of the crest on the way to the lungs. Was the crest an adaptation? If so, for what? Turner lists a variety of responses that have been tabled at one time or another for the functions of hadrosaur crests: they were weapons; they were anchors for a short trunk; they housed an olfactory system that detected predators. If we focus only on data that relate to the rough structure of the hadrosaur crest, then it seems there are very many design hypotheses that will make sense of the structure. Some will be ruled out – the crest could not have been a snorkel, for there are no holes in the crest that would permit air to be drawn in to the lungs. But many others remain, including the delightful hypothesis that *Corythosaurus*'s crest was more SCUBA than snorkel – a short-term air tank that would have enabled the dinosaur to forage underwater.

It is not only adaptationist hypotheses that face the problem of underdetermination. Maybe hadrosaur crests are not adaptations at all – perhaps a cranial crest is a side effect of selection working on some other aspect of hadrosaur anatomy, with no special function of its own. If we are to assert this we need evidence; otherwise this is a just-so story, too, albeit not an adaptationist one. Nonetheless, the historical nature of some biological hypotheses does present them, as a class, with especially acute problems of underdetermination compared with those of nonhistorical sciences. If we have two incompatible hypotheses about hadrosaur crests, we should try hard to find some data that might discriminate between them. The problem for biology is that sometimes we can say what data we need, but these data are inaccessible. Are hadrosaur crests air tanks, predator detectors, or something else? We could tell quite easily if we could observe hadrosaurs in action, but that avenue is (and always has been) closed off to us.

For the remainder of this section I will follow Turner in discussing the discovery of adaptation in the more general context of *inference to the best explanation*. Within science and without, we often infer that a proposition is true on the grounds that if it were true, it would be the best explanation of our data (Lipton 2004). Such inferences feature throughout the *Origin of Species*, for example. Time and again, Darwin says that his theory should be accepted because it offers a better explanation of diverse sets of facts than do its competitors.

To support inference to the best explanation is not to say merely that once we have a hypothesis that would make our data probable, we should believe that hypothesis. This would entitle us, for example, to infer any combination of past ecological circumstances, no matter how outlandish, just so long as they make likely the sparse structural data that we have to hand relating to fossil remains. For an explanation to be good, hence for it to command our assent, it must meet further constraints, and we can use these further constraints to illuminate some of the ways we might discriminate between alternative adaptive hypotheses, including hypotheses about *Corythosaurus*'s crest.

If an explanatory hypothesis makes the data very probable, we say that the hypothesis has high *likelihood*. Immediately, it will be clear that likelihood does not handle underdetermination problems well. If several incompatible hypotheses all entail the data, then they all have equal likelihoods, so likelihood cannot be used to choose among them. And likelihood certainly is not the only thing we look for in a good explanation (although it is important). Suppose our datum is the fact that Sam has won the National Lottery. The hypothesis that Sam entered the lottery fair and square makes this datum exceptionally improbable. The hypothesis that Sam has a brother who rigged the machines makes our datum far more probable. But we do not assume that every lottery winner is a cheat, even though such hypotheses have far greater likelihoods than the hypotheses that these individuals are legitimate winners. Is there evidence that Sam has a brother? Is there any reason to think his brother has control over the lottery machine? Without further evidence in favour of these assumptions the fact that they constitute an explanatory hypothesis with a high likelihood does not count strongly in favour of the truth of that hypothesis. Similarly, unless we have evidence in favour of the assumptions laid out in some selection hypothesis, the high likelihood of that hypothesis does not count strongly in its favour. The first lesson, then, is that as well as offering adaptive hypotheses that 'make evolutionary sense' of the traits we are interested in, we need to test the assumptions of those hypotheses directly before we put faith in them. Is there any reason to think that hadrosaurs foraged in water, so SCUBA apparatus would have been useful to them? Perhaps the crest is a musical instrument or resonator (a hypothesis much favoured today), but is there reason to think that hadrosaur hearing was good enough for the noises the crest produced to have been audible? Did hadrosaurs live in groups, in which auditory signals were therefore valuable? The answers we give to these questions can help to rule some hypotheses in, and others out.

A second way to bolster our explanatory inferences is to see whether they conform to any general pattern. Suppose we have observed lottery fixing in many cases that resemble Sam's own victory. This counts as further evidence in favour of the claim that Sam's victory was also fixed. Similarly, we can bolster selective hypotheses by looking to see whether the hypothesis of adaptation in question conforms to any general pattern. This is where the so-called comparative method can be powerful. Consider Darwin's example of the bald head of the vulture. An engineering analysis of what a bald head might be good for, coupled with knowledge of the feeding habits of the vulture, could lead us to the hypothesis that the bald head is an adaptation 'for wallowing in putridity' (1964, 197). This, after all, makes evolutionary sense of the trait's appearance. Darwin points out, however, that 'we should be very cautious in drawing any such inference, when we see that the skin on the head of the clean-feeding male turkey is likewise naked' (ibid.).

Suppose we conclude, on the basis of the fact that male turkeys have bald heads, that some common ancestor of turkeys and vultures had a bald head, and that both turkeys and vultures have inherited this trait. That would undermine the claim that the bald head of the vulture has been shaped by selection to enable it to wallow in putridity, providing we think the common ancestor was clean-feeding as the turkey is. But it is consistent with this scenario that the bald head of the vulture contributes to fitness by enabling the vulture to wallow in putridity, and even that there has been maintaining selection for this effect in the recent past that explains the continued presence of the vulture's bald head. So appeals of this kind to the traits of related species in the refutation of adaptation claims makes best sense if one understands adaptation neither in the manner of Reeve and Sherman, nor in the manner of Sober, but as Gould and Vrba understand it – as a trait shaped by selection for its function.

For a third way to strengthen our adaptive inferences, let us return to the lottery case. Suppose we think that Sam cheated in winning the lottery. This claim will lead us to make certain predictions about what else we should observe. One prediction that issues from our hypothesis might be that someone Sam knows will have been in the London area (where the lottery machines are located) on the week-end when the lottery was held. If this prediction turns out to be right, then it provides only the weakest evidence in favour of our hypothesis. On the other hand, we might make a more detailed prediction of the same type: a member of Sam's family will have been lurking in the television studio where the machines are kept at the precise time of the lottery. If CCTV evidence shows this pre-diction to be true, then we will be far more confident of our hypothesis. In general, we should be more confident of an explana-tory hypothesis if its fine-grained predictions turn out true than if its more hand-waving predictions turn out true. Here, then, is a final way, championed in a series of papers by Orzack and Sober (e.g., Orzack and Sober 1994), for us to increase our confidence in the truth of plausible adaptive explanations. We should demand that our hypotheses yield rigorously quantified predictions. Such predictions might include detailed engineering specifications of what structure some trait ought to have, on the assumption that it has a certain function. This, roughly speaking, is Rudwick's (1964) *Paradigm Method* for determining the function of fossils. But Orzack and Sober point out that we can also make quantitative predictions at the level of the population, by specifying the exact frequency of trait distributions we should expect to find under some functional hypothesis.

6.   ADAPTATION AND ADAPTATIONISM

Although I have concentrated for most of this essay on adaptation, I want to close with some words about *adaptationism*. The problems we have looked at so far concern what adaptations are, how we should explain their appearance, and how we can tell when we have found one. The varieties of adaptationism all assert, in one way or another, that adaptation is of special importance for biology. Needless to say, that is an ambiguous claim, and various writers have made efforts to distinguish the varieties of adaptationism, producing estimates from two forms through to seven (e.g., Godfrey-Smith 2001, Lewens 2002, Lewens forthcoming).

It will serve our purposes well enough here to distinguish four types of adaptationism:

> *Empirical adaptationism* – most traits are adaptations. Nature is, in some sense, 'well designed'.
>
> *Methodological adaptationism* – regardless of the actual level of design quality in nature, or the prevalence of adaptations, the best way to investigate nature is to assume that all traits are adaptations.
>
> *Explanatory adaptationism* – the proper business, or the most important or interesting business, of evolutionary biology is the explanation of adaptation.
>
> *Epistemological adaptationism* – biologists have methods that are good enough to establish fairly decisively the truth and falsehood of most hypotheses about adaptation.

I include the fourth in my list of forms of adaptationism only for historical reasons. One of the favourite criticisms of antiadaptationists has been that hypotheses of adaptation are frequently 'just-so stories'. We saw in the last section how evolutionary historical hypotheses of all kinds – not just the hypotheses about adaptation – can be subject to underdetermination problems, and we looked at ways to enrich our inferential practices when trying to overcome these problems. Even so, antiadaptationists have historically been sceptical of the methods by which we test hypotheses of adaptation; conversely, a strand of adaptationism has incorporated a greater optimism about such methods. I now set this fourth form of adaptationism aside.

Let me say a little about the other three kinds of adaptationism, beginning with empirical adaptationism. I phrased this as the view that nature is well designed. But what does that mean? There are several options for how to explain this. At the strong end of the spectrum we might take adaptationism to be the view that all traits are the best ones possible. Even here we run into problems (what is the meaning of 'possible'?), but we are likely to dismiss this version of the hypothesis as false, even obviously so. After all, much of the evidence against intelligent design uses the imperfection of organic traits for their apparent purposes as evidence. At the weak end of the spectrum is the view that natural selection has been involved in some way in the history of most traits. This threatens to turn out trivially true. An effort to give an interesting reading to the hypothesis of adaptationism comes from Sober and Orzack: 'Natural selection has been the only important cause of most of the phenotypic traits found in most species' (Sober 1998, 72). What this means is that the fittest available phenotypes are always the only ones present in populations. Take a population of moths, in which some are well camouflaged, others are poorly camouflaged, and none has (or has ever had) chemical secretions that kill birds instantly. If the poorly camouflaged moths end up the only ones in the population, then this result counts against adaptationism, for a better available variant did not reach fixation. If moths with deadly chemicals fail to reach fixation, then this does not count against adaptationism, for such moths were never available to be selected from.

On Orzack and Sober's view, the adaptationist hypothesis could be true even if the range of available variation for selection to act on turns out to be very highly constrained. Indeed, adaptationism would be true at a possible world where the laws of nature dictate that only two kinds of entity exist, both make copies of themselves, and the type that is more fecund replaces the less fecund type. Yet there may be nothing much like good design at this world – no 'organs of extreme perfection'. This makes clear the difficulty of finding a reading of the adaptationist hypothesis that satisfies our intuition that if adaptationism is true, then natural selection is a powerful force. Certainly in the dull world I just described selection seems to have very little power.

Adaptationism is sometimes better understood as a heuristic rather than an empirical hypothesis. It is a recommendation for how to go about investigating nature in a fruitful manner. Once again, we could catalogue many variant forms of methodological adaptationism ranging from the boringly sensible advice that sometimes it is useful to test hypotheses of the form 'X is an adaptation for E', through to the implausible insistence that nothing that is of value in biology can be discovered unless one tests hypothesis of the form 'X is an adaptation for E'. What is important to note is that versions of methodological adaptationism can be useful even if one thinks that adaptation is not ubiquitous. The adaptationist heuristic is useful when the failure to establish an adaptationist hypothesis suggests to us that selection is not responsible for the aspect of organic form we are investigating.

Explanatory adaptationism tells us that, in some sense or another, the proper business of evolutionary biology is the study of adaptations (for this reason I have elsewhere called it 'Disciplinary Adaptationism' [Lewens forthcoming], but for the sake of clarity I will stick with Godfrey-Smith's terms here). John Maynard Smith is an adaptationist of this kind, as is Richard Dawkins, who writes that 'large quantities of evolutionary change may be non-adaptive, in which case these alternative theories may well be important in parts of evolution, but only in the boring parts' (Dawkins 1986, 303).

Sterelny and Griffiths (1999, 228) worry that there may be some kind of vacuity in explanatory adaptationism. The problem stems, once again, from the widely accepted definition of adaptation as 'trait selected for some function'. If explanatory adaptationism includes the claim that adaptation is always explained by selection, then this aspect of the position is, indeed, trivial under this definition. But this does not make the general claim that biologists ought to busy themselves with adaptations vacuous. The standard definition of adaptation leaves open a variety of questions that one might try to answer, none of which is trivial, and not all of which are about adaptation. Which are the adaptations? Which traits are adaptations for what? Which traits are functionless? Which organisms are descended from which others? An exhortation to focus on questions of the first two types is not vacuous.

The fact that evolutionary biology contains questions that are not about adaptation saves explanatory adaptationism from vacuity, but this only generates new problems for the position. How could we argue against someone who says that an evolutionary biologist's proper business is to determine genealogical relationships among species, regardless of which traits might be adaptations for what? One might reply by saying that explaining how selection shapes traits for their functions is important because it is necessary in the struggle against creationism. These 'well-designed' traits are just the ones that, absent a good selection explanation, will be used as evidence by intelligent design theorists. But first, this would only establish adaptation as the most important part of biology if we agreed that intelligent design was so threatening that its defeat should be the discipline's primary goal. And even if we were to inflate intelligent design in this way, it still would not undermine the importance of tracing lines of descent between species. This project, too, is centrally important in undermining creationist arguments.

We arrive back at the beginning, then, with natural theology. Explanatory adaptationism is merely a statement of explanatory interests, interests that we should not feel compelled to share, but whose salience is derived from Anglo-American biology's roots in natural theology. This conclusion is ironic: far from expressing enmity between modern biology and natural theology, explanatory adaptationism is testimony to the fellowship between the two traditions.

2   Population Genetics

1.   INTRODUCTION

Population genetics is a subfield of evolutionary biology that aims to represent mathematically the changes in the genetic variation of populations (specifically, sexually reproducing populations with Mendelian heredity) over time. The mathematical models of population genetics provide a theoretical basis for experimental studies of laboratory populations and studies of natural populations. Our primary focus in this essay is on population genetics theory itself, rather than its applications, although towards the end of the essay we give some discussion of the latter.

Population genetics attempts to measure the influence of the causes of evolution, namely, mutation, migration, natural selection, and random genetic drift,[1] by understanding the way those causes change the genetics of populations. But how does it accomplish this goal? We begin in Section 2 with a brief historical outline of the origins of population genetics. In Section 3, we sketch the model theoretic structure of population genetics, providing the flavor of the ways in which population genetics theory might be understood as incorporating causes. In Sections 4 and 5 we discuss two specific problems concerning the relationship between population genetics and evolutionary causes, namely, the problem of conceptually distinguishing natural selection from random genetic drift, and the problem of interpreting fitness. In Section 6, we briefly discuss the methodology and key epistemological problems faced by population geneticists in uncovering the causes of evolution. Section 7 of the essay contains concluding remarks.

We are focusing on the issue of causality in population genetics because we take this issue to be at the core of many of the contemporary philosophical debates in population genetics. However, it should be noted that population genetics raises other philosophical issues that this essay will not address. To give two examples, there has been much debate over the questions of whether population genetics describes any scientific laws (see, e.g., Ruse 1977; Beatty 1995; Waters 1998), and whether the models of population genetics have been reconciled with the "semantic conception" of theories (see, e.g., Beatty 1981, Lloyd 1988).

2. ORIGINS OF POPULATION GENETICS AND THE EVOLUTIONARY SYNTHESIS

In the early part of the twentieth century, Gregor Mendel's experimental work on pea plants was commonly perceived to be at odds with Darwinian natural selection; the former, it was argued, was evidence for discontinuous evolution (involving large changes from parent to offspring), whereas the latter required continuous evolution (involving small gradual changes from parent to offspring). However, no later than 1932, the field of theoretical population genetics emerged as a reconciliation between Mendelism and Darwinism. Most biologists at the time accepted the fact of evolution, or Darwin's "descent with modification," but Darwin's idea of natural selection as a cause of that modification was controversial. Indeed, lacking were any generally accepted accounts of genetic variation in populations (is the variation continuous or discontinuous?) or evolutionary change (is change gradual or saltational?), and an understanding of the appropriate use of statistical methods for studying these. The emergence of theoretical population genetics, which addressed all of these issues, is typically associated with the work of R. A. Fisher, Sewall Wright, and J. B. S. Haldane. The foundational works that ushered in theoretical population genetics are Fisher's (1930) *The Genetical Theory of Natural Selection*, Wright's (1931) "Evolution in Mendelian Populations," and Haldane's (1932) *The Causes of Evolution*. What follows is a brief summary of the views of these three biologists.

Fisher (1890–1962) was an English biologist trained in mathematics at Cambridge University. In addition to his unsurpassed contributions to statistics, his initial contributions to evolutionary biology predate those of the other two theorists. Fisher's aim in

The Genetical Theory of Natural Selection was formally and mathematically to demonstrate how the "vague" concept of natural selection (as it was then considered) could possibly work. And he does this by considering the theory of natural selection against the principles of Mendelian inheritance on an analogy with the mathematical techniques of statistical mechanics. Fisher's view is typically understood as follows: *Evolution is driven primarily by natural selection, or mass selection, at low levels acting on the average effects of single allele changes (of weak effect) at single loci independent of all other loci.*

Wright (1895–1988) was an American biologist trained by William E. Castle at Harvard University in physiological genetics. Prior to his groundbreaking research in evolutionary theory, which he carried out at the Universities of Chicago and Wisconsin, Madison, he worked as a staff scientist for the U.S. Department of Agriculture. Wright's views changed greatly over the years, but what Wright was looking for in his 1931 essay "Evolution in Mendelian Populations" were the ideal conditions for evolution to occur, given specific assumptions about the relationship between Mendelian heredity and the adaptive value of gene complexes. Ideal conditions, for Wright, are those conditions that produce the fastest rate of evolution to the highest "adaptive peak." Wright believed that these conditions required that populations be subdivided and semi-isolated, and that selection, along with random genetic drift and migration, operated in a "shifting balance" of phases. Wright's Shifting Balance Theory can be summarized: Evolution proceeds via a shifting balance process through three phases: Phase I, *Random genetic drift causes subpopulations semi-isolated within the global population to lose fitness;* Phase II, *Mass selection on complex genetic interaction systems raises the fitness of those subpopulations;* Phase III, *Interdemic selection then raises the fitness of the large or global population.*

Haldane (1892–1964) was an English biologist trained in mathematics, classics, and philosophy at Oxford University. Haldane began his work on evolutionary problems in 1922 with theoretical or mathematical inquiries into the consequences of natural selection in Mendelian populations. Haldane's 1932 *The Causes of Evolution* is an original and important contribution to the origins of theoretical population genetics with its critical analysis of extant empirical work against the background of his own and Fisher's and Wright's theoretical work. The appendix to *Causes* is a critical discussion of his own and Fisher's and Wright's achievements regarding the reconciliation of Mendelism with Darwinism. Haldane agrees with portions of both Fisher's and Wright's evolutionary theories. For instance, Haldane supported Wright's emphasis on epistasis and migration, and he supported Fisher's view on the importance of natural selection over random genetic drift. Haldane further thought that natural selection would proceed rapidly in large populations, an idea neither Fisher nor Wright believed. It is easy and fairly common to view Haldane as a popularizer of population genetics among biologists generally. But this is a mistake: Much of Haldane's work in the 1920s adumbrates ideas found in the work of Fisher and Wright.

The mathematical reconciliation of Mendelism with Darwinism achieved by Fisher, Wright, and Haldane began the historical period of evolutionary biology called the "Evolutionary Synthesis" (also known as the "Modern Synthesis"). Their theoretical achievements combined with early experimental work by such luminaries as Theodosius Dobzhansky (1937) set the stage for the integration of previously divergent fields such as paleontology, zoology, botany, systematics, and genetics. To be sure, there was considerable disagreement among the architects of the Synthesis. Fisher and Wright in particular were engaged in an initially friendly controversy that rapidly became heated, from 1929 until Fisher's death in 1962, over how to interpret their mathematical theories. By the 1950s, as Stephen Jay Gould (1983) points out, the Synthesis would "harden," emphasizing natural selection as the most significant evolutionary cause.

To a large extent, the applications of contemporary population genetics are deeply rooted in the achievements of the period between 1918 and 1960. Indeed, there is, for instance, persistent controversy over the relative significance of Fisher's and Wright's population genetics theories (Skipper 2002). Since 1960, application of molecular techniques to evolutionary problems has led to revisions in the interpretation of some of the basic assumptions of population genetics theory as well as of evolution at the molecular level. Moreover, advances in microbiology and developmental biology have led to challenges to the explanatory scope of population genetics.[2]

3.   CONTEMPORARY POPULATION GENETICS

The models of contemporary population genetics exemplify the generalized reconciliation between Mendelism and Darwinism using the now well-entrenched statistical methods introduced by the architects of the field. As such, population genetics defines "evolution" as change in gene frequencies, or more strictly, any change in the frequency of alleles within a population from one generation to the next. Differently put, population genetics aims to account for the dynamics of genetic variation in populations. And it does so by attempting to uncover the patterns of those dynamics via the causes of evolution, namely, mutation, migration, natural selection, and random genetic drift. Our goal in this section is to provide the reader with a general sense of what the models of contempory population genetics are like; consequently, our discussion must take a slight technical turn.

Population genetics begins its task by specifying the conditions under which gene frequencies remain *unchanged* from one generation to the next: the conditions under which evolution is *not* occurring. These conditions are captured by the foundational principle of population genetics called the "Hardy-Weinberg Principle." The Principle begins with a set of assumptions about the genetic system, mating system, and population structure: Assume a randomly interbreeding, large (mathematically, infinite) population of diploid organisms with one genetic locus and two alleles. In fact, these assumptions are fundamental to most standard presentations of population genetics. Given these basic assumptions, the Hardy-Weinberg Principle states that in the absence of evolutionary causes, that is, mutation, migration, natural selection, and random genetic drift, the gene frequencies of the population will remain unchanged from one generation to the next; the population will be in "Hardy-Weinberg equilibrium."[3] In Hardy-Weinberg equilibrium, when the two allele frequencies are equal, the distribution of genotype frequencies will map on to the Mendelian 3:1 phenotypic ratio.

The mathematical relation between the allele frequencies and the genotype frequencies is

$$AA : p^2 \qquad Aa : 2pq \qquad aa : q^2$$

where p^2, $2pq$, and q^2 are the frequencies of the genotypes AA, Aa, and aa in zygotes of any generation; p and q are the allele frequencies of A and a in gametes of the previous generation; and $p + q = 1$. The chance that all possible combinations of alleles will occur randomly is $(p + q)^2 = 1$ so that we arrive at the famous equation describing the Hardy-Weinberg Principle, $p^2 + 2pq + q^2 = 1$. As long as the basic assumptions hold in the absence of the evolutionary causes, the allele frequencies p and q will remain constant and genotype frequencies will be in accord with the equation; in other words, there is no evolutionary change in a population in Hardy-Weinberg equilibrium.

Understanding evolution as change in gene frequencies, then, is understanding the ways in which populations deviate from Hardy-Weinberg equilibrium. Population geneticists may begin with assumptions about the genetic system, mating system, and population structure, and then proceed to modify the mathematical representation of the Hardy-Weinberg Principle by adding parameters for mutation, migration, natural selection, and random genetic drift.

Consider a simple case. First, consider that the preceding assumptions concerning genetic system, mating system, and population structure hold. Second, assume we want to understand how natural selection may cause a population to deviate from Hardy-Weinberg equilibrium; specifically, we want to understand a simple case of viability selection. The frequency of the genotypes in our population before selection is given by the Hardy-Weinberg equilibrium equation, $p^2 + 2pq + q^2 = 1$. Since we want to understand how natural selection causes a deviation from that equilibrium, we modify the equation to include a parameter that captures the "selective pressure" on the genotypes, or in other words, the probability of survivorship of the genotype. This parameter is called "fitness" (w) and is usually measured relatively so that the fitness of one genotype is expressed relative to another genotype; the genotype that is the standard of comparison is assigned a fitness value of 1.

Given the fitness parameter, if the frequencies of the genotypes AA, Aa, aa before selection are p^2, $2pq$, and q^2, respectively, then the frequencies of the genotypes after selection are $p^2 w_{AA}$, $2pq w_{Aa}$, and $q^2 w_{aa}$ by incorporating the fitnesses of the genotypes in the computation of their postselection frequencies. Indeed, the sum of the frequencies of the genotypes after selection equals the average fitness for the population, that is, $p^2 w_{AA} + 2pqw_{Aa} + q^2 w_{aa} = \bar{w}$. And we have as the mathematical relation between the allele frequencies and the genotype frequencies

$$AA : \frac{p^2 w_{AA}}{\bar{w}} \qquad Aa : \frac{2pqw_{Aa}}{\bar{w}} \qquad aa : \frac{q^2 w_{aa}}{\bar{w}}$$

We may then compute the postselection frequencies of A and a, which are designated as p' and q', respectively:

$$p' = \frac{p^2 w_{AA} + pqw_{Aa}}{\bar{w}} \qquad q' = \frac{pqw_{Aa} + q^2 w_{aa}}{\bar{w}}$$

From these equations, the outcome of selection can be deduced: For instance, if $p > p'$, where $p = [(p^2 + 2pq) / 2]$, then selection is acting to decrease the frequency of allele A in the next generation.

The previous example is not intended to provide a primer on the statistical methods of population genetics let alone an understanding of them. Rather, it is intended to provide the flavor of the way in which evolution as change in gene frequencies is approached using a version of those tools: Starting from a mathematical statement about the distribution of allele frequencies in the absence of evolutionary causes, one may understand the ways in which those causes change that distribution by modifying the mathematical statement with parameters measuring the influence of those causes. Indeed, mutation, migration, multiple modes of selection, and random genetic drift are treated in more or less the same ways, that is, by modifying and extending the basic mathematical statement of the Hardy-Weinberg Principle. Moreover, the basic tools roughly introduced here can be expanded to cover evolution for alternative assumptions regarding the genetic system, mating system, and population structure. Further, the theoretical apparatus can be made more powerful and expressive by introducing models that allow population geneticists to represent the probabilities of a range of possible results, rather than simply predicting a single result as in the model described. (Biologists call models that predict one specific value *deterministic* models; this should not be confused with the Laplacean or philosophical sense of determinism, which generally refers to a property of the world rather than a property of a model. Models that provide a probability distribution for a range of results are called *stochastic* models. Deterministic and stochastic models will be discussed further in the next section.)

4.   POPULATION GENETICS THEORY AND EVOLUTIONARY CAUSES

As Michael Ruse has documented, Charles Darwin construed natural selection as a cause (or, more precisely, a *vera causa*) in order to conform to the predominant philosophies of science of his time (Ruse 1979, chap. 7). Contemporary population genetics, as we have seen, incorporates not just natural selection, but also mutation, migration, and random genetic drift. Is natural selection still construed as a cause? And are the other phenomena causes as well?

As we have seen, population geneticists define "evolution" as "change in gene frequencies." For selection, drift, mutation, and migration to be causes of evolution, they must be able to bring about such changes – at least theoretically, if not in reality as well. Unfortunately, ever since David Hume's skeptical challenge to cause-effect relationships, philosophers have been unable to agree on a definition of "cause," or even whether we can legitimately infer the existence of causes at all. Assuming, however, that there are such things as causes and that we can develop a satisfactory account of causation, it seems fair to say that in the context of population genetics, mutation, migration, selection, and drift are causes of evolution. For example, it is easy to see how mutation within a population will lead to a change in gene frequencies within that population. Similarly, migration into or away from a population also yields a change in gene frequencies in the population. (Selection and drift will be discussed further later.)

The implications of the population geneticist's construal of evolution are threefold. First, the commonly held notion that evolution and natural selection are the same is false. Second, with four possible causes to consider, the equations can become quite complex. This is because, unlike in the simplified previous scenarios, these causes can act in combination, as is implied by Richard Lewontin's suggestion that

population genetic theory is a descriptive theory that provides the mapping of causal processes as genetic outcomes. It says, '*if* mutation rates are such and such, *if* the mating pattern is such a one, *if* there are five genes affecting the character with the following norms of reaction, *then* the trajectory of the population in time, or the equilibrium state, or the steady state distribution of gene frequencies will be such and such.' (Lewontin 1985, 10)[4]

Third, even though the causes can act in concert, they are considered to be *distinct* causes. The most difficult case of distinguishing between the causes of evolution is that encountered in distinguishing selection from drift. Thus, we pay special attention to that case here. However, our discussion here will of necessity be relatively brief; for further discussion, see Millstein (2002) and earlier works (Beatty 1984, Hodge 1987).

The problem in distinguishing selection from drift arises at least in part as a result of ambiguities in the models of population genetics. This will require an exploration of three different aspects of population genetics. We will argue that it is a mistake to characterize selection and drift in terms of the first two of these aspects; the proper characterization of selection and drift derives from the third aspect.

Consider first the model of natural selection discussed in the previous section. Although the fitness value (the w in the equations) is generally understood to be a probability, namely, the probability of survivorship of the genotype, the equations themselves will not generate a range of possible future genotype frequencies. Rather, they will generate one specific genotype frequency for each of the genotypes. That is, the model of natural selection is "deterministic," in the sense described previously.

On the other hand, according to the standard presentation, models become stochastic and generate a range of possible genotype frequencies, when – and, according to some authors, only when (see, e.g., Brandon 2005; but cf. Millstein 2005) – the assumption of infinite population size is relaxed. To understand this, imagine an urn filled with red and green balls where balls are sampled without respect to color. If a large sample of balls were taken, we would expect the frequencies of colored balls in the sample to be very close to the frequencies in the urn. On the other hand, if we only take a small sample of colored balls, our sample may very well have different proportions of colored balls than the urn does. In the same way, if our population (the "sample" that is taken with each generation) is infinite, then we expect (with a very high probability) that descendant generations will have genotype frequencies very close to those of the parent generations. However, if the population size is finite, then the sample may not be representative; that is, the genotype frequencies of the descendant generation may diverge considerably from that of the parents. But in which direction will they diverge? For example, will the frequency of AAs increase or decrease? And by how much? We cannot say for certain; we can only predict the *probability* of various divergences, just as we would not be able to say for certain whether a small sample of balls would have a smaller or greater (or equal) percentage of green balls as compared to the urn, only the probabilities of drawing various numbers of green balls. So, the introduction of finite population size yields a stochastic model, that is, a model that generates a probability distribution of future outcomes.

If one were to try to understand what selection and drift are from a literal interpretation of these models, one might be tempted to conclude that natural selection is the achievement of the predictions of the models. That is, one might be tempted to conclude that natural selection occurs when genotype frequencies are exactly those that the fitness values lead us to expect. One might be further tempted to label the introduction of finite population size into the models as the introduction of drift; again, reading literally from the models, drift then becomes any deviation from the expectations of selection. On this view, selection is deterministic, but drift is stochastic (in the senses described). This is, in fact, one way of distinguishing selection from drift, but, as we shall argue later, it is not a very good way. The point to notice now is that on this interpretation of population genetics, selection and drift are distinguished by the *outcomes* that are produced (agreement with fitness predictions and divergence from fitness differences, respectively).

Now let us consider a second aspect of population genetics. In discussing whether selection or drift predominates in a particular population, biologists will sometimes rely on the following "rule of thumb": natural selection has prevailed if $4N_e s \gg 1$, whereas random drift has prevailed if $4N_e s \ll 1$, where N_e is the effective population size (i.e., the number of individuals in a population who contribute offspring to the next generation) and s is the selection coefficient (Futuyma 1986, 173). In other words, when the effective population is large and/or the selection coefficient is high, selection tends to prevail. When the effective population size is small and/or the selection coefficient is low, random drift tends to prevail.

If you were to take this second aspect of population genetics on its face, you probably would come to a very different conclusion than before. Namely, you would conclude that natural selection and random drift are not entirely distinct; instead, it would appear that selection and drift are on a continuum. With a low selection coefficient and a small population size, you have drift, but increase the selection coefficient and/or the population size and eventually you will have selection. With an intermediate selection coefficient and an intermediate population size, however, it is unclear on this view whether the population is undergoing selection or drift. Although it might not appear so, the conclusion that there is a continuum between selection and drift is also reached by a consideration of outcomes. The question is, which contributes more to the genotype frequency produced – the achievement of fitness expectations or the deviation from them? The idea is that when there are a low selection coefficient and small population size, the effects of drift (the effects of sampling) swamp the effects of selection, but when there are a high selection coefficient and large population size, the effects of selection swamp the effects of drift.

Thus, the literal reading of these two aspects of population genetics has led to a conundrum; two different aspects of the models of population genetics yield different conclusions about whether drift and selection are distinct concepts. On the first view, they *are* distinct concepts; selection is the achievement of fitness expectations and drift is any deviation from that expectation. On the second view, the two concepts are *not* distinct; rather, there is a continuum between drift and selection.

There is, however, a third alternative, which takes an altogether different approach. This is the approach that one of us has endorsed (e.g., Millstein 2002). Rather than literally interpreting the models in isolation, we derive our concepts from *phenomena* that the models are intended to represent.

Interestingly enough, the presentation of the phenomenon of natural selection in population genetics textbooks generally does not deviate much from Darwin's own presentation. In order for selection to occur in a population, 1) there must be heritable variation among individuals, 2) the variation must confer a differential ability to survive and reproduce in the given environment, and 3) more offspring are produced than can survive in the given environment (i.e., there is, to use Darwin's phrase, a "struggle for existence"). But from these conditions Darwin did not conclude, as the selection equations discussed seem to imply, that organisms having advantageous variations would necessarily be the ones with greater reproductive success. Instead, Darwin claimed, "if variations useful to any organic being do occur, assuredly individuals thus characterized will have the *best chance* of being preserved in the struggle for life" ([1859] 1964, 127; italics added; see also, e.g., pp. 61, 81). In other words, we expect that the fittest organisms will be the most successful, but that does not always happen; perhaps, for example, the fittest organism fails to find food or is crushed by a falling boulder. In fact, unless one were whiggishly to claim that Darwin, in acknowledging that the expected may not happen, had a notion of drift, one is left with the conclusion that the phenomenon that Darwin called natural selection – arguably, the same phenomenon that the models are attempting to represent – is not "deterministic" at all, but rather "stochastic."

What phenomena, then, are the drift models purportedly representing? There are at least seven different kinds of drift phenomena (Millstein 2002). Here, we mention only two: indiscriminate gamete sampling and indiscriminate parent sampling (see Beatty 1984). Gamete sampling is the process in which some – but not all – gametes are successfully united in zygotes, whereas parent sampling is the process in which some – but not all – organisms successfully reproduce and become parents. But there is a tempting rejoinder, which goes something like the following: "But *why* is it that some gametes become joined together in zygotes and others do not? Perhaps the successful gametes are fitter in some way; perhaps the sperm swim faster or the eggs are more robust. And *why* is it that some organisms survive to become parents when others do not? Again, perhaps they are just fitter." This rejoinder puts its finger on the difference between the phenomenon of selection and the phenomenon of drift. *If* some gametes *were* fitter, or *if* the individuals *were* fitter, then we would *not* be describing drift at all; we would be describing selection (i.e., *discriminate* sampling). The point behind discussing drift is that there may *not* be any fitness differences (although there may be physical differences that do not confer any fitness benefits), and yet some gametes or individuals may still be more successful than others. To use Hartl and Clark's example, imagine shellfish that "produce vast numbers of pelagic larvae that drift about in the sea" (1989, 70). Although Hartl and Clark do not elaborate, the image is of virtually identical larvae, subject to the vagaries of tides and predators (i.e., *indiscriminate* sampling).

Thus, examination of the phenomena, prior to any representation by the models, yields a third way of understanding the difference between selection and drift. Selection, on this view, is a discriminate sampling process in which physical differences between biological entities (gametes, organisms, etc.) are causally *relevant* to differences in reproductive success. Drift, on the other hand, is an indiscriminate sampling process in which physical differences between biological entities are causally *irrelevant* to differences in reproductive success.[5] Note that unlike in the first two attempts to spell out the difference between selection and drift, the distinction is made by identifying selection and drift as different types of *processes* rather than different *outcomes*. In other words, selection and drift are different kinds of causal processes. In using the term "causal process," we mean to suggest that selection and drift are *physical* processes occurring in nature and in the laboratory; furthermore, they are to be distinguished from pseudoprocesses such as the movement of a shadow (Salmon 1984). Finally, in using the term "causal process" to describe selection and drift, we mean to suggest that selection and drift consist of a series of states occurring *through time*, where the states are generated causally.[6] The *outcomes* of these processes, on the other hand, refer to one state (e.g., the genotype frequencies of a population) at a particular point in time.

Not only are selection and drift different kinds of causal processes, they are different kinds of causes, both of which can lead to evolution. Considering selection first, if the individuals whose variations confer on them a greater ability to survive and reproduce do in fact reproduce in greater numbers than individuals who lack these variations, then the gene frequencies of the second generation have changed from those of the preceding generation. Now there exists a greater proportion of individuals with "advantageous" variations; evolution has occurred. Natural selection has caused evolution. But if there is a change in the proportions of types from one generation to the next, but that change is *not* due to physical differences between individuals, then it is drift that has caused evolution. Each is a different cause because each is a different kind of causal process.

There are at least three advantages to this view over the other two.

First, we think it is a mistake to interpret models literally. The models of population genetics are highly idealized metamathematical structures that, at best, are understood as bearing a similarity relationship to the real world systems they describe. And determining precisely how to understand the extent of similarity between the models and the real world is no easy task (Wimsatt 1980a). At any rate, the models were developed subsequent to understanding the phenomena they are trying to capture. It seems backward, therefore, to try to understand the phenomena via the models.

Second, confusion arises in distinguishing selection and drift in large part because population geneticists sometimes speak of selection and drift as causal *processes* (as in the quote from Lewontin), yet at other times they speak of selection and drift as *outcomes*, or effects (thus, e.g., drift is sometimes referred to as the "Sewall Wright Effect").[7] A moment's reflection will show, however, that population geneticists cannot have it both ways. If, for example, selection is identified with its outcomes, then selection occurs when organisms having a greater ability to reproduce as compared to their conspecifics do in fact enjoy greater reproductive success. However, this is just evolution, that is, a change in gene frequencies from one generation to the next. It would not make sense for selection in this sense to be a *cause* of evolution; selection, considered as an outcome, is one *form* of evolution. But, as we discussed earlier, biologists commonly construe selection as a cause of evolution. This makes sense on the view of selection as a process, but not on the view of selection as an outcome.

And third, further confusion arises because the outcomes of the different processes often cannot be distinguished. To see this, first consider a population in which physical differences between organisms do *not* confer any differences in survival or reproductive ability (a population undergoing drift, on our account), so that the relative values of different types may fluctuate from generation to generation. Now, consider a second population, in which physical differences between organisms *do* confer differences in survival and reproductive ability (a population undergoing selection, on our account). Suppose that the environment of the second population is fluctuating. Because of the fluctuating environment, different types may be favored in different generations, producing a fluctuation of types over the generations that produces a pattern that is indistinguishable from that of drift. If selection and drift are understood purely as outcomes, then either both the populations are undergoing drift or both the populations are undergoing selection. We are skeptical that any biologist would, when presented with this scenario, actually take either of these positions, because the two populations are biologically very different. Characterizing drift and selection as processes instead of outcomes captures, rather than glosses over, that difference.

We should emphasize that the claim here is that the *concepts* of natural selection and random genetic drift can be distinguished from one another, and that that distinction should be based on the kinds of processes they are, and not on the kinds of outcomes, as a literal interpretation of the models would suggest. However, that is not to suggest that selection and drift can be easily distinguished *empirically*. That, unfortunately, is a much more complicated problem, which we present in a more general fashion in the following.

5.   CAUSES, BUT NOT ALL OF THE CAUSES

Even though population genetics models the causes of evolution, it is not clear that population genetics tells a complete causal story of evolution. That is, the question arises as to whether there are causes involved in the process of evolution that are not captured by the models of population genetics. Here we will focus on just one area: the concept of fitness. There is an extensive body of literature on the concept of fitness, especially on the propensity interpretation of fitness (Brandon 1978, Mills and Beatty 1979). However, although we acknowledge our intellectual debt to the propensity interpretation, we do not intend our discussion here to be a defense of this or any other interpretation of fitness. Rather, in this section we seek only to explore some issues of causality that the concept of fitness raises.

Earlier, we argued that natural selection is a cause of evolution.[8] Yet natural selection is a causal process in two senses: it is itself a cause (of evolution), and it is made up of causes. It is to this latter sense, the causality *within* the process, that we now turn.

Prima facie, it makes sense to invoke fitness as a causal concept in a colloquial description of natural selection. After all, what makes one organism fitter than another organism in a given environment are its physical characteristics. And those physical characteristics cause the organism to have superior survival and reproductive success in the given environment. That is, the organism's superior fitness, under auspicious circumstances, causes its superior reproductive success. The identification of "fitness" as a cause of reproductive success seems so trivial as to be tautological.

However, as Elliott Sober (1984) has pointed out, the situation is not as simple as the colloquial story makes it appear. After all, in population genetics, fitness (the w in the equations described) does not represent just one physical trait. That is, the fitness of a finch is not just the shape and size of its beak, not just its ability to avoid predators and disease, and not just its ability to find a mate and reproduce; its fitness is *all* of those things together. Sober argues that whereas any of those elements individually may cause or prevent reproductive success, they generally do not *all* cause or prevent reproductive success in any given instance. Thus, according to Sober, fitness is "causally inert."

It should be noted, however, that population genetics models typically subdivide fitness traits into two broad components: an organism's ability to survive (viability) and an organism's ability to reproduce (fecundity). Yet this gross distinction does not address the different kinds of abilities that fitness encompasses. Consider fecundity. In the models of population genetics, fecundity encompasses at least three different kinds of abilities: 1) an organism's ability to produce various numbers of offspring, 2) an organism's ability to attract mates, and 3) an organism's ability to fight for potential mates.

For Darwin, on the other hand, the latter two abilities were in a distinct category. In fact, Darwin considered these traits to be the basis of *sexual selection*, which for Darwin was a different type of cause from natural selection. This is because natural selection, which would involve viability as well as fecundity in the first sense, would tend to produce organisms that are adapted to their environments: finches with beaks of a certain shape and size, for example. Sexual selection, on the other hand, would primarily involve fecundity in the second or third senses, which would mean that the traits that were produced would not tend to be those that were adaptive.

Instead, we find sexual selection yielding traits like the inefficient-but-beautiful tail of a peacock (as a result of the second sense of fecundity) or antlers of a male deer (as a result of the third sense of fecundity). If sexual selection and natural selection truly are different kinds of causes, as Darwin thought, then they are causes that population genetics models do not distinguish between (in the sense that they are not treated any differently – although of course sexual selection is explicitly discussed in most population genetics textbooks), and this is in part a result of not distinguishing among different types of fecundity.

Further complicating matters is the fact that, as discussed, population genetics uses *relative* fitness. That is, it is not an individual's ability to survive and reproduce that appears in the equation, but rather, its – or, more precisely, its genotype's – ability to survive and reproduce as compared to the abilities of others in the population.[9] Of course, relative fitnesses are calculated from genotype fitnesses. But this is not simple computational convenience; the relative fitnesses are what really matter for the evolution of the population. If there are no fitness differences among genotypes, there will be no selection, and a population in which one genotype is twice as fit as its conspecific will evolve differently than one in which it is three times as fit. And yet, is relative fitness a causal concept? If it is, it does not seem to be a causal concept that adheres to any one individual. So, interpreting the causality of relative fitness is challenging.

Thus, it would appear either that population genetics fails to capture fully the actual causes that operate in a population undergoing selection, or that biologists and philosophers have yet to provide an adequate interpretation of fitness. Given the numerous alternatives in the literature, the latter would seem to be the consensus view.

6.   HOW DO WE FIND THE CAUSES?

Thus far we have interpreted population genetics as a causal theory from the point of view of the theory, its structure. Our view may be expanded by considering how population genetics theories are related to real populations of organisms. What we want to explore, then, is how population geneticists credential empirical causal claims made about their models.

Elisabeth Lloyd (1988) has constructed a simple, broad framework for the confirmation of evolutionary and ecological models. On her view, empirical claims, or hypotheses, that some model is similar to some natural system are confirmed by way of (C1) fit between model and data, (C2) independent support for aspects of the model, and (C3) variety of evidence. Fit between a model and data is just evidence that demonstrates a matching between the model and the data. Since population genetics theories are embedded with numerous assumptions, for example, that mating is random or that populations are large, any independent empirical support for those assumptions will increase the confirmational standing of a model. Most important is that there are a variety of types of support for a model, that is, a variety of instances of fit and a variety of instances of independent support for any assumptions. On Lloyd's view, standard statistical techniques for analyzing data, common across the biosciences, are used to analyze (C1)–(C3).

One of us (Skipper 2004) has argued that a constellation of experimental strategies forms the basis of Lloyd's confirmation framework. The idea here is that fit, independent support, and variety of evidence are all driven by experimental methodology. Thus, for instance, fit between model and data is at bottom driven by the practical procedures and techniques used to collect the data. There are a number of strategies, including (E1) experimental checks and calibration, (E2) reproducing artifacts known to be present, (E3) intervention, (E4) independent confirmation using different experiments, (E5) elimination of error, (E6) using the results to argue for their validity, (E7) using an independently well-corroborated theory of the phenomenon to explain the results, (E8) using an apparatus based on a well-corroborated theory, and (E9) using statistical arguments (cf. Rudge 1998). And they are used to justify experimental claims across the three main classes of experiment, that is, natural experiments, or observations of evolution in action, field experiments, controlled manipulations of populations in the wild, and laboratory experiments, highly controlled manipulations of populations in the laboratory (Diamond 1986).[10]

As an illustration, consider the famous case of the Scarlet Tiger moth, *Panaxia dominula*, perhaps (at more than sixty years) the longest-running field study in ecological genetics, first studied by R. A. Fisher and the famed ecological geneticist E. B. Ford (Fisher and

Ford 1947). Fisher and Ford carried out a field experiment using the novel (at the time) capture and release protocol to determine whether natural selection or random genetic drift caused the fluctuations in the *medionigra* gene of the moth, responsible for a specific wing coloration phenotype. By capturing moths from an Oxfordshire field, marking them with a dab of paint, and recapturing them, Fisher and Ford could, over time, track the fluctuations in the *medionigra* gene via its phenotype, a particular wing coloration pattern. From data collected between 1939 and 1946, Fisher and Ford performed statistical analyses (E9) that supported a fit with their selectionist model (C1) over a random genetic drift model. Note that Fisher and Ford inferred that selection controlled the fate of the *medionigra* gene and not that they had direct observational evidence of selection acting on the moths with this phenotype. Their statistical argument (E9) was that the fluctuations in the gene from year to year were too great to be due to drift and, so they must have been caused by selection.

Over the last 10–15 years, further field and laboratory experiments have revealed flaws in Fisher and Ford's experimental procedures, flaws that have been carried through the sixty years of field work on *Panaxia* (e.g., Goulson and Owen 1997). Fisher and Ford, as well as subsequent experimenters, failed to account for temperature fluctuations in the moths' environment during their experiment. The interpretation of Fisher and Ford's results hinges on performing this experimental check (E1), since temperature fluctuations affect the expression of the *medionigra* gene in the moth during the larval stage, turning the wing color darker in either of the extremes of cold or hot. Affected moths with the *medionigra* gene would look more like the dominant form, f. *dominula*, and, thus, would be scored as such. The capture-mark-release census data would be skewed, and, thus, the fit between the selectionist model and the data would be called into question. Because the check was not performed, a large portion of the experimental results on *Panaxia* are ambiguous. Ironically, Sewall Wright (1948) pointed out that Fisher and Ford's (1947) "argument by elimination" of drift to selection was not as strong as they believed. But Wright's critiques, now vindicated, were ignored.

The example of the Scarlet Tiger moth highlights the way in which we view the way population geneticists (and ecological geneticists) credential empirical or experimental claims about evolutionary causes. The example also highlights the ways in which such claims can go wrong. The preceding has only been a sketch, but we think it is a plausible one. What, ultimately, can be claimed about the strength of such claims? Lewontin (2000b) makes plain that the epistemological landscape of population genetics is a continuum from a "maximal inferential program" and a "minimal deductive program." The minimal program is tantamount to theoretical population genetics, or the program of providing the network of relationships between evolutionary causes and their outcomes at the genetic level. The maximal program aims to give a correct account of evolutionary causes that have led to any and all observed patterns of genetic variation in natural populations. The maximal program is epistemologically unrealistic because it requires that scientists know the apparently unknowable: all of the biological and natural historical details of any arbitrarily chosen species. The minimal program is entirely analytic, having "no truly epistemological problems, only questions of methodological ingenuity" (Lewontin 2000b, 200). Somewhere between the two extremes is what Lewontin thinks is epistemic reality in population genetics. Indeed, Lewontin claims, "the best to which population geneticists can aspire is a formal structure that sets the limits of allowable explanation and a set of existentially modified claims about what has actually happened in the real history of organisms" (Lewontin 2000b, 213). We agree.

7. CONCLUSION

The problematic of population genetics is to account for the dynamics of genetic variation in natural populations via the causes of evolution, namely, mutation, migration, natural selection, and random genetic drift. The present essay has very briefly surveyed the historical development of population genetics, the current model theoretic structure of population genetics, key conceptual problems in understanding important evolutionary causes, and the problem of ferreting out those causes via theoretical and experimental work. We have here managed to pick away only at the very tip of the iceberg; there is much philosophical work to be done on the key theme of this essay – population genetics as a causal theory – as well as on other problems of the field. Moreover, it is important to note that while population genetics aims to understand the causes of evolution, it is generally understood as not contributing significantly or at all to our understanding of other evolutionary causes, such as the causes of speciation and extinction. That is, there is more to evolutionary studies than population genetics, in spite of the dramatic problems and progress of the field.

ACKNOWLEDGMENTS

Thanks to John Beatty, Michael Dietrich, Jon Hodge, David Hull, Anya Plutynski, and Michael Ruse for helpful comments and discussion on an earlier draft of this essay. Remaining infelicities are our own.

NOTES

1. Biologists refer variously to "random genetic drift," "genetic drift," "random drift," or simply, "drift." These are all the same phenomenon.
2. See, in particular, Chapter 18 (Laubichler) of this volume on the challenge issued from "evo-devo."
3. There is some question here about what counts as an "assumption" and what counts as a "cause," as well as concerning which assumptions are "basic." Here we ignore these complexities and simply echo the most common way that population genetics models are presented.
4. Lewontin's point here is actually to show the limitations of population genetics, an issue that will be discussed later; his quote also hints at an even greater complexity to the causality of population genetics than is presented here.
5. John Beatty has suggested to us (personal communication) that perhaps indiscriminate sampling phenomena ought to be seen as *causes* of drift rather than drift itself. On this view, drift should be seen as the analog to *evolution by natural selection*, not the analog to natural selection itself; the analog to selection (discriminate sampling) would be simply indiscriminate sampling. Here we must acknowledge that biologists do sometimes refer to indiscriminate sampling as drift and at other times refer to the effects (or outcomes) of indiscriminate sampling as drift. So, if we want to prevent confusion, we have to decide whether drift refers to the causal process or the outcome. Since biologists do sometimes speak of drift and selection as alternatives, and since they are both in a broad sense treated as causes of evolution in population genetics textbooks (where it is common to list selection,

drift, mutation, and migration as causes), we argue that it makes sense to have drift refer to the causal process. That is, it makes sense to treat drift in the same way that selection is treated. However, it is probably true that in some sense it does not matter whether you call the causal process "drift" or the outcome "drift," *as long as it is clear when one is referring to the causal process and when one is referring to the outcome* (Millstein 2002).

6. Even though indiscriminate sampling is characterized as a process whereby physical differences between biological entities are causally *irrelevant* to differences in reproductive success, it is still a causal process in this sense. For example, in gamete sampling, the uniting of gametes to form zygotes over time consists of many states that are the result of underlying causes.

7. One might easily object to the attribution of this phenomenon to Sewall Wright; our point here is only to emphasize how common it is for drift to be called an effect.

8. Even here a problem of interpretation presents itself: at what level (gene, organism, group, species) does this cause operate? But see Chapter 3 (Lloyd), this volume.

9. Using the fitness of genotypes rather than the fitness of individuals may be another way in which population genetics does not capture a complete causal picture – assuming, that is, that the causal stories of individuals are part of the evolutionary story, which is not entirely clear.

10. Diamond introduces his own experimental strategies as well.

3   Units and Levels of Selection

The theory of evolution by natural selection is, perhaps, the crowning intellectual achievement of the biological sciences. There is, however, considerable debate about which entity or entities are selected and what it is that fits them for that role. In this chapter I aim to clarify what is at issue in these debates by identifying several distinct, though often confused, concerns and then identifying how the debates on what constitute the units of selection depend to a significant degree on which of these different questions a thinker regards as central. Chief among these distinctions are replicators versus interactors as well as who benefits from a process of evolution by selection, that is, who benefits in the long run from a selection process and who gets the benefit of possessing adaptations that result from a selection process. Because Richard Dawkins is the primary source of several of the confusions addressed in this essay, I treat his work at some length.

1.   INTRODUCTION

For more than twenty-five years, certain participants in the "units of selection" debates have argued that more than one issue is at stake. Richard Dawkins (1978, 1982a), for example, introduced "replicator" and "vehicle" to stand for different roles in the evolutionary process. He proceeded to argue that the units of selection debates should not be about vehicles, as they had formerly, but about replicators. David Hull (1980) in his influential article "Individuality and Selection" suggested that Dawkins's "replicator" subsumes two quite distinct functional roles and broke them up into "replicator" and "interactor." Robert Brandon (1982), arguing that the force of

Hull's distinction had been underappreciated, analyzed the units of selection controversies further, claiming that the question about interactors should more accurately be called the "levels of selection" debate to distinguish it from the dispute about replicators, which he allowed to keep the "units of debate" title.

The purpose of this chapter is to delineate further the various questions pursued under the rubric of "units and levels of selection." This analysis is not meant to resolve any of the conflicts about which research questions are most worth pursuing; moreover, I make no attempt to decide which of the questions or combinations of questions discussed ought to be considered "the" units of selection question.

2.   FOUR BASIC QUESTIONS

With respect to the controversies that surround the units and levels of selection question, four basic questions can be delineated as distinct and separable. As will be demonstrated in Section 3, these questions are often used in combination to represent the units of selection problem. But let us begin by clarifying terms (see Lloyd 1992, 2001).

The term *replicator*, originally introduced by Dawkins but since modified by Hull, is used to refer to any entity of which copies are made. Dawkins (1982a, 47) classifies replicators using two orthogonal distinctions. A "germ-line" replicator, as distinct from a "dead-end" replicator, is "the potential ancestor of an indefinitely long line of descendant replicators" (1982a, 47). For instance, DNA in a chicken's egg is a germ-line replicator, whereas that in a chicken's wing is a dead-end replicator. An "active" replicator is "a replicator that has some causal influence on its own probability of being propagated," whereas a "passive" replicator is never transcribed and has no phenotypic expression whatsoever. Dawkins (1982a, 47) is especially interested in *active germ-line replicators*, "since adaptations 'for' their preservation are expected to fill the world and to characterize living organisms."

Dawkins (1982b, 295) also introduced the term *vehicle*, which he defines as "any relatively discrete entity ... which houses replicators, and which can be regarded as a machine programmed to preserve and propagate the replicators that ride inside it." According to

Dawkins (1982a, 62), most replicators' phenotypic effects are represented in vehicles, which are themselves the proximate targets of natural selection.

Hull (1980, 318), in his introduction of the term *interactor*, observes that Dawkins's theory has replicators interacting with their environments in two distinct ways: they produce copies of themselves, and they influence their own survival and the survival of their copies through the production of secondary products that ultimately have phenotypic expression. Hull (1980, 318) suggests *interactor* for the entities that function in this second process. An interactor denotes that entity that interacts, as a cohesive whole, directly with its environment in such a way that replication is differential – in other words, an entity on which selection acts directly. The process of evolution by natural selection is "a process in which the differential extinction and proliferation of interactors cause the differential perpetuation of the replicators that produced them" (Hull 1980, 318; see also Brandon 1982, 317–18).

Hull also introduced the concept of "evolvers," which are the entities that evolve as a result of selection on interactors: these are usually what Hull (1980, 327) calls *lineages*. So far, no one has directly claimed that evolvers are units of selection. They can be seen, however, to be playing a role in considering the question of who owns an adaptation and who benefits from evolution by selection, which we will consider in Sections 2.3 and 2.4.

2.1   The Interactor Question

In its traditional guise, the interactor question is, What units are being actively selected in a process of natural selection? As such, this question is involved in the oldest forms of the units of selection debates (Darwin 1859, Haldane 1932, Wright 1945). In his classic review article, Lewontin (1970, 7) contrasts the levels of selection, "especially as regards their efficiency as causers of evolutionary change." Similarly, Slobodkin and Rapaport (1974, 184) assumed that a unit of selection is something that "responds to selective forces as a unit – whether or not this corresponds to a spatially localized deme, family, or population."

Questions about interactors focus on the description of the selection process itself, that is, on the interaction between an entity, that entity's traits and environment, and how this interaction produces evolution; they do not focus on the outcome of this process (see Wade 1977; Vrba and Gould 1986). The interaction between some interactor at a certain level and its environment is assumed to be mediated by "traits" that affect the interactor's expected survival and reproductive success. Here, the interactor is possibly at any level of biological organization, including a group, a kin-group, an organism, a gamete, a chromosome, or a gene. Some portion of the expected fitness of the interactor is directly correlated with the value of the trait in question. The expected fitness of the interactor is commonly expressed in terms of genotypic fitness parameters, that is, in terms of the fitness of combinations of replicators; hence, interactor success is most often reflected in and counted through replicator success. Several methods are available for expressing such a correlation between interactor trait and (genotypic or genic) fitness, including partial regression, variances, and covariances.

In fact, much of the interactor debate has been played out through the construction of mathematical genetic models – with the exception of Wade's (1978, 1980) and some of Wilson and Colwell's (1981) work on female-biased sex ratios (see especially Griesemer and Wade 1988). The point of building such models is to determine what kinds of selection, operating at which levels, may be effective in producing evolutionary change.

It is widely held, for instance, that the conditions under which group selection can effect evolutionary change are quite stringent and rare. Typically, group selection is seen to require small group size, low migration rate, and extinction of entire demes. Some modelers, however, disagree that these stringent conditions are necessary. Matessi and Jayakar (1976, 384), for example, show that in the evolution of altruism by group selection, very small groups may not be necessary, contra Maynard Smith (1964). Wade and McCauley (1980, 811) also argue that small effective deme size is not a necessary prerequisite to the operation of group selection. Similarly, Boorman (1978, 1909) shows that strong extinction pressure on demes is not necessary. And finally, Uyenoyama (1979) develops a group selection model that violates all three of the "necessary" condition usually cited.

That different researchers reach such disparate conclusions about the efficacy of group selection occurs partly because they are using different models with different parameter values. Wade (1978) highlighted several assumptions, routinely used in group selection models, that biased the results of these models against the efficacy of group selection. For example, he noted that many group selection models use a specific mechanism of migration; it is assumed that the migrating individuals mix completely, forming a "migrant pool" from which migrants are assigned to populations randomly. All populations are assumed to contribute migrants to a common pool from which colonists are drawn at random. Under this approach, which is used in all models of group selection prior to 1978, small sample size is needed to get a large genetic variance between populations (Wade 1978, 110; see discussion in Okasha 2003).

If, in contrast, migration occurs by means of large populations, higher heritability of traits and a more representative sampling of the parent population will result. Each propagate is made up of individuals derived from a single population, and there is no mixing of colonists from the different populations during propagule formation. On the basis of Slatkin and Wade's (1978, 3531) analysis, much more between-population genetic variance can be maintained with the propagule model. They conclude that by using propagule pools as the assumption about colonization, one can greatly expand the set of parameter values for which group selection can be effective.

Another aspect of this debate that has received a great deal of consideration concerns the mathematical tools necessary for identifying when a particular level of biological organization meets the criteria for being an interactor. Examples of suggested techniques within the philosophical community include Bandon's use of Salmon's notion of screening off and the work by Wimsatt (1980, 1981) and Lloyd ([1988] 1994) on the additivity approach (see Sarkar 1994 and Godfrey-Smith 1992 for criticisms of this last approach, and Griesemer and Wade 1988, and Okasha 2004a for defenses of it). Biologists have also suggested a variety of statistical techniques for addressing this issue. See, for example, the work of Arnold and Fristrup (1982), Heisler and Damuth (1987), and Wade (1985), respectively.

Overall, while many of the suggested techniques have had strengths, no one approach to this aspect of the interactor question has been generally accepted, and indeed it remains the subject of debate in biological circles (Okasha 2004b, c). Discussions of these issues within philosophy have been muted of late as a result of the influence of genic pluralism, which regards the entire interactor debate as a mistake.

Note that the "interactor question" does not involve attributing adaptations or benefits to the interactors, or indeed, to any candidate unit of selection. Interaction at a particular level involves only the presence of a trait at that level with a special relation to genic or genotypic expected success that is not reducible to interactions at a lower level. A claim about interaction indicates only that there is an evolutionarily significant event occurring at the level in question; it says nothing about the existence of adaptations at that level. As we shall see, the most common error made in interpreting many of the interactor-based approaches is that the presence of an interactor at a level is taken to imply that the interactor is also a manifestor of an adaptation at that level.

2.2   The Replicator Question

The focus of discussions about replicators concerns just which organic entities actually meet the definition of replicator. Answering this question obviously turns on what one takes the definition of replicator to be. In this connection Hull's contribution turned out to be central. Starting from Dawkins's view, Hull (1980, 318) refined and restricted the meaning of "replicator," which he defined as "an entity that passes on its structure directly in replication." The terms *replicator* and *interactor* will be used in Hull's sense in the rest of this essay.

Hull's definition of replicator corresponds more closely than Dawkins's to a long-standing debate in genetics about how large or small a fragment of a genome ought to count as a replicating unit – something that is copied, and that can be treated separately in evolutionary theory (see especially Lewontin 1970). This debate revolves critically around the issue of linkage disequilibrium and led Lewontin, most prominently, to advocate the usage of parameters referring to the entire genome rather than to allele and genotypic frequencies in genetical models. The basic point is that with much linkage disequilibrium, individual genes cannot be considered as replicators because they do not behave as separate units during reproduction. Although this debate remains pertinent to the choice of state space of genetical models, it has been eclipsed by concerns about interactors in evolutionary genetics.

This is not to suggest that the replicator question has been solved. Work on the replicator question is part of a rich and continuing research program; it is simply no longer a large part of the units debates. That this parting of ways took place is largely due to the fact that evolutionists and philosophers working on the units problems tacitly adopted Dawkins's suggestion that the replicator, whatever it turned out to be, be called the 'gene' (see Section 3.3). This move neatly removes the replicator question from consideration. Exactly why this move should have met with near-universal acceptance is to some extent historical. However, the fact that the intellectual tools (largely mathematical models) of the participants in the units debates were better suited to dealing with aspects of that debate other than the replicator question, which requires mainly bio-chemical investigation, surely contributed to this outcome.

There is a very important class of exceptions to this general abandonment of the replicator question. Susan Oyama, Paul Griffiths, and Russell Gray have been leading thinkers in formulating a radical alternative to the interactor/replicator dichotomy known as Developmental Systems Theory (Oyama 1985; Griffiths and Gray 1994, 1997; Oyama, Griffiths, and Gray 2001). Here the evolving unit is understood to be the developing system as a whole, privileging neither the replicator nor the interactor. James Griesemer (2000) has originated a profound reconceptualization of the evolution by selection process and has rejected the role of replicator as misconceived. He proposes in its place the role of "reproducer," which focuses on the material transference of genetic and other matter from generation to generation. The reproducer plays a central role, along with a hierarchy of interactors, in his much-awaited book on the evolutionary process.

2.3   The Beneficiary Question

Who benefits from a process of evolution by selection? There are two predominant interpretations of this question: Who benefits ultimately in the long term, from the evolution by selection process? And who gets the benefit of possessing adaptations as a result of a selection process? Take the first of these, the issue of the ultimate beneficiary.

There are two obvious answers to this question – two different ways of characterizing the long-term survivors and beneficiaries of the evolution by selection process. One might say that the species or lineages (Hull's evolvers) are the ultimate beneficiaries of the evolutionary process. Alternatively, one might say that the lineages characterized on the genic level, that is, the surviving alleles, are the relevant long-term beneficiaries. I have not located any authors holding the first view, but, for Dawkins, the latter interpretation is the *primary fact* about evolution. To arrive at this conclusion, Dawkins adds the requirement of agency to the notion of beneficiary (see Hampe and Morgan 1988). For Dawkins, a beneficiary, by definition, does not simply passively accrue credit in the long term; it must function as the initiator of a causal pathway. Under this definition, the replicator is causally responsible for all of the various effects that arise further down the biochemical or phenotypic pathway, irrespective of which entities might reap the long-term rewards.

A second and quite distinct version of the beneficiary question involves the notion of adaptation. The evolution by selection process may be said to "benefit" a particular level of entity under selection, through producing adaptations at that level (Williams 1966, Maynard Smith 1976, Vrba 1984, Eldredge 1985). On this approach, the level of entity actively selected (the interactor) benefits from evolution by selection at that level through its acquisition of adaptations.

It is crucial to distinguish the question concerning the level at which adaptations evolve from the question about the identity of the ultimate beneficiaries of that selection process. One can think – and Dawkins does – that organisms have adaptations without thinking that organisms are the "ultimate beneficiaries" of the selection process. This sense of "beneficiary" that concerns adaptations will be treated as a separate issue, discussed in the next section.

2.4 The Manifestor-of-Adaptation Question

At what level do adaptations occur? Or, as Sober (1984, 204) puts this question, "When a population evolves by natural selection, what, if anything, is the entity that does the adapting?" As mentioned previously, the presence of adaptations at a given level is sometimes taken to be a requirement for something to be a unit of selection. Wright (1980), in an absolutely crucial observation, distinguished group selection for "group advantage" from group selection per se. In other words, he claimed that the combination of the interactor question with the question of what entity had adaptations had created a great deal of confusion in the units of selection debates in general.

Some, if not most, of this confusion is a result of a very important but neglected duality in the meaning of "adaptation." Sometimes "adaptation" is taken to signify any trait at all that is a direct result of a selection process at that level. In this view, any trait that arises directly from a selection process is claimed to be, by definition, an adaptation. Sometimes, on the other hand, the term "adaptation" is reserved for traits that are "good for" their owners, that is, those that provide a "better fit" with the environment and that intuitively satisfy some notion of "good engineering." These two meanings of adaptation, the *selection-product* and *engineering* definitions, respectively, are distinct, and in some cases, incompatible.

Williams, in his extremely influential book *Adaptation and Natural Selection* (1966), advocated an engineering definition of adaptation. He believed that it was possible to have evolutionary change result from direct selection favoring a trait without having to consider that changed trait as an adaptation. Consider, for example, his discussion of Waddington's (1956) genetic assimilation experiments. Williams (1966, 70–81) interprets the results of Waddington's experiments in which latent genetic variability was made to express itself phenotypically because of an environmental pressure (see the lucid discussion in Sober 1984, 199–201).

Williams (1966, 75–78) considers the question of whether the bithorax condition (resulting from direct artificial selection on that trait) should be seen as an adaptive trait, and his answer is that it should not. Williams instead sees the bithorax condition as "a disruption ... of development," a failure of the organism to respond. Hence, Williams drives a wedge between the notion of a trait that is a direct product of a selection process and a trait that fits his stronger engineering definition of an adaptation (see Gould and Lewontin 1979; Sober 1984, 201; cf. Dobzhansky 1956).[1]

In sum, when asking whether a given level of entity possesses adaptations, it is necessary to state not only the level of selection in question but also which notion of adaptation – either *selection-product* or *engineering* – is being used. This distinction between the two meanings of adaptation also turns out to be pivotal in the debates about the efficacy of higher levels of selection, as we will see in Sections 3.1 and 3.2.

2.5   Summary

In this section, four distinct questions have been described that appear under the rubric of "the units of selection" problem: What is the interactor? What is the replicator? What is the beneficiary? And what entity manifests any adaptations resulting from evolution by selection? There is a serious ambiguity in the meaning of "adaptation"; which meaning is in play has had deep consequences for both the group selection debates and the species selection debates. Commenting on this analysis, John Maynard Smith (2001, 1497) wrote in *Evolution*, Lloyd (2001) argues, "correctly I believe, that much of the confusion has arisen because the same terms have been used with different meanings by different authors ... [but] I fear that the confusions she mentions will not easily be ended." In Section 3, this taxonomy of questions is used to sort out some of the most influential positions in three debates: group selection (3.1), species selection (3.2), and genic selection (3.3).

3.   AN ANATOMY OF THE DEBATES

3.1   Group Selection

George Williams's (1966) famous near-deathblow to group panselectionism was, oddly enough, about benefit. He was interested in cases in which there was selection among groups and the groups as a whole *benefited from* organism-level traits (including behaviors) that seemed disadvantageous to the organism. (Similarly, for Maynard Smith [1964].) Williams argued that the presence of a benefit to the group was not sufficient to establish the presence of group selection. He did this by showing that a group benefit was not necessarily a group adaptation. (Hence, Williams is here using the term benefit to signify the manifestation of an adaptation at the group level.) His assumption was that a genuine group selection process results in the evolution of a group-level trait – a real adaptation – that serves a design purpose for the group. The mere existence, however, of traits that benefit the group is not enough to show that they are adaptations; in order to be an adaptation, under Williams's (1966) view, the trait must be an *engineering* adaptation that evolved by natural selection. Williams argued that group benefits do not, in general, exist *because* they benefit the group; that is, they do not have the appropriate causal history.

Implicit in Williams's discussion is the assumption that being a unit of selection at the group level requires two things: (1) having the group as an interactor and (2) having a group-level engineering-type adaptation. That is, Williams combines two different questions, the interactor question and the manifestor-of-adaptation question, and calls this combined set *the* unit of selection question. These requirements for "group selection" make perfect sense given that Williams's prime target was Vero Wynne-Edwards, who promoted a view of group selection that incorporated this same two-pronged definition of a unit of selection.

This combined requirement of engineering group-level adaptation in addition to the existence of an interactor at the group level is a very popular version of the necessary conditions for being a unit of selection within the group selection debates. David Hull (1980, 325) claims that the group selection issue hinges on "whether entities more inclusive than organisms exhibit adaptations." John Cassidy (1978, 582) states that the unit of selection is determined by "who or what is best understood as the possessor and beneficiary of the trait." Similarly, Eldredge (1985, 108) requires adaptations for an entity to count as a unit of selection, as does Vrba (1983, 1984).

Maynard Smith (1976, 282) also ties the engineering notion of adaptation into the version of the units of selection question he would like to consider. In an argument separating group and kin selection, Maynard Smith concludes that group selection is favored by small group size, low migration rates, and rapid extinction of groups infected with a selfish allele and that "the ultimate test of the group selection hypothesis will be whether populations having these characteristics tend to show 'self-sacrificing' or 'prudent' behavior more commonly than those which do not." This means that the presence of group selection or the effectiveness of group selection is to be measured by the existence of nonadaptive behavior of individual organisms along with the presence of a corresponding group-level adaptation. Therefore, Maynard Smith does require a group-level adaptation from groups to count as units of selection.

As with Williams, it is significant that he assumes the *engineering* notion of adaptation rather than the weaker *selection-product* notion. As Maynard Smith (1976, 278) puts it, "an explanation in terms of group advantage should always be explicit, and always calls for some justification in terms of the frequency of group extinction."

In contrast to the preceding authors, Sewall Wright (1929, 1931) separated the interactor and manifestor-of-adaptation questions in his group selection models. He distinguishes between what he calls "intergroup selection," that is, interdemic selection in his shifting balance process, and "group selection for group advantage." Wright (1980, 840) cites Haldane (1932) as the originator of the term "altruist" to denote a phenotype "that contributes to group advantage at the expense of disadvantage to itself." Wright (1980, 841) connects this debate to Wynne-Edwards, whom he characterizes as asserting the evolutionary importance of "group selection for group advantage." He argues that Hamilton's kin selection model is "very different" from "group selection for the uniform advantage of a group." Hamilton himself concurred in a little-known paper from 1975 (Hamilton 1996, vol. 1, 337).

Wright (1980, 841) takes Maynard Smith, Williams, and Dawkins to task for mistakenly thinking that because they have successfully criticized group selection for group advantage, they can conclude that "natural selection is practically wholly genic." Wright (1980, 841) argues that "none of them discussed group selection for organismic advantage to individuals, the dynamic factor in the shifting balance process, although this process, based on irreversible local peak-shifts is not fragile at all, in contrast with the fairly obvious fragility of group selection for group advantage, which they considered worthy of extensive discussion before rejection."

This is a fair criticism of Maynard Smith, Williams, and Dawkins. According to Wright, the problem is that these authors failed to distinguish between two questions: the interactor question and the manifestor-of-adaptation question. Wright's interdemic group selection model involves groups only as interactors, not as manifestors of group-level adaptations. Further, he is interested only in the effect the groups have on organismic adaptedness and expected reproductive success. More recently, modelers following Sewall Wright's interest in structured populations have created a new set of genetical models that are also called "group selection" models and in which the questions of group adaptations and group benefit play little or no role.

For a period spanning two decades, however, Maynard Smith, Williams, and Dawkins did not acknowledge that the position they attacked, namely, Wynne-Edwards's, is significantly different from other available approaches to group selection, such as that of Wright, Wade, Wilson, Uyenoyama, Feldman, and Lewontin. Ultimately, however, both Williams and Maynard Smith recognized the significance of the distinction between the interactor question and the manifestor-of-an-adaptation question. As a result, Williams (1985, 7–8) wrote, "If some populations of species are doing better than others at persistence and reproduction, and if such differences are caused in part by genetic differences, this selection at the population level must play a role in the evolution of the species," while concluding that group selection "is unimportant for the origin and maintenance of adaptation."

Shortly thereafter, Maynard Smith (1987, 123) made an extraordinary concession.

There has been some semantic confusion about the phrase "group selection," for which I may be partly responsible. For me, the debate about levels of selection was initiated by Wynne-Edwards' book. He argued that there are group-level adaptations ... which inform individuals of the size of the population so that they can adjust their breeding for the good of the population. He was clear that such adaptations could evolve *only* if populations were units of selection. ... Perhaps unfortunately, he referred to the process as "group selection." As a consequence, for me and for many others who engaged in this debate, the phrase came to imply that groups were sufficiently isolated from one another reproductively to act as units of evolution, and not merely that selection acted on groups.

The importance of this debate lay in the fact that group-adaptationist thinking was at that time widespread among biologists. It was therefore important to establish that there is no reason to expect groups to evolve traits ensuring their own survival unless they are sufficiently isolated for like to beget like. ... When Wilson (1975) introduced his trait-group model, I was for a long time bewildered by his wish to treat it as a case of group selection and doubly so by the fact that his original model ... had interesting results only when the members of the group were genetically related, a process I had been calling kin selection for ten years. I think that these semantic difficulties are now largely over. (1987, 123)

Dawkins (1989a) also seems to have rediscovered the evolutionary efficacy of higher-level selection processes in an article on artificial life. In this article, he is primarily concerned with modeling the course of selection processes, and he offers a species-level selection interpretation for an aggregate species-level trait. Still, he seems not to have recognized the connection between this evolutionary dynamic and the controversies surrounding group selection because in his second edition of *The Selfish Gene* (Dawkins 1989b) he had yet to accept the distinction made so clearly by Wright in 1980. This was in spite of the fact that by 1987, the importance of distinguishing between evolution by selection processes and any engineering adaptations produced by these processes had been acknowledged by the workers Dawkins claimed to be following most closely, Williams and Maynard Smith.

The most recent significant entry into these debates is Elliott Sober and David Sloan Wilson's *Unto Others*, which they published in 1998. In this work Sober and Wilson develop a case for group selection based on the need to account for the existence of biological altruism. Biological altruism is any behavior that benefits another organism at some cost to the actor. Such behavior must always reduce the actor's fitness, but it may, as Sober and Wilson (following the work of Haldane and Wright) show, increase the fitness of certain groups within a structured population. While the biological modeling in *Unto Others* was not new, the book did call the issues involved in the group selection debates to the attention of the larger philosophic community.

3.2   Species Selection

Ambiguities about the definition of a unit of selection have also snarled the debate about selection processes at the species level. The combining of the interactor question and the manifestor-of-adaptation question (in the engineering sense) led to the rejection of research aimed at considering the role of species as interactors, *simpliciter*, in evolution. Once it is understood that species-level interactors may or may not possess design-type adaptations, it becomes possible to distinguish two research questions: Do species function as interactors, playing an active and significant role in evolution by selection? And does the evolution of species-level interactors produce species-level engineering adaptations and, if so, how often?

For most of the history of the species selection debate, these questions have been lumped together; asking whether species could be units of selection meant asking whether they fulfilled *both* the interactor and manifestor-of-adaptation roles. For example, Vrba (1984) used Maynard Smith's treatment of the evolution of altruism as a touchstone in her definition of species selection. Maynard Smith (1976) argued that kin selection could cause the spread of altruistic genes but that it should not be called group selection. Again, this was because the groups were not considered to possess design-type adaptations themselves. Vrba (1984, 319) agreed that the spread of altruism should not be considered a case of group selection because "there is no group adaptation involved; altruism is not emergent at the group level" (Maynard Smith gives different reasons for his rejection). This amounts to assuming that there must be group benefit in the sense of a design-type group-level adaptation in order to say that group selection can occur. Vrba's (1983, 388) view was that evolution by selection is not happening at a given level unless there is a benefit or engineering adaptation at that level. She explicitly equates units of selection with the existence of an interactor plus adaptation at that level. Furthermore, it seems that she has adopted the stronger engineering definition of adaptation.

Eldredge (1985, 134, 196) also argues that species selection does not happen unless there are species-level adaptations. Eldredge (1985, 133) rejects certain cases as higher-level *selection processes* overall because "frequencies of the properties of lower-level individuals which are part of a high-level individual simply do not make convincing higher-level adaptations." Vrba, Eldredge, and Gould all defined a unit of selection as requiring an emergent, adaptive property. This amounts to asking a combination of the interactor and manifestor-of-adaptation questions.

But consider the lineagewide trait of variability. Treating species as interactors has a long tradition (Thoday 1953, Dobzhansky 1956, Lewontin 1958). If species are conceived as interactors (and not necessarily manifestors-of-adaptations), then the notion of species selection is not vulnerable to Williams's original anti-group-selection objections. As Williams (1992, 27) remarks, "the answer to these difficulties must be found in Lloyd's idea that higher levels of selection depend, not on emergent characters, but on any and all emergent fitnesses." The old idea was that lineages with certain properties of being able to respond to environmental stresses would be selected for, that the trait of variability itself would be selected for, and that it would spread in the population of populations. In other words, lineages were treated as interactors.

The earlier researchers spoke loosely of adaptations whereby adaptations were treated in the weak sense as equivalent simply to the outcome of selection processes (at any level). They were explicitly *not* concerned with the effect of species selection on organismic level traits but with the effect on species-level characters such as speciation rates, lineage-level survival, and extinction rates of species. Lloyd and Gould (1993) argue that this sort of case represents a perfectly good form of species selection even though some balk at the thought that variability would then be considered, under a weak definition, a species-level adaptation (cf. Lloyd [1988] 1994).

Vrba (1989) also eventually recognized the advantages of keeping the interactor question separate from a requirement for an engineering-type adaptation. In her more recent review article, she has dropped her former requirement that in order for species to be units of selection, they must possess species-level adaptations. Ultimately, her current definition of species selection is in conformity with a simple interactor interpretation of a unit of selection (cf. Damuth and Heisler 1988; Lloyd [1988] 1994).

It is easy to see how the two-pronged definition of a unit of selection – as interactor and manifestor-of-adaptation – held sway for so long in the species selection debates. After all, it dominated much of the group selection debates until just recently. Some of the confusion and conflict over higher-level units of selection arose because of a historical contingency – Wynne-Edwards's implicit definition of a unit of selection and the responses it provoked.

3.3 Genic Selection: The Originators

One may understandably think that Dawkins is interested in the replicator question because he claims that the unit of selection ought to be the replicator. This would be a mistake. Dawkins is interested primarily in a specific ontological issue about benefit. He is asking a special version of the beneficiary question, and his answer to that question dictates his answers to the other three questions flying under the rubric of the "units of selection."

Briefly, Dawkins argues that because replicators are the only entities that "survive" the evolutionary process, they must be the beneficiaries. What happens in the process of evolution by natural selection happens *for their sake*, for their benefit. Hence, interactors interact for the replicators' benefit, and adaptations belong to the replicators. Replicators are the only entities with real agency as initiators of causal chains that lead to the phenotypes; hence, they accrue the credit and are the real units of selection.

Dawkins's version of the units of selection question amounts to a combination of the beneficiary question plus the manifestor-of-adaptation question. There is little evidence that he thinks he is answering the predominant interactor question; rather, he argues that people who focus on interactors are laboring under a misunderstanding of evolutionary theory. One reason he thinks this might be that he takes as his opponents those who hold a combination of the interactor plus manifestor-of-adaptations definition of a unit of selection (e.g., Wynne-Edwards). Unfortunately, Dawkins ignores those who are pursuing the interactor question alone; these researchers are not vulnerable to the criticisms he poses against the combined interactor-adaptation view.

Dawkins (1982b, 113–16) believes that interactors, which he calls "vehicles," are not relevant to *the* units of selection problem. The *real* units of selection, he argues, should be replicators, "the units that actually survive or fail to survive." Organisms or groups as "vehicles" may be seen as the unit of function in the selection process, but they should not, he argues, be seen as the units of selection because the characteristics they acquire are not passed on (Dawkins 1982b, 99). Here, Dawkins is following Williams's (1966, 109) line. Genotypes have limited lives and fail to reproduce themselves because they are destroyed in every generation by meiosis and recombination in sexually reproducing species; they are only temporary. Hence, genes are the only units that survive in the selection process. The gene (replicator) is the real unit because it is an "indivisible fragment"; it is "potentially immortal" (Williams 1966, 23–24; Dawkins 1982b, 97).

The issue, for Dawkins (1982b, 82), is whether, "when we talk about a unit of selection, we ought to mean a vehicle at all, or a replicator." He clearly distinguishes the dispute he would like to generate from the group-versus-organismic selection controversy, which he characterizes as a disagreement "about the rival claims of two suggested kinds of vehicles." In his view, replicator selection should be seen as an alternative framework for both organismic and group selection models.

There are two mistakes that Dawkins is *not* making. First, he does not deny that interactors are involved in the evolutionary process. He emphasizes that it is not necessary, under his view, to believe that replicators are directly "visible" to selection forces. Dawkins (1982b, 176) has recognized from the beginning that his question is completely distinct from the interactor question. He remarks, in fact, that the debate about group versus organismic selection is "a factual dispute about the level at which selection is most effective in nature," whereas his own point is "about what we ought to mean when we talk about a unit of selection." Dawkins (1982a, 46–47) also states that genes or other replicators do not "literally face the cutting edge of natural selection. It is their phenotypic effects that are the proximal subjects of selection."

Second, Dawkins does not specify how large a chunk of the genome he will allow as a replicator; there is no commitment to the notion that single genes are the only possible replicators. He argues that if Lewontin, Franklin, Slatkin, and others are right, his view will not be affected (see Section 2.2). If linkage disequilibrium is very strong, then the "effective replicator will be a very large chunk of DNA" (Dawkins 1982b, 89). We can conclude from this that Dawkins is not interested in the replicator question at all; his claim here is that his framework can accommodate any of its possible answers.

On what basis, then, does Dawkins reject the question about interactors? I think the answer lies in the particular question in which he is most interested, namely, what is "the nature of the entity *for whose benefit adaptations* may be said to exist?" On the face of it, it is certainly conceivable that one might identify the beneficiary of the adaptations as – in some cases, anyway – the individual organism or group that exhibits the phenotypic trait taken to be the adaptation. In fact, Williams (1966) seems to have done just that in his discussion of group selection. But Dawkins 1982a, 60) rejects this move, introducing an *additional* qualification to be fulfilled by a unit of selection; it must be "the unit that actually survives or fails to survive." Because organisms, groups, and even genomes are destroyed during selection and reproduction, the answer to the survival question must be the replicator. Strictly speaking, this is false; it is copies of the replicators that survive. He therefore must mean replicators in some sense of information and not as biological entities (see Hampe and Morgan 1988).

But there is still a problem. Although Dawkins (1982a, 60) concludes, "there should be no controversy over replicators versus vehicles. Replicator survival and vehicle selection are two aspects of the same process," he does not just leave the vehicle selection debate alone. Instead, he argues that we do not need the concept of discrete vehicles at all. The important point is that, on Dawkins's analysis, the fact that replicators are the only survivors of the evolution-by-selection process automatically answers also the question of who owns the adaptations. He claims that adaptations must be seen as being designed for the good of the active-gene-line replicator for the simple reason that replicators are the only entities around long enough to enjoy them over the course of natural selection.

Dawkins (1982b, 114) acknowledges that the phenotype is "the all important instrument of replicator preservation," and that genes' phenotypic effects are organized into organisms (that thereby might benefit from them in their lifetimes). But because only the active germ-line replicators survive, they are the true *locus of adaptations* (Dawkins 1982b, 113; emphasis added). The other things that benefit over the short term (e.g., organisms with adaptive traits) are merely the tools of the real survivors, the real owners. Hence, Dawkins rejects the vehicle approach partly because he identifies it with the manifestor-of-adaptation approach, which he has answered by definition, in terms of the long-term beneficiary.

The second key aspect of Dawkins's views on interactors is that he seems to want to do away with them entirely. Dawkins (1982b, 116) is aware that the vehicle concept is "fundamental to the predominant orthodox approach to natural selection." Nevertheless, he rejects this approach in *The Extended Phenotype*, claiming that the "main purpose of this book is to draw attention to the weaknesses of the whole vehicle concept" (Dawkins 1982b, 115). But his "vehicle" approach is not equivalent to "the interactor question"; it encompasses a much more restricted approach.

In particular, when Dawkins (1982b, 5, 55) argues against "the vehicle concept," he is only arguing against the desirability of seeing the individual organism as the one and only possible vehicle. His target is explicitly those who hold what he calls the "Central Theorem," which says that *individual organisms should be seen as maximizing their own inclusive fitness*. Dawkins's arguments are indeed damaging to the Central Theorem, but they are ineffective against other approaches that define units of selection as interactors.

One way to interpret the Central Theorem is that it implies that the individual organism is always the beneficiary of any selection process. Dawkins seems to mean by "beneficiary" both the manifestor-of-adaptation and that which survives to reap the rewards of the evolutionary process. He argues, rightly and persuasively, I think, that it does not make sense always to consider the individual organism to be the beneficiary of a selection process.

But it is crucial to see that Dawkins (1982b, 189) is not arguing against the importance of the interactor question in general, but rather against a particular definition of a unit of selection. The view he is criticizing assumes that the individual organism is the interactor, *and* the beneficiary, *and* the manifestor-of-adaptation. Consider his main argument against the utility of considering vehicles: the primary reason to abandon thinking about vehicles is that it confuses people. But look at his examples; their point is that it is inappropriate always to ask how an organism's behavior benefits that organism's inclusive fitness. We should ask instead, says Dawkins (1982b, 80), "whose inclusive fitness the behavior is benefiting." He states that his purpose in this book is to show that "theoretical dangers attend the assumption that adaptations are for the good of ... the individual organism" (Dawkins 1982b, 91).

So, Dawkins is quite clear about what he means by the "vehicle selection approach"; it always assumes that the organism is the beneficiary of its accrued inclusive fitness. Dawkins advances powerful arguments against the assumption that the organism is always the interactor cum beneficiary cum manifestor-of-adaptations. This approach is clearly not equivalent to the approach to units of selection characterized as the interactor approach. Unfortunately, Dawkins extends his conclusions to these other approaches, which he has, in fact, not addressed. Dawkins's lack of consideration of the interactor definition of a unit of selection leads to two grave problems with his views.

One problem is that he has a tendency to interpret all group selectionist claims as being about beneficiaries and manifestors-of-adaptations as well as interactors. This is a serious misreading of authors who are pursuing the interactor question alone. Consider, for example, Dawkins's (1982b, 85) argument that groups should not be considered units of selection:

To the extent that active germ-line replicators benefit from the survival of the group of individuals in which they sit, over and above the [effects of individual traits and altruism], we may expect to see adaptations for the preservation of the group. But all these adaptations will exist, fundamentally, through differential replicator survival. The basic beneficiary of any adaptation is the active germ-line replicator.

Notice that Dawkins begins by admitting that groups can function as interactors, and even that group selection may effectively produce group-level adaptations. The argument that groups should not be considered real units of selection amounts to the claim that the groups are not the ultimate beneficiaries. To counteract the intuition that the groups do, of course, benefit, in some sense, from the adaptations, Dawkins uses the terms "fundamentally" and "basic," thus signaling what he considers the most important level. Even if a group-level trait is affecting a change in gene frequencies, "it is still genes that are regarded as the replicators which actually survive (or fail to survive) as a consequence of the (vehicle) selection process" (Dawkins 1982b, 115). Thus, the replicator is the unit of selection, because it is the beneficiary, and the real owner of all adaptations that exist.

Saying all this does not, however, address the fact that other researchers investigating group selection are asking the interactor question and sometimes also the manifestor-of-adaptation question, rather than Dawkins's special version of the (ultimate) beneficiary question. He gives no additional reason to reject these other questions as legitimate; he simply reasserts the superiority of his own preferred unit of selection. In sum, Dawkins has identified three criteria as necessary for something to be a unit of selection: it must be a replicator; it must be the most basic beneficiary of the selection process; and it is automatically the ultimate manifestor-of-adaptation through being the beneficiary.

NOTE

1. Note that Williams says that "natural selection would produce or maintain adaptation as a matter of definition" (1966, 25; cf. Mayr 1976). This comment conflicts with the conclusions Williams draws in his discussion of Waddington; however, Williams later retracts his bithorax analysis (1985). Williams is committed to an engineering definition of adaptation (personal communication 1989).

4 What's Wrong with the Emergentist Statistical Interpretation of Natural Selection and Random Drift?

Population-level theories of evolution – the stock and trade of population genetics – are statistical theories par excellence. But what accounts for the statistical character of population-level phenomena? One view is that the population-level statistics are a product of, are generated by, probabilities that attach to the individuals in the population. On this conception, population-level phenomena are explained by individual-level probabilities and their population-level combinations. Another view, which arguably goes back to Fisher (1930) but has been defended recently,[1] is that the population-level statistics are sui generis, that they somehow emerge from the underlying deterministic behavior of the individuals composing the population. Walsh, Lewens, and Ariew (2002) label this the *statistical interpretation*. We are not willing to give them that term, since everyone will admit that the population-level theories of evolution are statistical, so we will call this the *emergentist statistical interpretation* (ESI). Our goals are to show that (1) this interpretation is based on gross factual errors concerning the practice of evolutionary biology, concerning both what is done and what can be done; (2) its adoption would entail giving up on most of the explanatory and predictive (i.e., scientific) projects of evolutionary biology; and finally (3) a rival interpretation, which we will label the *propensity statistical interpretation* (PSI), succeeds exactly where the emergentist interpretation fails.

1.   PROPENSITY AND EMERGENTIST INTERPRETATIONS

The propensity interpretation of fitness was introduced into the philosophical literature in 1978 (see Brandon 1978; also see Mills and Beatty 1979). The prime motivation was to make room for an explanatory theory of natural selection, which is tantamount to solving the so-called tautology problem. This problem arises from a casual inspection of the phrase "survival of the fittest," followed by the question of what defines the fittest. If the answer is those that reproduce the most, then it seems we are explaining a phenomenon, differential reproduction, in terms of itself, which is no explanation at all.

Brandon's approach was to think of fitness (or adaptedness) as a disposition. Just as it is not explanatorily empty to cite the water solubility of salt in explaining the behavior of a particular sample of salt when placed in water, so too it is not explanatorily empty to cite differences in adaptedness to a common environment when explaining a particular case of differential reproduction. Of course, in the case of water solubility we want, and indeed have, a deeper explanation of that disposition – a general explanation given in terms of molecular bonding.

The case of fitness differs in two important ways from that of water solubility. First, ceteris paribus, water-soluble substances dissolve when placed in water, period. That is, although we qualify the claim with a ceteris paribus clause – we want to exclude cases such as that when the water is frozen, or already saturated, and so on – the claim itself is not probabilistic. But we think chance can intervene in real biological populations so that higher fitness and higher levels of reproductive success can be dissociated. On the propensity interpretation, fitness (or adaptedness) is an explicitly probabilistic concept.[2] Thus it is a probabilistic propensity. Second, unlike in the case of water solubility, there is no general underlying explanation of differential fitness (see Brandon 1978 or 1990, 13–25; also see Rosenberg 1978, 1985; and Sober 1984). The underlying causal basis of fitness differences can be uncovered by detailed study of particular populations in particular selective environments, but it will not be general. It is our impression that this interpretation of fitness is widely accepted in both the philosophical and biological communities. But some of its broader implications are probably not appreciated.

If the propensity interpretation is correct, then population-level probabilities are derivable from individual-level probabilities in a familiar way. For example, if a coin and tossing device yields a probability of heads of .5, then we can calculate the probability of various results in an ensemble of tosses, say four, by the laws of probability theory. For those of you who have trouble thinking of coin tossing as genuinely stochastic, substitute the following example. Oxygen-15 has a half-life of two minutes. Take four atoms of that isotope. What is the probability that exactly two of them will decay during a two-minute time interval? To answer that question we do the exact same calculation as in the coin example and get the same answer. The probability of that outcome is .375. So the propensity interpretation of fitness yields a familiar and natural way of understanding the population-level probabilities that are essential to evolutionary theory. The only sticking point here is that, taken literally, the propensity interpretation of fitness is committed to the fundamental indeterminacy of the lives, deaths, and ultimately reproductive successes of individual organisms (see Brandon and Carson 1996). Some people find that a difficult ontological commitment.

So perhaps one motivation for the emergentist statistical interpretation is that it is not committed to the indeterminacy of individual lives and deaths. Indeed it seems that the inspiration for this interpretation is the relation between statistical thermodynamics and Newtonian mechanics. Here, *supposedly*, the underlying mechanics of the molecules in a gas are deterministic, but at the macrolevel we get the explicitly probabilistic second law of thermodynamics.[3] Perhaps then there is an analogue of the emergentist statistical interpretation in physics. But we do not think one should be much impressed by that, since no one understands the relationship between mechanics and statistical thermodynamics (Sklar 1999). Looking to physics will not help us understand the emergence of population-level probabilities.

Another possible motivation for the emergentist interpretation is that we are stuck with it. That is, we can clearly see that there are population-level probabilities governing the evolutionary trajectories of populations, but we have *no access* to the individual-level probabilities postulated by the propensity interpretation – either because they do not exist, or because we have epistemic limitations. We will deal with this possibility in the next section.

Finally, one might find the emergentist interpretation attractive primarily because one thinks that evolutionary theory deals solely with population-level probabilities, and therefore has no need for individual-level probabilities.[4] So ontological parsimony suggests we do without them.[5] This point will be dealt with in Section 3.

2.   THE ARGUMENT FOR THE EMERGENTIST
     STATISTICAL INTERPRETATION

Basing fitness on type- or population-level effects has precedent, especially among biologists. Fisher (1930) took the fitness of a type to be the objective representation of that type in the next generation. Similarly, a standard evolutionary biology textbook (Futuyma 1986) defines fitness as "the average contribution of one allele or genotype to the next generation or succeeding generations, compared with that of other alleles or genotypes."

Despite drawing support from a number of biologists, until recently this position has received little support from philosophers. In our discussion of the ESI, we will focus on recent defenses of this view by Walsh et al. (2002) and Matthen and Ariew (2002). Walsh et al. ask whether evolutionary theory is a statistical theory or a dynamical theory. A statistical theory is phenomenological, not causal, and a dynamical theory is a theory of forces, à la Newtonian mechanics. Thus their question is more or less equivalent to this: is evolutionary theory like the kinetic theory of gases or Newtonian mechanics? This seems an impoverished range of options. Why should it be relevantly similar to either one? Are those the only two types of scientific theories? Although they present us with a false choice the logic of their argument is clear: Sober's (1984) description of evolutionary theory as a theory of forces is, they claim, wrong; therefore the emergentist statistical interpretation is correct.[6]

Sober's description of evolutionary theory as a theory of forces has flaws, some more serious than others. As Endler (1986) has pointed out, there are a number of disanalogies between natural selection and the concept of force in physics.[7] But this is a quibble compared to the most important problem with Sober's analogy, which is the fact that selection and drift are not opposing forces, but rather two copossible outcomes of the same process – the process of sampling from a population where the probabilities of being sampled for each member of the population do not all equal 1 or 0 (see Brandon 2005). That is a serious flaw in Sober's account, but it does not mean we have to settle for a purely phenomenological account.

Why do Walsh et al. think that it does? Given the impoverished range of options they present, logic seems to force this choice on them. But we think there is more to it than that.

In a related article Matthen and Ariew (2002) present another argument for what amounts to the same conclusion. Again Sober's comparison of evolutionary theory to Newtonian mechanics is the target. Matthen and Ariew (2002, 67) argue that different components of fitness are not comparable, and in particular that there is nothing like vector addition that would allow us to combine different "forces" of selection. So although we know that, everything else being equal, it is best to produce the minority sex in a population with a skewed sex ratio:

we have no way of calculating whether a given sex-selection strategy interacts with a given parental-care-strategy, and how the fitness produced by variants of these strategies combine. This inability to add the "forces" of fitness is even more pronounced when the source laws are in unrelated domains. Suppose a certain species undertakes parental care, is resistant to malaria, and is somewhat weak but very quick. How do these fitness factors add up? We have no idea at all. The theory of probability has no general way to deal with such questions. (2002, 67)

(This last sentence of this quote is quite odd. Why should the theory of probability tell us how different components of fitness interact in biology? Should that not be a matter of biology?)

The conclusion of this is the following:

The disanalogy is that, while force affords Newtonian mechanics the means to compare and add up the consequences of these diverse causes, fitness does not add up or resolve. *This is why population geneticists are forced to estimate fitness by measuring population change.* (2002, 68, emphasis added)

The logic of this argument is, we think, clear enough. Its conclusion is false, and we want to focus on that. But let us briefly examine the major premise.

First, population genetic models regularly do combine different factors of evolutionary change in straightforward ways. One could, for instance, write down a simple model that tracks the evolution of two alleles, A and a, in a haploid population with discrete generations. Here the frequency of A in generation 2 is simply the product of the fitness of A, w_A, and its frequency in generation 1, p_1. Thus (where p_2 is the frequency of A in generation 2):

$$p_2 = w_A p_1$$

and similarly for change in a (where q_i is the frequency of a in generation i):

$$q_2 = w_a q_1.$$

Given this very simple model we can easily add the effects of migration and mutation (where μ is the mutation rate from A to a, υ is the mutation rate from a to A, and m_{1A} is the rate of loss of A due to emigration, m_{2A} the gain in A due to immigration, m_{1a} the rate of loss of a due to emigration, and m_{2a} the rate of gain of a due to immigration):

$$p_2 = w_A p_1 + p_1(1 - \mu) + (1 - p_1)\upsilon - m_{1A} + m_{2A}$$
$$q_2 = w_a q_1 + q_1(1 - \upsilon) + (1 - q_1)\mu - m_{1a} + m_{2a}$$

The frequency of A in generation 2, p_2, equals $w_A p_1$ plus the mutation rate from a to A, minus the mutation rate from A to a, minus the emigration rate of A, plus the immigration rate of A. Mutation, migration, and selection are fully comparable. This, of course, is not a discovery by us but is simply elementary population genetics. Thus, if Matthen and Ariew's claim were that different factors of evolutionary change, such as selection and mutation, are not comparable, their claim would be contrary to standard practice in population genetics and would be wrong.

But that is not their claim; rather they claim that different components of fitness are not comparable. Again this claim seems to be contradicted by standard population genetics. As Michod (1999, 12) points out, "Almost all models of natural selection involve some kind of fitness decomposition in one form or another." Perhaps the examples most familiar to philosophers are group selection models for the evolution of altruism. In such models different components of fitness are separated in fitness equations. The following equations are representative:

$$w_s = 1 + bx$$
$$w_a = 1 + bx - c$$

where w_s is the fitness of a selfish type and w_a the fitness of an altruist, c is the cost of altruism, b is the benefit, and x is the number of altruists within the particular group; c represents the component of selection due to the selective disadvantage of altruism within a group, while differences in the value of x, and therefore of bx, represent the component of selection due to the selective advantage of groups with a larger number of altruists. Thus, in this case at least, different components of selection are comparable, just as selection, mutation, and migration are fully comparable. And so, it would seem, the major premise of Matthen and Ariew's argument is simply wrong.

We think it is wrong, but we are not sure that the preceding example fully addresses their point. Their point, we think, is that we have no *general* theory that would allow us to compare different components of fitness, that is, nothing like vector addition in Newtonian mechanics. In their example, we have no *theory* that allows us to combine the components of selection due to sex-ratio differences and those due to parental-care differences. This interpretation may explain their odd remark about the failure of probability theory to provide a framework for such a comparison. So our group selection example would be atypical in that in this case we do have an explicit theory of how to compare the individual- and group-selection components. If this interpretation of their remarks is correct, then we agree with them, but then the conclusion of their argument does not follow.

It is hardly surprising that we have no general theory that would allow us to predict the fitness of every possible combination of every character state. Any such theory we develop is likely to be local and post hoc. In a population where various sex-ratio strategies are extant as are various parental-care strategies we can *measure* the fitnesses of the extant combinations. With sufficient study we may be able to offer an ecological explanation of these fitness values. But the resulting generalizations do not derive from any general theory.

The fitness values of various types are among the basic parameters of models in population genetics. They are like other basic parameters, such as mutation rates, migration rates, and effective population size, in that they need to be measured empirically and cannot be predicted from some general theory. There is here a difference from, and a similarity to, Newtonian mechanics. The difference is the locality of these parameter values – that they apply to particular populations in particular environments – and the resultant need to remeasure them time and again (see Brandon 1994). The similarity is that the basic parameters of Newtonian physics need to be empirically measured as well, for example, the value of G, the gravitational constant.

But have we not just conceded Matthen and Ariew's point, namely, that fitness must be measured in terms of its consequences? No. First, we can develop, through detailed ecological investigations, local theories of organism-environment interactions that would allow us to measure fitness indirectly. Second, and much more importantly, when looking at the effects of fitness we do not have to look at evolutionary, or transgenerational, change.[8] We can, and biologists often do, look at something else.

3.   METHODS FOR DETECTING SELECTION

John Endler (1986) in his comprehensive overview of studies of natural selection in the wild describes ten methods for detecting natural selection. For present purposes we do not need an analysis at that fine a grain, although we would recommend his account to any philosopher who would pronounce on how biologists *must* measure natural selection. A simpler classification results from first distinguishing between those methods that detect selection in terms of its effects versus those that detect selection in terms of its causes. Let us label the second category *CF*, for *c*auses of *f*itness. The first category needs to be further subdivided. The first subdivision – methods that detect selection in terms of *e*volutionary *c*onsequences – will be labeled *EC*. We will label the second subdivision – methods that detect selection by *d*irect *m*easurement of (parts of) the process of natural selection, that is, measurement of differential survivorship, mating ability, fertility, fecundity, and so forth – *DM*.

Studies using method *CF* are difficult in that they require detailed knowledge of organism-environment relations. As we will see they are rare. But it is important that they are not impossible (Lewontin 1978; Brandon 1990, chap. 1). They are not. However, on the basis of our knowledge of evolutionary biology, we would say that we are close to being able to apply method *CF*, usually, if not inevitably, through repeated applications of method *DM*. We will return to this shortly.

Method *EC* is the method that Matthen and Ariew (2002) claim biologists *must* use.[9] What is common to all the cases we lump under *EC* is that patterns of variation, either extant (horizontal) or over time (vertical), are used to compare models of selection to a null model of no selection. The models may be informal and qualitative (Endler's I–III and some cases of V), or they may be formal and quantitative (Endler's IV and some cases of V). But the essential feature of all of these cases is that the past or present existence of selection is *inferred* from data that eliminate the null (no-selection) hypothesis. A few examples will clarify just how *EC* works.

If one observes a consistent correlation between some environmental feature and character state, then one can hypothesize that these environmental differences lead to different selective environments that result in the observed distribution of character states. For example, one might, as Kettlewell (1955, 1956) did, observe a correlation between the darkness of tree trunks (due to industrial pollution) and the frequency of the dark morph in *Biston betularia*. The selection hypothesis is that in woods affected by pollution the dark morph is selectively favored over the light form, and vice versa, in nonpolluted woods.[10] Of course, Kettlewell did not stop with this hypothesis; he went on to demonstrate experimentally, using method *DM*, that selection was indeed operating in accordance with the selection hypothesis. We think all will agree that this was a good thing. Although the observed patterns of variation were not consistent with the null hypothesis (which in this case would be that the different color morphs were distributed randomly about the different areas Kettlewell studied), they are consistent with still other hypotheses. For instance, the hypothesis that air pollution has a developmental effect on moths that darkens wing color is not eliminated by the observed patterns of variation. This sort of problem seems to be quite general when the models in question are informal and qualitative. If, as in the preceding case, there are multiple nonselection hypotheses, then the elimination of one of them will not automatically support the hypothesis of selection.

This is less of a problem, but still a problem, when the models are quantitative. To illustrate how *EC* works with such models we will describe two examples, the first a simple "toy" example, and the second a genuinely interesting piece of contemporary biological research.

Consider a simple example of heterozygote superiority. Suppose that there are two alleles, *A* and *a*, at a locus and that the locus is in linkage equilibrium with all other loci. The fitness of *Aa* is normalized to 1, and selection coefficients are assigned to the two homozygotic genotypes. A simple population genetic model shows how a population satisfying the preceding will settle into an equilibrium. The equilibrium frequencies of *A* and *a* can be mathematically derived given the values of the selection coefficients associated with the two homozygotes. That is, the fitness values mathematically determine the equilibrium frequencies of the two alleles. But the relevant equations work in both directions, so given the equilibrium frequencies we can determine the fitness values. How could we use this fact to infer not just the existence of selection, but the quantitative strength of selection?

If we observed stable allele frequencies at the locus over a number of generations, then we could show that the likelihood of the null hypothesis (in this case that the alleles are selectively neutral) is considerably lower than that of the selection hypothesis (see Brandon 2005 and Brandon and Nijhout forthcoming). And given the support of the selection hypothesis we could then go on to infer, in the manner outlined, the fitness values of the three genotypes. Of course these inferences are based on the assumptions mentioned, so one's confidence in the inferences should be proportional to one's confidence in the truth, or approximate truth, of the assumptions.

Consider this slightly different scenario. In this case we have no access to vertical data, but we can observe strong selection against one of the two homozygotes (e.g., we observe the negative effects of sickle cell anemia associated with a known homozygotic genotype at the hemoglobin locus). *If* we assume that the population is at equilibrium, then we can again eliminate the null hypothesis and estimate the fitness values of the other two genotypes.

The preceding "toy" examples are meant to illustrate clearly the inferential character of the *EC* method. Let us now turn to the work of Marty Kreitman and others, which we consider to be the most interesting use of *EC* method.[11] Kreitman has developed an elegant method for determining where in the genome selection is acting, with drift being the explicit null hypothesis. Because of the redundancy of the genetic code, substitutions in the third position of a codon often produce synonymous codons (i.e., codons that code for the same amino acid). Given this fact, one can compare the behavior of the first two codons with that of the third. If selection is acting at the relevant genomic region, the first two positions (substitutions in which will not produce synonymous codons) should behave differently than the third. In contrast, if selection is not acting on the region, then the first two positions should be as free to drift as the third, and thus no difference is expected in the behavior of the third position. This is exciting work and is certainly a powerful way of investigating the selectionist/neutralist debate. But notice that this sort of work tells us nothing about the "why" of selection. It offers no ecological explanation of selection.

According to Darwinian theory, small differences in organisms can result in differences in various abilities and capacities, such as the ability to survive, the ability to attract mates, fertility, or fecundity. Although Matthen and Ariew (2002) complain that we have no way to combine or compare these different capacities, nature surely does, since at the end of the day, at the end of the generation, they combine to produce a given level of reproductive success. And this capacity is exactly what the propensity interpretation of fitness defines. As empirical biologists we should not be surprised that there is no general *theory* about how these various capacities combine to produce fitness, but there is a general *method* to investigate this. It is called *fitness component analysis* (see Endler 1986, 84–86, for discussion and references). Basically, we want to sample the population under study at as many life history stages as possible. The idea is that different capacities will manifest themselves at different stages of life history and we can then see *empirically* how they combine. The ideal, though rarely attainable, would be to observe every member of the population throughout its entire lifetime. In practice, biologists typically look at the *effects* of a small number of these capacities, getting a *direct measure* of some component of fitness. This is the method we are labeling *DM*.

Let us first describe a couple of familiar studies that have used the *DM* method; then we want to explore some important philosophical differences between it and the *EC* method.

The most famous studies of natural selection in the wild are those conducted by H. B. D. Kettlewell (1955, 1956). As we have already seen, the correlational data supported (perhaps only weakly) the hypothesis that selection was responsible for the increase in the melanic form of *Biston betularia* in woods downwind of large industrial areas. But Kettlewell did a series of experiments to support that hypothesis more strongly, and in his best known experiments he marked individual moths, released them into the woods, and then recaptured them several days later. He then compared the relative frequency of the two color morphs in the recaptured group to that in the released group. In the experiments conducted in woods downwind of industrial areas, he observed an increase in the relative frequency of the dark form in the recaptured group compared to the released group. On the basis of many auxiliary studies, he attributed this change in frequency to selection (by birds), and thus had a measure of *one component of fitness* of the two morphs in that environment.

As a measure of lifetime fitness, the sort of fitness that ultimately matters for evolution, Kettlewell's study is incomplete. It tells us nothing about how the two forms perform in the larval stage; it tells us nothing about any differences that might exist between them in mating ability, fertility, fecundity, and so on. Its power to explain the existing patterns of variation depends on the truth of the assumption that the two forms are more or less equivalent with respect to their other components of fitness. But given the fit between the observed selection differentials and the observed patterns of variation, that assumption is not at all unreasonable. And, most importantly, Kettlewell's work is a direct demonstration of the existence of selection in the areas he studied, during the life history stage that he studied.

Less familiar, but more complete, are the studies of the evolution of heavy metal tolerance in grasses conducted by Janis Antonovics and others (Antonovics, Bradshaw, and Turner 1971). Here fitnesses were measured more directly and more completely. As is that of Kettlewell's moths, this too is a study of adaptation to an environmental perturbation. Mining activities produce soils with high levels of heavy metals. These metals are typically toxic to most plants. When this contaminated soil is piled by the side of a mine, there is often a sharp boundary between metal-contaminated soils and normal soils. This dramatic difference in a factor in the external environment leads to a dramatic difference in selective environments. Antonovics was able to show this by monitoring the lives of large numbers of individuals, both in the contaminated soil and in noncontaminated soil. Looking at differences in lifetime survivorship the strength of selection was measured in both selective environments. Genetically tolerant plants were strongly selected over nontolerant types in the contaminated soil, and vice versa in the normal soil. Surprisingly, selection was so strong that genetically differentiated subpopulations were produced over very short distances in spite of considerable gene flow between them.

Antonovics's measures of fitness values were more complete than Kettlewell's in that he was able to look at lifetime survivorship, rather than survivorship during a small portion of the life cycle. They were more direct in that differential deaths were actually observed, rather than inferred from differences between released and recaptured groups. But this still is a measure of a component of fitness, not complete fitness. For instance, there was no attempt to measure potential differences in fertility and fecundity. But given the strength of selection observed, and the remarkable genetic differentiation associated with the different selective environments, we can be reasonably confident that Antonovics's measures of fitnesses captured a crucial part of the causal story.

As we said previously, the ideal application of the *DM* method would be to observe every stage of the life histories of the organisms in the population under study, and so to measure every component of fitness and to then see how, in this particular situation, those components combine to produce overall fitness. There is absolutely no philosophical or conceptual difficulty in doing this. The difficulties are of a practical nature. And so biologists using the *DM* method almost always measure some component or components of fitness. Such measures provide good (but not complete)[12] explanations of evolutionary change to the extent that the measured components of fitness dominate the unmeasured components.

One might think that the difference between the *EC* and *DM* methods is rather minor, that they both detect fitness values in terms of effects, and that the only real difference is that the *EC* method looks for effects over an evolutionary time scale while the *DM* method looks at effects of traits such as differential mating ability that are observable over the time scale of a single generation. That is a difference, and an important difference, but it is not the only difference.

Before discussing the less obvious differences between these two methods, let us briefly comment on the major consequence of the preceding difference. It is not minor. Since *EC* looks at trans-generational effects, it necessarily confounds the effects of the ecological process of selection (which is basically what *DM* studies) with multiple effects of the genetic system. For instance, in cases of heterozygote superiority, once the population reaches equilibrium there is no evolutionary change. Thus there are no *EC* effects. With the appropriate vertical data (many generations with no change at that locus) or horizontal data (the exact same system found in a number of related species) the *EC* method could eliminate the null hypothesis of neutrality, but it could not estimate the selection differentials. Without those data the *EC* method cannot even differentiate a case of strong selection, say that both homozygotes are lethal, from no selection. Combining the results of two separable processes results in a loss of information. And, unfortunately for the *EC* approach, that information is crucial to evolutionary explanation and prediction. We will return to this shortly.

At least as important philosophically is a less obvious difference between the two methods. Although we will need to add a little nuance to this, the *EC* method is model based and inferential; the *DM* method is not model based and is appropriately described as *measurement*.

Let us focus on how we can come to know fitness values by means of these methods. It should be clear from our discussion of *EC* examples that data about patterns of variation are used to support a model of selection, from which fitness values can be estimated. Without the model, there could be no estimation of fitness values, since the data are simply patterns of horizontal and/or vertical variation. Philosophically speaking, the inference to fitness values in these cases is *abductive*.

Now we do not want to claim that in using the *DM* method no inferences are made. Remember that Kettlewell inferred that the differences in relative frequency of the two forms in the recaptured class compared to the released group reflected differential predation. He, in fact, had a lot of evidence to back up that inference. We do not think, however, we should describe that evidence as a model. Antonovics's more direct measurements avoided that particular inference. We are not particularly concerned here with the curse of post-Kuhnian philosophy of science – the view that all observation is theory laden. Whatever one thinks about that, we would hope one would still be able to distinguish measurements of some quantity in nature from model-based inferences. Surely much *DM* work is done because the investigator is interested in some hypothesis or model. And one can always describe any parameter measurement as a hypothesis test (see Brandon 1994). But that possibility does not mean that that is the most perspicuous description. The *DM* method is appropriately named; it is a method of parameter measurement, one that is more or less direct.

4. APPLYING THE METHODS OF SELECTION DETECTION

When Matthen and Ariew claim that "population geneticists are forced to estimate fitness by measuring population change," we take this to be both a descriptive and a prescriptive claim. If biologists are forced to use the *EC* method, then they do use it, and presumably use it exclusively. That is descriptive. Furthermore if they must estimate fitness that way, then they ought to do it that way. That is prescriptive.

Our own prescriptive views may well have come through in the last section, but for the record, let us be explicit. We think that it is important that it be possible to apply the *CF* method, in principle at least. But we think that the only way we will have the biological knowledge required to apply that method is through repeated applications of the *DM* method. In this way the *DM* method has priority over the *CF*. We have also said that some really interesting work has been done by using the *EC* method. Were we in charge, we would certainly fund more of it. But the *EC* method is really a method of last resort. In Krietman's studies it is used because he is looking over vast expanses of evolutionary time and is looking at genomic regions where the function is often unknown. It would be impossible to apply the *DM* method here. But we can imagine no situation in which both the *DM* and *CF* methods were applicable and the *CF* method preferable. In this sense, the *DM* method is a better way of doing evolutionary biology.

Prescriptive disputes are not easily settled, though we hope we have given some good reasons in support of our views. But the descriptive implications of Matthen and Ariew's claim can be easily dismissed. There are real data here.

Table 5.1 in Endler (1986) lists all of the published demonstrations of natural selection in the wild that Endler could find. Surely he missed one or two, but this is by far the most comprehensive survey of the literature in existence.[13] Endler lists the studies by species and then by the traits studied in that species. For instance, in *Homo sapiens* there are entries for tooth size, birth weight and gestation time, height, body shape, and haemoglobin S. For each species trait he records the method(s) of demonstration. We mapped Endler's ten methods onto our three methods as follows: Endler's I–V are our *EC*, his VI–VIII are our *DM*, and his IX–X equate to our *CF*. When Endler listed more than one method we counted more than one of our methods *if* his listed methods crossed our categories (as happened only once). The results are as follows:

Method	Number of studies
EC	1
DM	172
CF	2

I think it is fair to conclude from this that Matthen and Ariew's descriptive claim is false. And if their prescriptive claim is true, then evolutionary biologists are certainly not behaving as they ought.

5.   CONCLUSIONS

If cavers were to race from one cave entrance to another, the winner would surely owe her success to such characteristics as her ability to navigate, her swiftness at making vertical ascents and descents, and her ability to squeeze through narrow apertures. The aboveground observer will recognize that these skills are necessary but would not be able to say how the caver's skills combined to lead her to victory. Those who argue for the ESI apparently think that biologists are in the same epistemological cul-de-sac.

Those who argue for the PSI would argue that the case of the cavers is disanalogous to natural selection in two ways. First, unlike the aboveground observer, scientists are able to observe not just the *outcome* of natural selection, but also the *process*. This would be like being able to track the cavers and see, for example, how many navigational errors they make or how fast they are able to make ascents and descents. From these data we could (1) *explain* why the winner won the race and (2) *predict* how the cavers would fare in different caves. Analogously, as we saw earlier, we can use *CF* to measure the variety of factors that lead to the success of an individual organism or type of organism. We can use these data to explain the success of organisms and predict how they would fare in different environments.

The second disanalogy is that unlike the singular cave race, biological phenomena are repeatable. Even if we could not observe how the cavers act underground, we could race them in a number of different caves, some without alternate routes (to eliminate navigational errors) and others with many, some with few vertical drops and others with many. Through this comparison, we could see which cavers fare well in which kinds of caves. This would allow us to learn which individual cavers (or caver type) do well in which kind of cave. We could use these data to predict which caver would win in a particular cave race and to explain why the winner won and the loser lost. Analogously, we could use *DM* to see how different organisms fare in different environments. We could even clone organisms and raise the clones in a diversity of environments. This would give us data to understand how different components of the organism's fitness combine to prove successful in a particular environment. As we saw in the previous section, *DM* is commonly employed by biologists.

In sum, the arguments for the ESI have been thoroughly refuted. This is a good thing. It is incapable of explaining differential reproduction, a key part of the process of evolution by natural selection. It can merely posit the existence of population-level statistical distributions – they emerge in mysterious ways. And it can make no sense of the way biologists actually measure fitness in the wild. In contrast, the PSI does these things easily and naturally. The ESI does have the advantage of allowing one to hang on to a philosophical prejudice, namely, that phenomena at the level of individual organisms are deterministic. But, as we have seen, hanging on to this particular prejudice is quite costly.

NOTES

1. Walsh, Lewens, and Ariew (2002), and Matthen and Ariew (2002); see also Sterelny and Kitcher (1988).
2. In 1978 Brandon defined adaptedness as follows: for an organism O in environment E there is a range of possible offspring numbers, Q_1^{OE}, $Q_2^{OE}, \ldots, Q_n^{OE}$, and for each number there is an associated probability, $P(Q_I^{OE})$. The adaptedness of O in E, $A(O, E)$, then is the *expected value* of O's reproductive success in E. That is, $A(O, E) = \Sigma\ P(Q_I^{OE})Q_I^{OE}$. Later, drawing on the work of John Gillespie (1973, 1974, 1977), it was discovered that this expected value needed to be discounted by some function of the variance in offspring number (see Brandon 1990, 18–20). For further discussion of this point, see Beatty and Finsen (1989) and Sober (2001).
3. We emphasize the word "supposedly" because it seems to us that our confidence that the underlying mechanics is really deterministic should be much lower than our confidence in the second law.
4. Sterelney and Kitcher (1988, 345) argue that "evolutionary theory, like statistical mechanics, has no use for such a fine grain of description: the aim is make clear the central tendencies in the history of evolving populations."
5. We will not be able to deal with the general issue of ontological commitment here, but let us simply assert our view that parsimony is not an ontological virtue; rather accuracy is. The world either is or is not a simple place. Our job is to describe it as it is, not as we wish it were.
6. See Stephens (2004) for a recent endorsement of the Newtonian option.
7. But see Brandon (2005).
8. Matthen and Ariew (2002) do not fully understand the implications of their position. As they define it *"predictive fitness* (as we shall call it) is a statistical measure of evolutionary change, the *expected* rate of increase (normalized relative to others) of a gene, a trait, or an organism's representation in future generations." Thus in the conclusion quoted when they speak of population geneticists' being forced to estimate fitness by "measuring population change," we must interpret "population change" as transgenerational change. Their conception of fitness is not novel, it is called Fisherian fitness. It is unsuited for explanatory purposes (see Brandon 1990, chap. 1; Ramsey 2006). Unfortunately for Matthen and Ariew, they do realize that they are committed to this (see footnote 30, 74).

9.   Walsh and colleagues (2002) are less explicit on this point, but it seems they also are committed to this view.

10. The vice versa hypothesis is not really necessary here but was part of Kettlewell's explicit experimental research. See Brandon (1999) and Rudge (1999).

11. See Yang and Bielawski (2000), Bamshad and Wooding (2003), and Hamblin, Thompson, and DiRienzo (2002).

12. See Brandon (1990, chap. 5) for an account of ideally complete adaptation explanations.

13. Of course, it is now seventeen years old. For a more recent (1984–97) list of studies of natural selection in the wild, see Kingsolver et al. (2001).

5 Gene

The historian Raphael Falk has described the gene as a 'concept in tension' (Falk 2000) – an idea pulled this way and that by the differing demands of different kinds of biological work. Several authors have suggested that in the light of contemporary molecular biology 'gene' is no more than a handy term that acquires a precise meaning only in some specific scientific context in which it is used. Hence the best way to answer the question 'What is a gene?', and the only way to provide a truly *philosophical* answer to that question is to outline the diversity of conceptions of the gene and the reasons for this diversity. In this essay we draw on the extensive literature in the history of biology to explain how the concept has changed over time in response to the changing demands of the biosciences. In this section we have drawn primarily on the work of Raphael Falk (1986, 1991, 1995, 2000, 2001, 2005, in press), Michael Dietrich (2000a, 2000b), Robert Olby (1974, 1985), Petter Portin (1993), and Michael Morange (1998). When our historical claims are commonplaces that can be found in several of these sources we do not cite specific works in their support. We have also chosen not to explain basic genetic terminology, as this would have occupied much of the chapter. More specialized terms are explained when they cannot be avoided. In the final part of the essay we outline some of the conceptions of the gene current today. The seeds of change are implicit in many of those current conceptions and the future of the gene concept appears set to be at as turbulent as its past.

THE INSTRUMENTAL GENE

In the first three decades of genetic research the gene had a dual identity (Falk 1986, 2005). Genes, or Mendelian factors, were intervening variables defined by the Mendelian pattern of inheritance. From this perspective, the fact that some trait of an organism can be resolved into one or more Mendelian characters establishes definitively that there are genes for those characters. Indeed, it seems that at least some of the earliest Mendelians did not clearly distinguish between the Mendelian character itself and the Mendelian factor 'underlying' it. That distinction was made clear by Wilhelm Johannsen's introduction of the terms 'phenotype' and 'genotype' in 1909. But as well as intervening variables, genes were hypothetical material constituents of the cell whose physical transmission from parent to offspring causally explained the Mendelian pattern of inheritance. In his Nobel Prize acceptance speech Thomas Hunt Morgan, the father of classical genetics, noted, "There is not consensus of opinion amongst geneticists as to what genes are – whether they are real or purely fictitious – because at the level at which genetic experiments lie, it does not make the slightest difference whether the gene is a hypothetical unit, or whether the gene is a material particle" (1933, quoted in Falk 1986, 148). In our view, one of the clearest themes in the century-long evolution of the concept of the gene is the dialectic between these two conceptions of the gene, a structural conception anchored first in cytology and later in biochemistry, and a functional conception anchored in the observable results of hybridizations, at first between organisms and later directly between DNA molecules.

Recent scholarship has stressed the fact that 'classical genetics' was not merely a theory of heredity, but at least as importantly an experimental practice – 'genetic analysis' – in which the regularities postulated by the Mendelian theory of heredity were used to address other questions about the structure and function of living systems (Waters 2004; Falk in press). This experimental practice imposed strong constraints on the concept of the gene. In the earliest days of Mendelian genetics, William Castle's hybridization experiments with hooded rats challenged the discreteness and constancy of Mendelian factors. In those experiments alleles appeared to be 'contaminated' by the alleles they had shared a cell with in previous generations. The resulting debate exposed a circularity of argumentation: 'unit factors' (individual Mendelian genes) can only be identified by their effect on 'unit characters' (those that display a single, consistent Mendelian pattern of inheritance), but how can a unit character that is supposed to stand for a unit factor be delimited? This circularity was resolved by definition: Mendelizing traits are determined by a single gene, and non-Mendelizing traits are controlled by more than one gene. The instrumental gene is by definition a Mendelizing unit – it is there to do a job that depends on this stipulation. The visible, heritable characters of organisms must be interpreted in such a way as to permit genetic analysis of those traits. If a character does not correspond to a gene then it must be decomposed into simpler characters that do (later described as 'primary characters'). In the same spirit, quantitative traits, which vary continuously between individuals and thus cannot occur in Mendelian ratios, were treated as the effect of many hypothetical genes, each of which makes an equal and inseparable contribution to the character, giving rise to the discipline of quantitative genetics.

THE MATERIAL GENE

The Morgan school rapidly established the chromosomal theory of heredity, according to which genes are arranged in a linear fashion along the chromosomes that cytologists had observed in the cell nucleus. They were able to explain many deviations from the standard Mendelian pattern of inheritance in terms of the observable behavior of chromosomes. Most importantly, they were able to correlate closely the linkage maps generated by genetic analysis with observable changes in the structure of chromosomes, an achievement facilitated by the discovery of huge, polytenic chromosomes in the salivary glands of *Drosophila*. Linkage was thus both a (functional) measure of the probability that two genes would be inherited together and a (structural) fact about the relative position of visible bands on the salivary gland chromosomes. But despite these achievements, most members of the Morgan school did not concern themselves with the material nature of genes, both because this was not a question that could be pursued via genetic analysis and because the pursuit of genetic analysis did not require it to be answered.

"Molecular biology was born when geneticists, no longer satisfied with a quasi-abstract view of the role of genes, focused on the problem of the nature of genes and their mechanism of action" (Morange 1998, 2). Foremost among these was Herman J. Muller, a student of

Morgan's not satisfied by the purely instrumental notion of the gene as an unknown physical entity localized on chromosomes. For Muller these particulate, atomic entities were the basis, the 'secret' of life, and the essential entities on which the Darwinian process of evolution rests. In order to fulfill these functions genes needed to have the properties of autocatalysis (self-replication) to make them units of heredity, heterocatalysis to allow them to contribute to the phenotype, and mutability to create heritable variation. Muller set up a research programme to study the material nature of the gene and reveal the physical basis of these properties. In 1927 Muller discovered the mutagenic effect of x-rays and used this to make the first estimates of the physical size of an individual gene.

For our purposes, Muller's emphasis on the material gene is important because of his commitment to finding an epistemic pathway to the gene that bypassed the observed effect of the gene of the phenotype. When this commitment started to bear fruit it became possible to advance a concept of the gene that abandoned some of the commitments required if genes were to be epistemically accessible via genetic analysis. Features of the gene that previously could not be meaningfully called into question – and that were thus treated as definitional – became features that could be tested and potentially rejected.

The material nature of the gene was progressively revealed by the new discipline of biochemistry, which came into being in the interwar years. One aim of this discipline was to understand the synthesis of the agents of organic *specificity* – organic molecules that interact only with a very narrow class of other molecules and thus allow the very precise chemistry required by living systems. From the mid-1930s it became increasingly clear that the specificity of organic molecules is explained by *conformation* and *weak interactions* between molecules. The conformation of a molecule is its three-dimensional shape, which determines whether specific sites on molecules can come together. The interactions between those sites are much weaker than the covalent bonds of standard inorganic chemistry, so that interactions between molecules and the conformation of individual molecules can be altered by relatively low energies. These principles turned out to underlie the structure and functioning of all forms of life (Morange 1998, 15). The concept of specificity rapidly began to be applied to the relationship between genes and their products, as well as to the relationship between enzymes and their substrates.

If the activity of the cell is explained in terms of molecular specificity it is natural to suppose that the effects of genes on phenotypes are mediated by the production of biomolecules with appropriate specificity. Thus in 1941 the 'one gene–one enzyme' hypothesis, which helped to forge an experimental association between biochemistry and genetics, was born. George Beadle and Edward Tatum chose to attack the problem of gene action by genetic analysis of a known biochemical process. They produced and isolated mutant strains of the fungus *Neurospora* each unable to synthesize one of several chemicals involved in a single biosynthetic pathway. Genetic analysis of these mutants showed that each deficiency was the result of a mutation in a single gene. Only three years later Oswald T. Avery produced experimental evidence that genes were made of DNA. Looking back, his evidence seems compelling, but it needed another eight years and a different line of experiment for it to change the received 'protein model of the gene'. If the relationship between genes and enzymes was one of specificity, like the relationship between enzymes and their substrates, then it seemed unlikely that DNA could be responsible for 'genetic specificity'. The little that was known about DNA suggested it was an unspecific and monotonous molecule, perhaps with a structural role in the chromosome.

Historians have stressed the very substantial changes in approach produced by the influx of scientists trained in physics into biology during the 1940s. These changes moved genetics and biochemistry closer together and paved the way for the molecular conception of the gene that prevailed from the 1950s to the 1970s. One of these former physicists, Max Delbrück, was convinced that understanding the secret of life would require a physical approach and an organism as simple and pure as a bacterial virus – an organism so simple that it could be conceived as a naked gene. The bacteriophage appeared to have hardly more than the one key characteristic of life, self-replication, and was thus deemed perfect to study this property "without opening the biochemical 'black box'" (Morange 1998, 45). The 'phage group' around Delbrück, Salvador Luria, and Alfred Hershey helped to establish bacterial genetics and the prokaryotic age in genetic research.

DOING WITHOUT GENES?

The clash between the leading geneticist Richard Goldschmidt and his contemporaries in the 1940s and early 1950s provides further insight into the classical gene concept. The successes of the Morgan school in determining the linear order of genes on chromosomes allowed the discovery of 'position effects' in which a change in the relative position of genes on the chromosome is associated with a change in their phenotypic effects. This in turn raises questions concerning the nature of mutation. Today we define a mutation as any heritable change in the nucleotide sequence of a chromosome, which may occur either by the substitution of one nucleotide for another or by the translocation or inversion of a chromosome segment. In classical genetics, however, mutation was necessarily defined as a change in the *intrinsic* nature of an individual gene manifest in a heritable difference in phenotype. Mutations were thus distinguished from position effects, in which an intrinsically identical gene has a different effect because it has changed its location. Goldschmidt challenged this distinction. As there was no direct evidence that chromosomes have distinctive structural parts corresponding to individual genes, he suggested that 'mutations' and 'position effects' were simply smaller and larger changes in the structure of the chromosome. Because chromosomal changes on very different scales were known to have phenotypic effects, Goldschmidt argued that chromosomes probably contained a hierarchy of units of function. Famously, he denied that 'genes' exist, by which he meant that no unique structural unit corresponded to the unit of function of classical genetics. Although "Goldschmidt's efforts from 1940 to 1958 stand out as one of the first attempts to develop a theory which integrated models of genetic structure, genetic action, developmental processes and evolutionary dynamics" (Dietrich 2000a, 738), his views were completely unacceptable to most of his contemporaries. Effectively, Goldschmidt was insisting that both aspects of the dual identity of the classical gene converge on a single unit – the material gene must correspond to the instrumental unit of genetic analysis. Evidence to the contrary is thus evidence that there are no genes in the classical sense. Goldschmidt's contemporaries perhaps differed in that they were more hopeful that future discoveries would reveal a unique unit of genetic function at the molecular level. They certainly differed in their commitment to continuing existing lines of research and unwillingness to undertake the radical reorientation that Goldschmidt was suggesting.

'NEO-CLASSICAL' GENETICS AND THE MOLECULAR GENE

By the mid-1950s DNA was established as the genetic material, its structure had been analyzed by James Watson and Francis Crick (1953), and Crick had stated the 'Central Dogma' of molecular biology and its related 'sequence hypothesis' (1958): the linear sequence of nucleotides in a segment of a DNA molecule determines the linear sequence of nucleotides in an RNA molecule, which in turn determines the sequence of amino acids in a protein by 'informational specificity', that is, via the genetic code whose details were to be elucidated in the early 1960s. The same period saw a sea change in the gene concept itself, one that Petter Portin has labeled the transition from the 'classical' to the 'neo-classical' gene (Portin 1993). It may appear slightly confusing that the latter conception has also been labeled the 'classical molecular gene' (Neumann-Held 1998), but as Portin's 'neo-classicism' is precisely a molecularized classicism, the two names are complementary.

The new, molecular concept of the gene was the result of technical developments that allowed much more detailed maps of the chromosome ('fine structure mapping') and the interpretation of the results of this enhanced form of genetic analysis in the light of the new understanding of the material gene. The new conception departed from the classical in recognizing that the gene is not the fundamental unit of mutation or of genetic recombination. Recombination in classical genetics was the process in which alleles from two copies of a chromosome were combined on a single copy as a result of crossing over between homologous chromosome pairs during meiosis. Recombination was thus recombination of an allele of one gene with an allele of another gene, so that genes themselves were the minimal unit of recombination. Working with bacteriophage from 1954 to 1961 Seymour Benzer was able to increase the resolution of the 'cis-trans' or 'complementation' test so as to map out in detail the location of different mutations within the same gene and demonstrated conclusively that recombination can occur between different parts of a single gene. Two mutations are said to be in *cis*-position when they are on the same copy of a chromosome. They are in *trans*-position when one is on each of two homologous chromosomes. The logic of the *cis-trans* test depends on the fact that most mutations are recessive in the heterozygote. Hence, if an offspring derives a mutant allele of one gene from one parent and a mutant allele of another gene from the other parent, it should also receive a mutation-free, functional copy of each gene from the other parent and appear phenotypically normal. If, however, an offspring receives a different mutation from each parent, but they are in the same gene, then it will have no mutation-free copy of that gene and will be a phenotypic mutant. Thus, crossing two mutant lines to produce offspring with the two mutations in *trans*-position tests whether they are in the same gene. If, however, genetic recombination can occur within a single gene, then a small proportion of the offspring of a cross between carriers of two different mutant alleles of the same gene will receive a copy of the gene that recombines the undamaged portion from one mutant allele with the undamaged portion from the other mutant allele and is thus restored to normal function. Benzer used an analogue of the *cis-trans* test in bacteriophage to demonstrate that the gene as a functional unit defined by the *cis-trans* test (the 'cistron') can be represented as a linear recombination map of mutated sites. This acknowledgment led him to distinguish between units of recombination, 'recons', mutation, 'mutons', and genetic function, the 'cistron'.

Benzer's work could have been seen as a vindication of Goldschmidt and other skepticism about the unified, particulate gene (Holmes 2000; Falk 2005). But this was not how it was viewed by his contemporaries. Instead, the cistron was more or less immediately identified with the gene. From this followed the conventional gene concept of molecular biology. One reason the results were interpreted in this way was that the physical structure of the DNA molecule was now known and offered a natural interpretation for Benzer's findings. The unit of recombination and mutation is the single nucleotide, whilst the unit of genetic function (heterocatalysis) is the sequence of nucleotides from which a single RNA is transcribed, corresponding to a single protein, and thus vindicating the existing doctrine of 'one gene–one enzyme'.

CHALLENGES TO THE CLASSICAL MOLECULAR GENE CONCEPT

By the mid-1960s many scientists thought that the major problems of molecular genetics had been solved and were inclined to leave other investigators "to iron out the details" (Stent 1968). But the claim that 'what is true for *E. coli* is true for the elephant' turned out to be premature, and it seems unlikely that molecular geneticists will find themselves out of work anytime soon. According to the classical molecular conception a gene is a series of contiguous nucleotides whose sequence corresponds to the sequence of amino acids in a single polypeptide chain (one or more of which makes up a protein). It was soon realized that some genes code for functional RNAs that are not translated to a protein, but this fact is easily accommodated by the classical conception. As C. Kenneth Waters has stressed, the fundamental molecular gene concept is that of a DNA sequence that determines the structure of some *gene product* by linear correspondence (Waters 1994, 2000). The molecular gene is the 'image in the DNA' of the molecule whose biological activity is of interest to the experimenter (Rob D. Knight, pers. comm.). The classical molecular gene seemed to unite the two identities of the classical gene in a single natural unit. The functional definition of the gene that underlay genetic analysis and the structural definition of the material gene had turned out to be two ways to pick out the very same thing. Looked at more closely, however, the functional definition had been significantly revised so as to take account of findings about the material gene. In Muller's original vision genes reproduce themselves (autocatalysis), influence the phenotype (heterocatalysis), and mutate. The classical molecular gene, however, is not the unit of replication, which is the whole DNA molecule of which it is a part. Nor is it the unit of mutation. The only function with respect to which the molecular gene is the unit of function is that of contributing to the phenotype (Muller's heterocatalysis). So the functional role of the gene was revised to fit the molecular reality that had been uncovered. Furthermore, the concept of the gene was restricted to sequences that fulfilled this new functional role: not all segments of chromosomes that behave as Mendelian factors count as genes under the new conception. Untranscribed regulatory regions not immediately adjacent to the coding sequences they regulate can segregate independently of those coding sequences, and so can function as separate Mendelian factors, but they are not separate molecular genes. Nevertheless, the classical molecular gene was a highly successful example of the research strategy of identifying a functional role, searching for the mechanism that fulfills that role at a lower level of analysis, and using knowledge of that mechanism to refine understanding of function at the original (in this case phenotypic) level of analysis.

Since the 1970s, however, further investigation of the underlying structural unit has tended to undermine the idea that the revised functional role of the gene – determining the structure of a gene product – is filled by natural units of structure at the level of the DNA. The structures in the genome that play a genelike role need not be physically distinct: they can overlap one another or occur inside one another (in the same direction on the DNA molecule or in reverse). The relationship between structural genes and genelike functions is not one to one but many to many: some gene products are made from more than one structural gene and individual structural genes make multiple products. Finally, the sequence of elements in the gene product depends on much more than the sequence of nucleotides in the structural gene: different sequence elements can be repeated, scrambled, and reversed in the product, and the precise sequence of a gene product can reflect posttranscriptional and translational processing as well as the original DNA sequence. To put flesh on these bones we will briefly describe some of these mechanisms and give an example (Figure 5.1). (For more examples, details, and references, see Stotz and Griffiths 2004; Stotz, Bostanci, and Griffiths 2006; Stotz 2006.)

In eukaryotes (organisms whose cells have a nucleus and organelles, including fungi, plants, and animals) the DNA sequence is transcribed into a premessenger RNA (pre-mRNA) from which the final RNA transcript is processed by cutting out large noncoding sequences, called *introns*, and splicing together the remaining coding sequences, the *exons*. Biologists speak of alternative *cis*-splicing[1] when more than one mature mRNA transcript results from these processes through the cutting and joining of alternative exons. Adjacent genes are sometimes cotranscribed, that is, transcribed together to produce a *single* pre-mRNA that is then spliced. Splicing may also occur between a gene and an adjacent 'pseudogene' that

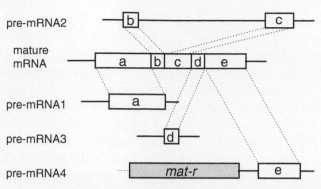

Figure 5.1 A contemporary molecular gene. Lines denote introns; boxes denote exons. Subunit 1 of the respiratory chain NADH dehydrogenase is encoded by the gene *nad1*, which in the mitochondrial genomes of flowering plants is fragmented into five coding segments that are scattered over at least 40 kb of DNA sequence and interspersed with other unrelated coding sequences. In wheat (illustrated) the five exons that together encode the polypeptide of 325 amino acids require one *cis*-splicing event (between the exons b/c) and three trans-splicing events (between exons a/b, c/d, and d/e) for assembly of the open reading frame. In addition, RNA editing is required, including a C to U substitution to create the initiation codon for this ORF. In some mosses and in mammals the ORF for NAD1 is an uninterrupted stretch of nuclear genomic DNA. Finally, in wheat, a separate ORF for a maturase enzyme (*mat-r*) is encoded in the intron upstream of exon e (Chapdelaine and Bonen 1991). For more examples, visit http://representinggenes.org.

would be incapable of producing a product on its own. Alternative gene products may also be derived from so-called overlapping genes. In these cases, the 'genes', in the sense of the 'open reading frames' (ORFs) that are transcribed into RNA, are not lined up like so many pearls on a string, but instead may overlap one another or even be completely contained one within another. While some cases of alternative splicing produce a range of proteins that are structurally related to one another, in other cases the products are quite different from each other (in which case they are often described as products of overlapping genes, rather than alternative splicing of the same gene). The degree of difference between the products depends on the extent of overlap between their exons, and on whether these shared sequences are read in the same reading frame. It is the precise nucleotide at which reading begins that determines which codons a DNA sequence contains. Starting at a different nucleotide is called 'frameshift', a phenomenon that would look like this in an English sentence: 'A gene is a flexible entity' becomes 'Age nei saf lex ibl een tit y'. But unlike any human language, a DNA sequence is always made up of meaningful 'three-letter words' (codons that specify an amino acid during translation) no matter where reading begins. This means that very different products can be read from the same sequence merely by frameshifting by one nucleotide. As well as alternative transcripts from a DNA sequence, multiple simultaneous transcripts can occur, as is the case of the parallel processing of functional noncoding RNAs (such as microRNAs) from the intronic regions of the premature transcript, which may be involved in the regulation of coding transcript of the same gene.

In the process of *trans*-splicing a final mRNA transcript is processed from two or more independently transcribed pre-mRNAs. Whilst the prefix *trans* might suggest that these pre-mRNAs are derived from DNA sequences far apart from each other, this is by no means always the case. In fact, two copies of the very same sequence can be spliced together this way, as can alternative exons in what would at first glance look like a 'normal' case of *cis*-splicing. Moreover, until very recently it was thought that only one strand of DNA is transcribed, but in fact DNA can be read both forwards and backwards by the cellular machinery, producing either different or matching (complementary) products. The latter case, in which exactly the same sequence is read in reverse, will result in an antisense transcript with likely regulatory function, possibly through silencing its complementary transcript. RNA editing is another mechanism of modification that can significantly diversify the 'transcriptome' or 'proteome' (the total complement of final transcripts or proteins in the cells of an organism). Whereas most other forms of posttranscriptional modifications of mRNA (capping, polyadenylation, and *cis*-splicing) retain the *correspondence* of the primary structure of coding sequence and gene product, RNA editing disturbs this correspondence by changing the primary sequence of mRNA after its transcription. The creation of 'cryptogenes' via RNA editing can potentially have radical effects on the final product, depending on whether editing changes the sense of the codon in which it occurs. While there are likely as many varieties of RNA editing as there are organisms, all belong to one of three known mechanisms: the site-specific *insertion* or *deletion* of one or several nucleotides, or nucleotide *substitution* (cytidine-to-uridine and adenosine-to-inosine deamination, uridine-to-cytidine transamination). Although we will not describe them here, other processes may occur before the final mRNA transcript is translated into a protein sequence or processed into a functional RNA. The relationship between DNA and gene product is indirect and mediated to an extent that was never anticipated when the basic mechanisms of transcription, RNA processing, and translation were clarified in the 1960s.

THE MODERN GENE

The 'modern gene' as Portin (1993) has termed it represents a further stage in the dialectic of structure and function described. The classical gene, primarily defined by the functional role it played in heredity, became identified with the structural gene revealed by early molecular biology, primarily through the study of prokaryotes and bacteriophage. As a result, the functional role of the gene was redefined as the determination, by linear correspondence, of the structure of a gene product. Further investigation of the manner in which a wider range of genomes generate a wider range of gene products has revealed that this functional role can be filled by diverse, highly flexible mechanisms at the level of the DNA itself: "We are currently left with a rather abstract, open and generalized concept of the gene, even though our comprehension of the structure and organization of the genetic material has greatly increased" (Portin 1993, 173). Goldschmidt's critique of the particulate gene has been explicitly revived in the light of our new understanding of genome structure and function:

The particulate gene has shaped thinking in the biological sciences over the past century. But attempts to translate such a complex concept into a discrete physical structure with clearly defined boundaries were always likely to be problematic, and now seem doomed to failure. Instead, the gene has become a flexible entity with borders that are defined by a combination of spatial organization and location, the ability to respond specifically to a particular set of cellular signals, and the relationship between expression patterns and the final phenotypic effect. (Dillon 2003, 457)

In a prescient paper twenty years ago, Raphael Falk reviewed what were then newly emerging challenges to the classical molecular gene and concluded:

Today the gene is not just *the* material unit or *the* instrumental unit of inheritance, but rather *a* unit, *a* segment that corresponds to *a* unit-function, as defined by the individual experimentalist's need. It is neither discrete – there are overlapping genes, nor continuous – there are introns within genes, nor does it have a constant location – there are transposons, nor a clearcut function – there are pseudogenes, not even constant sequences – there are consensus sequences, nor definite borderlines – there are variable sequences both 'upstream' and 'downstream'. (Falk 1986, 169)

Thus, as early as 1986 we were well on the way from the "well defined material entity back to an abstraction, a hypothetical construct, if not an intervening variable, devised by scientist for their needs" (Falk 1986, 160).

Focusing on the cutting edge of contemporary genomics can induce an extremely deflationary view of the gene. Some molecular biologists, realizing that the concepts of 'gene' transcription or 'gene' expression may not suffice to capture the variation in expressed genomic sequences, have proposed the more general term 'genome transcription' to allow for the incorporation of RNA transcripts that contain sequences outside the border of canonical genes. This view does not sit easily with the classical molecular conception of genes, which from the new perspective seem like "statistical peaks within a wider pattern of genome expression" (Finta and Zaphiropoulos 2001). One pragmatic, technological reason that today's biologists are prepared to consider such radical options is that the challenge of automated gene annotation has turned the apparently semantic issue of the definition of 'gene' into a pressing and practical one as the limitations of a purely structural, sequenced-based definition of the gene have become apparent. One influential recent review concludes that "one solution for annotating genes in sequenced genomes may be to return to the original definition of a gene – a sequence encoding a functional product – and use functional genomics to identify them" (Snyder and Gerstein 2003, 260).

The gene concept, however, plays a role in many other contexts besides the cutting edge of genomics (Stotz, Griffiths, and Knight 2004). We suggest, therefore, that there are at least three answers to the question 'What is a gene?', none of which can be neglected if we hope to depict the state of contemporary biology accurately. These are the traditional, instrumental gene; the postgenomic molecular gene; and the 'nominal gene'.

THE TRADITIONAL GENE

Biologists can and do still use genetic analysis – the analysis of the phenomenon of heredity by the analysis of the results of hybridization, either between organisms or directly between DNA molecules (Waters 2004; Falk in press). Genetic analysis remains a key tool in addressing broader biological questions. For these purposes the gene remains an intervening variable, defined by the inheritance patterns that it enables us to follow, and the difficulties of providing a univocal account of its identity as a material unit can be put to one side. The traditional gene concept is retained in much the same way in population genetics. In an important recent analysis, Lenny D. Moss introduces the term 'Gene-P' for something very like our 'traditional' gene (Moss 2003). The P stands variously for 'phenotype', 'prediction', and 'preformation' since these genes are identified in terms of their phenotypic effects, are used to predict the phenotypic results of hybridization, and reflect what Moss terms 'instrumental preformationism' – a strategic neglect of the ways in which the gene-phene relationship depends upon other factors. Moss contrasts his Gene-P to a materialistic concept of the gene that he calls 'Gene-D' (for 'development'). Genes-D are defined by their intrinsic chemical capacity to template for gene products. Here, we wish to distinguish two importantly different ways to conceptualize genes that fall within the general area of conceptual space that Moss labels Gene-D.[2]

THE POSTGENOMIC MOLECULAR GENE

We use the phrase 'postgenomic molecular gene' to refer to the entities that continue to play the functional role of the molecular gene – making gene products – in contemporary molecular biology. The postgenomic molecular gene concept embodies the continuing project of understanding how genome structure supports genome function, but with a deflationary picture of the gene as a structural unit. These genes are "things you can do with your genome" (Stotz, Bostanci, and Griffiths 2006): although the gene is still an 'image' in the DNA of the target molecule (the molecule whose activity we wish to understand) this image may be fragmented or distorted to such an extent that it cannot be discerned until functional genomics has revealed how these sequence elements are used in the broader genomic and cellular context. This conception of the gene remains a critical aspect of the epistemology of molecular bioscience simply because linear correspondence between molecules is fundamental to biologists' ability to identify and manipulate them, via technologies ranging from cDNA libraries to microarrays to RNA interference. But although it is important to know the 'gene for' some molecule in this sense, it does not matter very much whether that collection of sequence elements is a gene! To put it less paradoxically, the utility of knowing the DNA elements that underlie the production of the target molecule or its precursors does not at all depend on whether it is possible to give a univocal definition of the material gene. Finding the 'gene for' the molecule in this sense remains important even on the most deflationary, postgenomic view of the molecular gene.

NOMINAL GENE

The use of databases containing nucleotide sequences is well established. Codified as part of this process is a particular use of gene concepts on the basis of which one can identify various genes and count the number of genes in a given genome. ... I call genes, picked out in this way, nominal genes. A good way of parsing my argument is that nominal genes are a useful device for ensuring that our discourse is anchored in nucleotide sequences, but that nominal genes do not, and probably can not, pick out all, only, or exactly the genes that are intended in many other parts of genetic work. (Burian 2004b, 64–65)

It is hard to disagree that for many practical purposes genes are simply sequences that have been annotated as genes and whose annotation as such has been accepted by the scientific community. But, as Burian himself makes clear, this does not imply that the scientific community has a clear understanding of what makes a sequence a gene that needs only to be made explicit. Thomas Fogle has argued powerfully that this is not the case (Fogle 2001). The working concept of the gene, according to Fogle, is something like a stereotype or prototype: a sequence is a gene if it has enough similarities to other genes: for example, it contains an open reading frame, has one or more promoters, has one or more transcripts that are not too functionally diverse from one another, and so on. This is more or less a description of automated 'gene discovery' methods, and Fogle's suggestion is that the *concept* of the gene is no more principled or definitionlike than this. The various 'genelike' features are not weighted against one another in any principled, theory-driven way, but rather are weighted differently on different occasions in order to segment the DNA sequence into fairly traditional-looking 'genes', sometimes giving up on structural criteria to save functional ones (as in the example in Figure 5.1), at other times giving up on functional criteria to save structural ones (as in cotranscription of a gene and a 'pseudogene').

Fogle is quite critical of this state of affairs, arguing that combining structural and functional features into a single stereotype, what he calls the 'consensus' gene, hides both the diversity of DNA sequences that can perform the same function and the diverse functions of particular DNA sequences. Burian takes a more positive view, emphasising the value of simply having a shared collection of named sequences known or suspected to be involved in the production of gene products.

CONCLUSION

The gene began life as an intervening variable, defined functionally in terms of the Mendelian pattern of heredity in observable phenotypic characters. It rapidly acquired a second identity as a hypothetical material unit. A productive dialectic between investigations of the gene that identified it in each of these two ways concluded with the 'neo-classical' or 'classical molecular' conception of the gene. The functional role of the gene was redefined to exclude mutation and recombination, which became properties of the DNA in its own right, rather than of individual genes. The function of the gene became the determination of the structure of gene products via linear correspondence between molecules. This functional role was played by a natural class of units at the molecular level – the structurally defined molecular gene. Further investigation of a wider range of genomes and a wider range of gene products has thrown into doubt whether an adequate structural definition is possible – the structural basis upon which gene products are generated may be a very broad class of 'things you can do with your genome'. At this point it remains possible to think of genes in the traditional manner that dates back to the early twentieth century as intervening variable in the genetic analysis of phenotypes. It is also possible to think of them as the often complex collections of sequence elements that fill the functional role of the molecular gene ('postgenomic molecular genes'). Finally, it is possible to think of genes as simply those sequences whose similarity on various dimensions to stereotypical genes has led them to be annotated as genes and whose annotation as such has been accepted by the scientific community ('nominal genes').

ACKNOWLEDGMENTS

This material is based upon work supported by the National Science Foundation under Grants 0217567 and 0323496 and supplemental funding from the University of Pittsburgh. Griffiths's work on this chapter was supported by an Australian Research Council Federation Fellowship. Any opinions, findings, and conclusions or recommendations expressed in this material are those of the authors and do not necessarily reflect the views of the NSF.

NOTES

1. In contemporary usage, *cis-* elements are those transcribed together as parts of a single pre-mRNA whereas *trans-* elements are transcribed separately and united at some stage of posttranscriptional processing (*trans*-splicing). Thus *trans-* elements in the modern sense (*trans-* on mRNA) may be *cis-* located on the DNA.

2. Moss (pers. comm.) suggests that our 'postgenomic molecular' and 'nominal' material genes are perspectives on genes-D corresponding to what are, somewhat perversely, called 'forward' and 'reverse' genetics. The postgenomic molecular gene embodies the traditional, 'forward', strategy of locating the template resources corresponding to a known phenotype. The nominal gene is a template resource whose use we set out to understand.

6   Information in Biology

1.   INTRODUCTION

The concept of information has acquired a strikingly prominent role in contemporary biology. This trend is especially marked within genetics, but it has also become important in other areas, such as evolutionary theory and developmental biology, especially where these fields border on genetics. The most distinctive biological role for informational concepts, and the one that has generated the most discussion, is in the description of the relations between genes and the various structures and processes that genes play a role in causing. For many biologists, the causal role of genes should be understood in terms of their *carrying information* about their various products. That information might require the cooperation of various environmental factors before it can be "expressed," but the same can be said of other kinds of message.

An initial response might be to think that this mode of description is entirely anchored in a set of well-established facts about the role of DNA and RNA within protein synthesis, summarized in the familiar chart representing the "genetic code," mapping DNA base triplets to amino acids. However, informational enthusiasm in biology predates even a rudimentary understanding of these mechanisms (Schrodinger 1944). And more importantly, current applications of informational concepts extend far beyond anything that can receive an obvious justification in terms of the familiar facts about the specification of protein molecules by DNA. This includes

(i)   The description of whole-organism phenotypic traits (including complex behavioral traits) as specified or coded for by information contained in the genes;

(ii) The treatment of many causal processes within cells, and perhaps of the whole-organism developmental sequence, in terms of the execution of a *program* stored in the genes;

(iii) The idea that genes themselves, for the purpose of evolutionary theorizing, should be seen as, in some sense, "made" of information. From this point of view, information becomes a fundamental ingredient in the biological world.

There is no consensus about the proper form and status of these kinds of description, and the result has been the development of a foundational discussion within both biology and the philosophy of biology. Some have hailed the employment of informational concepts here as a crucial advance (Williams 1992). Others have seen almost every biological application of informational concepts as a serious error, one that distorts our understanding and contributes to lingering genetic determinism (Francis 2003). Most of the possible options between these extreme views have also been defended. These include various arguments that *some*, though not all, of the popular uses of informational concepts in biology are legitimate (Godfrey-Smith 2000, Griffiths 2001). They also include arguments that even the more tendentious uses of these concepts are legitimate so long as the concepts are applied consistently (Sterelny, Smith, and Dickison 1996, Jablonka 2002). Other philosophers and biologists regard the whole matter as a tempest in a teacup; they do not think that the development of an informational language for describing genes makes much of a difference to anything, as it is obviously a loose metaphorical usage that carries no real theoretical weight (Kitcher 2001).

The philosophical discussion has developed for two reasons. One is the general philosophical interest in abstract conceptual problems in particular areas of science – an interest in debates that seem resistant to empirical adjudication, but do not seem merely terminological. So some philosophical interest here is akin to more familar philosophical attention to such biological concepts as fitness, species, and natural selection. But the concept of information is not merely an ordinary theoretical concept within a particular part of science. It is also part of a family of concepts that has been the focus of intense study in several parts of philosophy, stretching back for centuries. "Information" itself does not have a long history in philosophy, but it is closely related to concepts that do, such as the concept of meaning, which is central to philosophy of language and much philosophy of mind. So philosophers are familiar with the kinds of puzzles that are generated by this family of "semantic" concepts. It is not that philosophers have developed a consensus theory that can be applied, in an off-the-shelf way, to new cases. But philosophers are intimately acquainted with many of the puzzles, twists and turns, red herrings, and trade-offs that arise in this area. So as information (and related concepts) have become more prominent in biology, some philosophers have thought that this is an area where they are qualified to help in the development of useful and coherent biological concepts.

This chapter has two main sections. The next section gives an outline of some of the arguments and options developed to date. The third section then develops some more novel ideas, which are presented in a cautious and exploratory way.

Before moving to the survey section, there are two other preliminary points to make. First, the topic of this essay is not the role of the concepts of information and representation in the parts of biology where they are most *obviously* relevant; the essay is not concerned with neuroscience, perception, language processing, and so on. The topic is the role of information (and its relatives) in parts of biology where its role is less obvious, such as the description of genes, hormones, and (to some extent) signaling systems at the cellular level. Second, in the early part of this discussion I will not put much emphasis on some of the finer distinctions between the concepts of information, representation, meaning, coding, and so on. As the discussion proceeds, distinctions between concepts within this family will become more important, but some of the subtle distinctions will be backgrounded initially.

2.   OUTLINE OF THE DEBATE

One common way to start organizing the problem is to make a distinction between two senses of "information," or two kinds of application of informational concepts. One of these is a weak or minimal sense, and the other is stronger and more controversial. In the weaker sense, informational connections between events or variables involve no more than ordinary correlations (or perhaps correlations that are "nonaccidental" in some physical sense involving causation or natural laws). This sense of information is associated with Claude Shannon (1948), who showed how the concept of information could be used to *quantify* facts about contingency and correlation in a useful way, initially for communication technology. For Shannon, anything is a *source* of information if it has a number of alternative states that might be realized on a particular occasion. And any other variable *carries information* about the source if its state is correlated with that of the source. This is a matter of degree; a signal carries more information about a source if its state is a better predictor of the source, less information if it is a worse predictor.

This way of thinking about contingency and correlation has turned out to be useful in many areas outside the original technological applications that Shannon had in mind, and genetics is one example. There are interesting questions that can be asked about this sense of information (Dretske 1981), but the initially important point is that when a biologist introduces information in this sense to a description of gene action or other processes, she is not introducing some new and special *kind* of relation or property. She is just adopting a particular quantitative framework for describing ordinary correlations or causal connections.

Consequently, philosophical discussions have sometimes set up the issue by saying that there is one kind of "information" appealed to in biology, Shannon's kind, that is unproblematic and does not require much philosophical attention. The term "causal" information is sometimes used to refer to this kind, though this term is not ideal. Whatever it is called, this kind of information exists whenever there are ordinary contingency and correlation. So we can say that genes contain information about the proteins they make, and also that genes contain information about the whole-organism phenotype. But when we say that, we are saying no more than what we are saying when we say that there is an informational connection between smoke and fire, or between tree rings and a tree's age. The more contentious question then becomes whether or not biology needs *another*, richer concept of information as well as Shannon's concept. Information in this richer sense is sometimes called "semantic" or "intentional" information.

What is the difference between them, and why might we think that biology needs to employ a richer concept? There is a range of differences between the two. First and most importantly, informational connections in the Shannon sense connect environmental conditions with biological traits in the same way that they connect genes and those traits. With respect to Shannon information, there is what Griffiths and Gray call a "parity" between the roles of environmental and genetic causes (Griffiths and Gray 1994, Griffiths 2001). In addition, information in the Shannon sense "flows" in both directions, as it involves no more than learning about the state of one variable by attending to another. So we can read off something about the phenotype from the state of the genes, but we can also learn something about the genes by attending to the phenotype.

Some talk about information in biology is consistent with these features of Shannon information, but some is not. It is usually thought that at least some applications of informational language to genes is supposed to ascribe to genes a special kind of causal property that is *not* ascribed to environmental conditions, even when they are causally important, and that is also unidirectional.

In addition, a message that carries "semantic information," it is often thought, has the capacity to misrepresent, as well as accurately represent, what it is about. There is a capacity for error. Shannon information does not have that feature; we cannot say that some variable carries false information about another, if we are using the original Shannon sense of the term. But biologists do apparently want to use language of that kind when talking about genes. Genes carry a message that is *supposed* to be expressed, whether or not it actually is expressed.

Once we take the alleged semantic properties of genes as seriously as this, some subtle questions arise. If genes are carrying a message in this sense, the message apparently has a prescriptive or imperative content, as opposed to a descriptive or indicative one. Genes contain instructions, not descriptions. Their "direction of fit" to their effects is such that if genes and phenotype do not match, what we have is a case of unfulfilled instructions rather than inaccurate descriptions.

Several philosophers and biologists have argued that much informational talk about genes uses a richer concept than Shannon's, but this concept can be given a naturalistic analysis. It is not a lapse back into unscientific teleological thinking. One way to proceed is to make use of a rich concept of biological *function*, in which the function of an entity derives from a history of natural selection (Sterelny, Smith, and Dickison 1996, Maynard Smith 2000, Shea forthcoming). This sort of move is familiar from the philosophy of mind, where similar problems arose in the explanation of the semantic properties of mental states. When an entity has been subject to and shaped by a history of natural selection, this can provide the grounding for a kind of purposive or normative description of the causal capacities of that entity. To use the standard example (Wright 1976), the function of a heart is to pump blood, not to make thumping sounds, because it is the former effect that has led hearts to be favored by natural selection. The hope is that a similar "teleofunctional" strategy might help make sense of the semantic properties of genes, and perhaps other biological structures with semantic properties.

There are several ways in which the details of such an account might be developed (Godfrey-Smith 1999), some focusing on the evolved functions of the genetic machinery as a *whole*, and others on the natural selection of *particular* genetic elements. All versions of this idea offered so far have problems of detail. One problem is that there is no overall connection between biological function and semantic properties; having a function in the rich historical sense is not generally sufficient for having semantic properties. Legs are for walking, but they do not represent walking. Enzymes are for catalyzing reactions, but they do not instruct this activity. There are things that legs and enzymes are *supposed* to do, but this does not make them into information carriers, in a rich beyond-Shannon sense. Why should it do so for genes?

Sterelny, Smith, and Dickison seem to think there is a quite intimate connection between evolutionary function and semantic properties in the case of biological structures that have been selected to play a causal role in *developmental* processes. They argue that genes, in virtue of these functional properties, represent the outcomes they are supposed to produce. They add, however, that any nongenetic factors that have a similar developmental role, and have been selected to play that role, also have semantic properties. So Sterelny, Smith, and Dickison want to ascribe very rich semantic properties to genes, but not only to genes. Some nongenetic factors have the same status.

Proposals that appeal to evolutionary design to "enrich" the informational properties of genes have problems of detail, but they also have attractive features. It is striking that John Maynard Smith, when he grappled with the status of his enthusiasm for informational concepts in biology, opted for something along these lines (2000). The resulting overall picture has good structural features. We would have a loose, uncontroversial Shannon sense of information that applies to all sorts of correlations, and an "overlay" of richer semantic properties in cases where we have the right kind of history of natural selection. Genes and a handful of nongenetic factors would have these properties; most environmental features that have a causal role in development would not. The neatness of the resulting picture provides, for some people, good reason to persevere with some account along these lines.

So far in this section I have mostly discussed the concept of information; there has not been much talk of "coding." And the ideas discussed so far do not put any emphasis on the special features of genetic mechanisms themselves, such as the combinatorial structure of the "genetic code." But surely these features of genetic mechanisms provide much of the underlying motivation for the introduction of semantic concepts into biology? It might seem so, but a lot of discussions have in effect treated this as an open question. As noted previously, the enthusiasm for semantic characterization of biological structures extends back before the genetic code was discovered. (See Kay 2000 for a detailed historical treatment.) But another line of thought in the literature, overlapping with the preceding ideas, has focused on the special features of genetic mechanisms, and on the idea of "genetic coding" as a contingent feature of these mechanisms.

Both Godfrey-Smith (2000) and Griffiths (2001) have argued that there is one highly restricted use of a fairly rich semantic language within genetics that is justified. This is the idea that genes "code for" the amino acid sequence of protein molecules, in virtue of the peculiar and contingent features of the "transcription and translation" mechanisms found within cells. Genes specify amino acid sequence via a templating process, which involves a regular mapping rule between two quite different kinds of molecules (nucleic acid bases and amino acids). This mapping rule is *combinatorial*, and apparently *arbitrary* (in a sense that is hard to make precise – Stegmann 2004).

The argument is that these features make gene expression into a causal process that has significant analogies to various paradigmatic symbolic phenomena, such as the use of natural language. Some have argued that this analogy becomes questionable once we move from the genetics of simple prokaryotic organisms (bacteria), to those in eukaryotic cells. This has been a theme of Sarkar's work (1996). Mainstream biology tends to regard the complications that arise in the case of eukaryotes as mere details, which do not compromise the basic picture we have of how gene expression works. An example is the editing and "splicing" of messenger RNA (mRNA) transcripts into a processed mRNA that is used in translation. This is a biologically important process, and it does make the DNA a much less straightforward predictor of amino acid sequence, but it can be argued that this does not much affect the crucial features of gene expression mechanisms that motivate the introduction of a symbolic or semantic mode of description.

So the argument in Godfrey-Smith (2000) and Griffiths (2001) is that there is one kind of informational or semantic property that genes and only genes have: coding for the amino acid sequences of protein molecules. But this relation "reaches" only as far as the amino acid sequence. It does not vindicate the idea that genes code for whole-organism phenotypes, let alone provide a basis for the wholesale use of informational or semantic language in biology. Genes can have a reliable causal role in the production of a whole-organism phenotype, of course. But if this causal relation is to be described in informational terms, then it is a matter of ordinary Shannon information, which applies to environmental factors as well.

In this section I have distinguished one line of thought that looks at Shannon information and its "enriched" relatives, and another line of thought that looks at the peculiar features of the mechanisms of gene expression, and the original narrow idea of a "genetic code." But the two lines of thought can be married in various ways. Maynard Smith, in response to problems with his teleofunctional account, appealed at one point to some special features of genetic mechanisms, including the apparent "arbitrariness" of the genetic code. This idea has been popular, but is hard to make precise. The key problem is that any causal relation can look "arbitrary" if it operates via many intervening links. There is nothing "arbitrary" about the proximal mechanisms by which a molecular binding event occurs. What makes the genetic code seem "arbitrary" is the fact that the mapping between base triplets and amino acids is mediated by contingent features of the sequences of transfer RNA (tRNA) molecules, and of the enzymes that bind amino acids to tRNA molecules. Because we often focus on the "long-distance" connection between DNA and protein and pay less attention to the intervening mechanisms, the causal relation appears arbitrary. If we picked out and focused on steps in any other biological cascade that are separated by three or four intervening links, the causal relation would look just as "arbitrary." Here it is also significant that the standard genetic code is turning out to have more systematic and nonaccidental structure than people had once supposed (Knight, Freeland, and Landweber 1999).

I will discuss three more topics, in a more self-contained way, to finish the survey. The first is the idea that genes contain a *program*, in a sense analogous to that in computer science (Mayr 1961, Moss 1992, Marcus 2004). This idea has not been discussed in such a concerted way by philosophers, though it is seen constantly in biological discussion. Here the focus is more on the control of *processes* by genes, as opposed to the specification of a particular *product*.

The "program" concept seems to be applied in biology in an especially broad and unconstrained way, often guided only by very vague analogies with computers and their workings. First, we might isolate a *very* broad usage, in which talk of programming seems merely aimed at referring to the intricate but orderly and well-coordinated nature of many basic processes in biological systems. Here, the most that talk of "programs" could be doing is indicating the role of evolutionary design. An example might be talk of "programmed cell death" in neuroscience, which is a very important process within neural development that could just as accurately be described as "orderly and adaptive cell-death in accordance with evolutionary design."

Second, however, we might isolate a sense in which talk of "programs" in biology is driven by a *close* analogy between some biological process and the *low-level* operation of modern computers. One crucial kind of causal process within cells is cascades of up- and down-regulation in genetic networks. One gene will make a product that binds to and hence down-regulates another gene, which is then prevented from making a product that up-regulates another – and so on. What we have here is a cascade of events that can sometimes be described in terms of Boolean relationships between variables. One event might follow only from the conjunction of another two, or from a disjunction of them. Down-regulation is a kind of negation, and there can be double and triple negations in a network. Gene regulation networks have a rich enough structure of this kind for it to make sense to think of them as engaged in a kind of computation. Computer chip "and-gates," neural "and-gates," and genetic "and-gates" have some genuine similarities. Most other biological processes, though just as much the product of evolutionary design, do not have a structure that motivates this sort of computational description. And once again we find, as in the case of "genetic coding," that the domain in which this computational language is well motivated, when applied to genes, is confined to the cellular level. Less elaborate cascades of this kind can also be found in the endocrinological (hormone-using) systems within the body. Here, too, informational language can seem naturally applicable and may be justified by a similar line of argument.

The second of the three topics I will discuss to finish this section is the link between informational description and genetic determinism. A number of critics have argued that the informational or semantic perspective on gene action fosters or encourages naive ideas about genetic determinism (Oyama 1985, Griffiths 2001). Others think that genetic determinism, when it is false, is an ordinary error about causal relations that has no particular link to the informational description of those relations. I side with the critics here who say that there is something definite about informational description of genes that encourages fallacies about genetic causation. The key point has been summarized by Griffiths. He notes that in complex systems, almost all causal factors are context dependent, and usually it is not hard to remember this. If we think in ordinary causal terms, it is straightforward to note that a genetic cause will have its normal effects only if accompanied by suitable environmental conditions, and an environmental cause will have its normal effect only if accompanied by suitable genetic conditions. (If the sensitivity on either side is high, then talk of "normal" effects itself may be misleading.) But, Griffiths suggests, the informational mode of describing genes (and other factors) fosters the *appearance* of context independence. "Genetic causation is interpreted deterministically because genes are thought to be a special kind of cause. Genes are instructions – they provide information – whilst other causal factors are merely material. ... A gay gene is an instruction to be gay even when [because of other factors] the person is straight" (2001, 395–96). So the idea is that the inferential habits and associations that tend to go along with the use of informational or semantic concepts lead us to think of genes as having an additional and subtle *kind* of extra causal specificity. These habits can have an effect even when people are willing overtly to accept context dependence of (most) causes in complex biological systems. Relatedly, the idea of internal genetic messages may also foster a tendency toward a kind of essentialist thinking; the meaning of the internal message tells us what the "true nature" of the organism is, regardless of whether this nature is actually manifested.

My final topic in this survey is the most strong and tendentious employment of informational language for genes, which arises in the context of evolutionary biology. It has been common for some time to say that, in the evolutionary context, we should think of a gene in terms of its sequence, which is preserved over many replication events, and not in terms of particular DNA molecules, which come and go (Dawkins 1976, 1986). The idea that sequence can be preserved across changes in the underlying molecules is certainly reasonable and important. But this message, important as it is, has been expressed in extreme and philosophically mysterious ways by some theorists. G. C. Williams (1992), for example, holds that because of these facts about the preservation of gene sequence across changes in molecules, we should think of information as a kind of fundamental ingredient of the universe, along with mass and energy, that exists in its own "domain." This makes the causal connections between the informational domain and the ordinary physical domain quite mysterious, and Williams himself finds this an important problem. But the appearance of a problem arises only from an unnecessary reification of information. We can instead say that what has been learned from work on the evolutionary features of genes is that various different physical objects can share their informational properties. These informational properties are explicable in terms of the lower-level physical properties of the objects, and the contexts in which the objects are embedded. Such a view does raise some further questions, but it does not introduce the idea of information as a separate "stuff" whose relations to ordinary physical things are tenuous and problematic.

The enthusiasm for a reified treatment of information can lead to other theoretical problems in biology. Some of the recent advocates of "intelligent design" creationism have tried to use the special and mysterious properties of information to mount anti-Darwinian arguments (Dembski 1997, criticized in Godfrey-Smith 2001). These arguments have no real force. Indeed, the resulting views tend to be *less* plausible than earlier versions of the argument from design, because even routine and low-level forms of evolution by natural selection, such as the evolution of drug resistance in bacteria, tend to be ruled out as impossible in principle. But the informational terminology in which the arguments are expressed lends them a spurious appearence of rigor.

3.   THE NEXT STEPS?

In this section I will cautiously introduce some ideas that approach the whole problem somewhat differently. I will motivate the change in tack by asking what appears to be an odd question. Is the informational or semantic description of genes *metaphorical*, or not? This should be an easy question to answer, but in fact seems to be surrounded by uncertainty. On the one hand, biologists sometimes say that the introduction of an informational framework was a crucial theoretical advance. This suggests that it is not at all a metaphor. If electrical charge, and entropy, were crucial theoretical advances in their day, it was not by being metaphors. But if one presses hard on what these informational properties are supposed to be, especially once we get beyond the simple idea of a combinatorial mapping from nucleic acids to amino acids, it is common to encounter a retreat to the idea of genetic information as a metaphor. It is not literally true that genes are programming development or representing the whole-organism phenotype, but this is a metaphor that has proved invaluable to biology.

Of course, we have to expect some vagueness here. And we can not expect biologists to be experts on the analysis of literal and nonliteral language. But what makes the situation odd is the fact that if someone tried carefully to adjudicate this question, he or she would run immediately into the fact that in the case of ascriptions of semantic properties, there is no clear and well-understood border between literal and metaphorical. There is not a clear and well-demarcated sense of what the literal domain is, to which metaphorical cases are being compared.

The same problem arises, to some extent, in cognitive science, which is often based on the idea that the mind/brain can be seen as a computer. Does this mean, there is such a thing as computation, and the brain literally does it? Or is the idea of neural computation something more like a metaphor? The abstract theory of computation, within mathematics, is not especially helpful for answering this question (Smith 2002).

In the case of computation in cognitive science, the question can be deflected initially by saying that computation is being treated as a "model" for the mind. But the term "model" is so ambiguous that this does not help much. Sometimes "model" means a provisional and cautiously defended theory. This does not help here because caution is not the issue. We want to know whether information processing, computation, representation, and so on, are real natural kinds that brain activity – and genetic activity – might be literal instances of, or whether some other story about the role of these concepts has to be told.

In the remainder of this section I will sketch one alternative story of this kind. It is designed to contrast with the simpler idea that informational properties are definite but elusive properties that genes either do or do not have. Instead, informational description of genes is motivated by a family of factors, which I will group into three categories. First, it is motivated by some real and uncontroversial features of genes and DNA themselves, which would not alone be sufficient to motivate an elaborate informational description. Second, the use of informational and semantic language introduces into biology a particular "causal schematism," derived from everyday contexts in which symbols are used. The schematism functions as a model, in a sense discussed in some recent philosophy of science.

Third, the informational framework reflects and reinforces a commitment to a way of demarcating the scientifically important features of genes and associated mechanisms. The framework foregrounds one set of properties and backgrounds another, and the properties of genes and other molecules that are being foregrounded are *sequence* properties, as opposed to all their other chemical properties. The result of this analysis is an account of the role of informational language in biology that is more focused on the entire disciplinary role of the informational framework and less on specific informational properties that might or might not be real.

I will say more about each of the three categories in turn. First, the informational framework is motivated, of course, by some real and uncontroversial features of genes and DNA themselves. Some of these were highlighted in the previous section, and they include the combinatorial structure and regularity of the mapping rule from nucleic acids to amino acids. But some motivation may also arise from a feature of DNA that is not so often remarked on in this context. This is the passivity, or comparative inertness, of DNA. Here we focus on some facts about what DNA does *not* do, as well as what it does.

The evolution of DNA as a repository of sequence information is often said to be due in part to its chemical stability. Origin of life work emphasizes the fact that RNA is a good initial replicator molecule because it has some enzymatic activity, but DNA is more stable once proteins have been developed for enzymatic work. And in modern cells, DNA does not *do* very much in chemical terms; almost all of its effects go via a particular indirect causal pathway by which DNA sequence is transcribed and translated. (The main exception to this is DNA's direct interaction with transcription factors, in gene regulation.) Proteins, as is always noted, do most of the actual chemical work in the cell. DNA specifies amino acid sequence and does not do much else. So to call DNA an "informational" molecule, in a modern context, is often a gesture toward what it does *not* do, as well as to what it does do.

My suggestion for a second set of motivations is more tendentious. It involves a general analysis of when and why people introduce semantic concepts (including information) into scientific and other explanatory contexts. The suggestion is that the use of these concepts is generally guided – not always consciously – by the postulation of an analogy between a particular everyday form of symbol use and the domain that the theorist is trying to understand. This analogy can be very partial, while still exerting influence on how the phenomena are described and understood.

How does the analogy work? A central aspect of everyday symbol use is that one object is used to "stand for" another. More precisely, a person guides behavior directed on one object or domain by attending to the state of another. This is the schematic core of everyday symbol use, and it shows up abstractly in many philosophical analyses of semantic phenomena (e.g., Millikan 1984), as well as in models of signaling games and the evolution of meaning (Skyrms 1996). This basic pattern is also installed in the basic picture that Shannon used in his theory of information: we have a source, and a signal whose state can be consulted to learn something about the source.

A central feature of this "causal schematism" is the distinction between some mechanism that reads or consumes the signal and the signal itself. In the genetic case, the idea that semantic description is guided by this model is quite helpful. First, we see that the basic cell-level machinery of transcription and translation is, in fact, a fairly good instance of the schematic structure in question. The ribosomal/tRNA machinery is, in effect, a reader or consumer of nucleic acid sequence, with the function of creating protein products that will have a variety of uses elsewhere in the cell. We also see that this realization of the causal schematism applies *only* at the cell level, at the level at which the transcription and translation apparatus shows up as a definite part of the machinery. So we see why it is true – if it is true, as I think it is – that the use of informational or semantic language in explaining how protein molecules are made is legitimate and well motivated, while the use of this language when talking about the role of genes in producing whole-organism phenotypes is not. Once we think in terms of the influence of analogy and a causal schematism here, we can also note a connection to the discussion of the comparative chemical "passivity" of DNA discussed earlier in this section. Paradigmatic cases of messages in everyday life are rather physically passive, too, having their significant effects only via their interpretation by a reader or consumer.

This second category of factors motivating informational description of genes involves a kind of model-based theorizing, in a sense that was developed for the analysis of very different parts of science (Giere 1988, Godfrey-Smith forthcoming, Weisberg forthcoming). The term "model" gives us some definite purchase here after all.

My third category involves a role for the informational framework that is not part of a causal hypothesis, a posited mechanism, or anything of that kind. Instead, it involves a commitment to a way of demarcating and categorizing an entire domain. Via the informational framework, a commitment is made to the importance of one set of properties and the unimportance of another. One set of properties of biological molecules is foregrounded, by introduction of a language that can naturally accomodate them, while another set of properties is backgrounded. What are being foregrounded are *sequence* properties, as objects of study, as opposed to all the other chemical properties of genes and associated biological molecules. The suggestion is that rather than attributing some particular causal powers to DNA sequence, the informational framework often functions to make sequences in general the primary focus of study. What results is a form of abstraction akin to that seen in statistical mechanics; there is a focus on a distinctive level of description and a particular set of statistical features of interactions between particles, abstracting away from lots of other properties (Griesemer 2005). The informational framework also brings with it a set of conceptual tools that are suited for the analysis of sequence properties, as opposed to other chemical properties. However, it should be added here that there are conspicuous uses of informational language in biology in contexts where sequence properties are not treated as central, for example in the description of hormonal signaling. In these cases I would emphasize the second of the three factors discussed in this section, the role of a causal schematism derived from public symbol use.

Suppose the actual patterns of use of informational language in genetics are in fact guided by factors like these, in a context-sensitive mixture. The use of the informational framework is guided by some real features of genetic mechanisms, but also by application of schematic causal model that guides many or most uses of a semantic language. It reflects and reinforces a general disciplinary focus on sequence properties as opposed to others. This would steer us away from the idea that there is some definite but hidden set of properties being posited by such language, which might or might not be real. I will leave somewhat open how this set of ideas relates to the more standard lines of thought outlined in the previous section. In some ways, the two can complement each other. In other ways, there is probably some tension.

ACKNOWLEDGMENTS

I am indebted to Arnon Levy and Nicholas Shea for comments and discussion of these issues.

7 Reductionism (and Antireductionism) in Biology

Accelerating developments in molecular biology since 1953 have strongly encouraged the advocacy of reductionism by a number of important biologists, including Crick, Monod, and E. O. Wilson, and strong opposition by equally prominent biologists, especially Lewontin, along with most philosophers of biology.

Reductionism is a metaphysical thesis, a claim about explanations, and a research program. The metaphysical thesis that reductionists advance (and antireductionists accept) is physicalism, the thesis that all facts, including the biological facts, are fixed by the physical and chemical facts; there are no nonphysical events, states, or processes, and so biological events, states, and processes are "nothing but" physical ones. This metaphysical thesis is one reductionists share with antireductionists. The reductionist argues that the metaphysical thesis has consequences for biological explanations: they need to be completed, corrected, made more precise, or otherwise deepened by more fundamental explanations in molecular biology. The antireductionist denies this inference, arguing that nonmolecular biological explanations are adequate and need no macromolecular correction, completion, or grounding. The research program that reductionists claim follows from the conclusion about explanations can be framed as the methodological moral that biologists should seek such macromolecular explanations. Antireductionists argue that such explanations are either or both unnecessary or unattainable. Reductionists argue that their view must be distinguished from eliminativism, the thesis that molecular biology not only provides the explanans (what does the explaining), but also describes all the facts to be explained (the explanantia). Antireductionists hold that reductionism must inevitably collapse into eliminativism.

1.   REDUCTIONISM – THE POSTPOSITIVIST PICTURE

The relationship of the rest of biology to molecular biology was originally envisioned by philosophers of science to vindicate a model of reduction associated with philosophers who had been positivists or their students (whence the label 'postpositivist'). In the locus classicus of reduction, Ernest Nagel's *Structure of Science* (1961), reduction is a form of intertheoretical explanation characterized by the deductive derivation of the laws of the reduced theory from the laws of the reducing theory. The deductive derivation requires that the concepts, categories, and explanatory properties, or natural kinds of the reduced theory, be captured in the reducing theory. To do so, terms of the narrower theory must be linked to concepts in the broader theory. As exponents of reduction such as Schaffner (1976) and Ruse (1976) noted, the most difficult and creative part of a reduction is establishing these connections, formulating bridge principles that link the concepts of the two theories. Thus, in particular, to reduce Mendelian to molecular genetics required the Nobel Prize–winning discoveries of Watson and Crick that identified the gene with DNA.

It was D. Hull who first noted (1976b) the difficulty of actually characterizing Mendelian properties by concepts drawn only from molecular biology. Most significantly, as the basic unit of phenotypic expression, mutation, and recombination, the gene could not be identified as either one DNA base (the smallest unit of mutation), or one stretch of DNA that constitutes a structural or regulatory gene, or the amount of DNA sequence minimally involved in recombination. Moreover, because of the redundancy of the genetic code, many different DNA sequences could code for the same gene. The relation between molecular DNA sequences and genes identified by their function was therefore "many-one" and "one-many": the same DNA sequence is implicated in many different genes and many different sequences can discharge the function of a single gene. Of course in the subsequent thirty or forty years matters made the relations between DNA and genes even more complicated: regulatory sequences and sites, introns and exons, posttranscriptional and translational modification, promoters, operons, open reading frames, junk DNA, repeats, transposons, RNA viruses, all these made completely intractable the problem of defining or characterizing either the gene in general, or even particular genes for particular traits, or for that matter even particular immediate gene products in terms of the structure of DNA molecules that compose them. There is of course no trouble identifying 'tokens' – particular bits of matter we can point to – of genes with particular 'tokens' of their molecular constituents. But token identities will not suffice for reduction, even if they are enough for physicalism to be true.

The second problem facing traditional reductionism in biology was the absence of laws (beyond those laws, if any, embodied in the theory of natural selection), either at the level of the reducing theory or the reduced theory or between them. Indeed, a good deal of the philosophy of biology's search for real nomological generalizations in biology has been motivated by the controversy about reduction; both its proponents and opponents needed real laws that they could show were either reducible or irreducible to other more fundamental laws, those of physical science. In particular, for Watson and Crick's achievement to vindicate traditional reductionism, there had to be laws of classical, Mendelian, or population genetics, which the identification of the gene as a sequence of DNA would enable us to derive from laws of molecular biology. But Mendel's laws are not laws: they began to be riddled with exceptions almost from the moment they were first rediscovered in the early 1900s: cross-over, linkage, mitotic drive, autosomal genes, and so forth. And when protected from exceptions by ceteris paribus clauses, they particularly defy derivation from more fundamental principles. (Other candidates for the status of nomological generalizations in biology, e.g., the Hardy-Weinberg "law" or Fischer's fundamental theorem, turn out to be either tautologies, or consequences of other equally controversial candidates for nomological status, such as the principle of natural selection.) Moreover, there were no distinctive nomological generalizations of molecular biology either, just the stocheometric equations of organic chemistry. As we shall see, the absence of laws bedevils both postpositivist reductionism and its denial.

That there are no laws of biology to be reduced to laws of molecular biology, and indeed that there are no laws of molecular biology, can be shown by the same considerations that explain why genes and DNA cannot satisfy reduction's criterion of connection.

The individuation of types in biology is almost always via function: to call something a wing, or a fin, or a gene is to identify it in terms of its function. But biological functions are naturally selected effects. And natural selection for adaptations – that is, environmentally appropriate effects – is blind to differences in physical structure that have the same or roughly similar effects. Natural selection "chooses" variants by *some of their effects*, those that fortuitously enhance survival and reproduction. When natural selection encourages variants to become packaged together into larger units, the adaptations become functions. Accordingly, the structural diversity of the tokens of a given Mendelian or classical or population biological gene or generally any functionally identified biological system or structure is inevitable. And no biological kind will be identical to any single molecular structure or manageably finite number of sets of structures.

To see why there can be no laws in biology, nor anything that could satisfy the criterion of connectability between an item functionally and structurally characterized, consider the form of a generalization about all Fs, where F is a functional term, like gene, or wing, or camouflage. The generalization will take the form $(x)[Fx \rightarrow Gx]$, a law about Fs and Gs. Gx will itself be either a structural predicate or a functional one: that is, either it will pick out Gs by some physical attribute common to them, or Gx will pick out Gs by descriptions of one of the effects that everything in the extension of Gx possesses. But if Fx is a functional kind, it will have been shaped by natural selection. Accordingly, there will be no single physical feature common to all items in the extension of Fx. It will be a physically heterogeneous class since its members have been selected for their effects. So Gx cannot be a structural predicate. Of course some structural feature may be shared by all of the members of Fx. But it will not be a biologically interesting one. Rather it will be a property shared with many other things – like mass or electrical resistance. These properties will have little or no explanatory role with respect to the behavior of members of the extension of Fx. For example, the generalization that "all endotherms are composed of confined quarks" does relate a structural property – quark confinement – to a functional one – endothermy – and is exceptionlessly true. But is not a law of biological interest. Gx cannot be a structural kind. Can Gx be another functional kind, like Fx, shared by all members of Fx? The existence of another functional property different from F that all items in the extension of the functional predicate Fx share is highly improbable. If Fx is a functional kind, then because of the blindness of selection to structure the members of the extension of Fx are physically diverse. As such, any two Fs have nonidentical (and usually quite different) sets of effects. Without a further effect common to all Fs, selection for effects cannot produce another selected effect; it cannot uniformly select all members of F for some further adaptation. Thus, there is no further function kind all Fs share in common. Whether functional or structural, there will be no predicate Gx that is linked in a law to Fx. Of course, the sort of connection reductionism requires between the Mendelian gene and the molecular gene will have to be an identity, $Fx = Gx$, which implies $(x)[Fx \rightarrow Gx]$. So, our conclusion precludes satisfaction of either of the two requirements of postpositivist reduction.

The unavoidable conclusion is that as far as the postpositivist model of intertheoretical reduction is concerned, neither of its conditions for reduction are satisfied by relations between nonmolecular and molecular biology.

2.   ANTIREDUCTIONISM VINDICATED?

If antireductionism were merely the denial that postpositivist reduction obtains among theories in biology, it would be obviously true. But antireductionism is not just a negative claim. It is the thesis that a) there are generalizations at the level of functional biology, b) these generalizations are explanatory, c) there are no further generalizations outside functional biology that explain the generalizations of functional biology or explain more fully what the generalizations of functional biology explain.

All three components of antireductionism are daunted by at least some of the same problems that vex reductionism: the lack of laws in functional biology and the problems facing an account of explanation in terms of derivation from laws. If there are no laws and/or explanation is not a matter of subsumption, then antireductionism is false too. But besides the false presuppositions antireductionism may share with reductionism, it has distinct problems of its own. Indeed, these problems stem from the very core of the antireductionist argument, the appeal to ultimate explanations underwritten by the theory of natural selection.

To see the distinctive problems for antireductionism, consider an example of putatively irreducible functional explanation advanced by antireductionists (due to Kitcher 1984, 1999). The example is the biologist's explanation of independent assortment of functional genes:

The *explanandum* is

> (G) Genes on different chromosomes, or sufficiently far apart on the same chromosome, assort independently.

According to Kitcher, the functional biologist proffers an *explanans* for (G), which we shall call (PS):

(PS) Consider the following kind of process, a *PS*-process (for *pairing* and *separation*). There are some basic entities that come in pairs. For each pair, there is a correspondence relation between the parts of one member of the pair and the parts of the other member. At the first stage of the process, the entities are placed in an *arena*. While they are in the arena, they can exchange segments, so that the parts of one member of a pair are replaced by the corresponding parts of the other members, and conversely. After exactly one round of exchanges, one and only one member of each pair is drawn from the arena and placed in the *winners box*. In any PS-process, the chances that small segments that belong to members of different pairs or that are sufficiently far apart on members of the same pair will be found in the winners box are independent of one another. (G) holds because the distribution of chromosomes to games at meiosis is a PS-process.

Kitcher writes, "This I submit is a full explanation of (G), an explanation that prescinds entirely from the stuff that genes are made of" (Kitcher, 1999, 199–200).

Consider why, according to the antireductionist, no molecular explanation of (PS) is possible. It is for the same reason no functional biological kind can be identified with those of molecular biology. Because the same functional role is almost always realized by a variety of structures, and because natural selection is blind to this variety, the full macromolecular explanation for (PS) or for (G) will have to advert to a range of physical systems that realize independent assortment in many different ways. These different ways will be an unmanageable disjunction of alternatives so great that we will not be able to recognize what they have in common, if indeed they do have something in common beyond the fact that each of them will generate (G). Even though we all agree that (G) obtains in virtue only of macromolecular facts, nevertheless, because of their number and heterogeneity, these facts will not explain (PS), still less supplant (PS)'s explanation of (G).

But this argument is unsatisfactory. Begin with (G). If the argument of the previous section is right, (G) is not a law at all. At most it reports particular facts about a spatiotemporally restricted kind, "chromosomes," of which there are only a finite number extant over a limited period at one spatiotemporal region (the Earth). Accordingly, (G) is not a law and so not a candidate for reduction, and the impossibility of reducing it to molecular biology is no objection against reductionism. More important, it is unclear what certifies (PS) – the account of PS-processes given – as explanatory, and what prevents the vast disjunction of macromolecular accounts of the underlying mechanism of meiosis from explaining (PS), or for that matter from explaining (G) (cf. Sober 1999).

The reductionist holds that, contrary to Kitcher's claim, there is something that the vast disjunction of macromolecular realizations of (PS) have in common that would enable the conjunction of them to explain (PS) fully to someone with a good enough memory for the details. Each was selected for because each implements a PS process and PS processes have been adaptive in the local environment of the Earth from about the onset of the sexually reproducing species onward. Since selection for implementing PS processes is blind to differences in macromolecular structures with the same or similar effects, there may turn out to be nothing else completely common and peculiar to all macromolecular implementations of meiosis besides their being selected for implementing PS processes. Of course, it may be that we never know enough of the "gory details" for a (disjunctive) macromolecular answer to the question of why (G) obtains. Similarly, we may never know enough for a macromolecular explanation of (PS) to be a complete answer to our question "Why do PS processes occur?" But this would be hollow victory for antireductionism, even if we grant the tendentious claim that we will never know enough for such explanations to succeed, for it relegates antireductionism to the status of a claim about biologists, not about biology. Such philosophical limitations on our epistemic powers have been repeatedly breeched in the history of science.

Antireductionists will need a different argument for the claim that neither (G) nor (PS) can be explained by the disjunction of macromolecular mechanisms that realize it, and for the claim that (PS) does explain (G) and (G) does explain individual cases of recombination. One argument for such a conclusion rests on the metaphysical thesis that there are no disjunctive properties or that if there are, such properties have no causal powers. Here is how the argument might proceed: the vast motley of alternative macromolecular mechanisms that realize (PS) have nothing in common. There is no property – and in particular no property with the causal power to bring about the truth of (G) – that they have in common. Physicalism (which all antireductionists party to this debate embrace) assures us that whenever PS obtains, some physical process or other obtains. Thus we can construct the identity (or at least the biconditional) that

$$(R) \ PS = P_1, \bigvee P_2 \bigvee \ldots \bigvee P_i, \bigvee \ldots \bigvee P_m,$$

where m is the number, a very large number, of all the ways macromolecular processes can realize PS processes.

Antireductionism cannot argue that (R) has no explanatory power, because it is merely a local fact about how PS processes are realized on Earth. For this is also true of (G) and antireductionists insist that (G) has explanatory power. A causal theory of explanation might rule out R as explaining PS on the ground that the disjunction, $P_1, \bigvee P_2 \bigvee \ldots \bigvee P_i, \bigvee \ldots \bigvee P_m,$ is not the *full* cause of PS processes. This might be either because it was incomplete – there is always the possibility that still another macromolecular realization of PS will arise – or because disjunctive properties just are not causes, have no causal powers, perhaps are not really properties at all. Thus, the antireductionist alleges, (PS) and (G) are the best and most complete explanations to which biology can aspire.

Consider the claim that (R) is not complete, either because some disjuncts have not occurred yet or perhaps that there are an indefinite number of possible macromolecular implementations for (SP). This in fact seems to be true, just in virtue of the fact that natural selection is continually searching the space of alternative adaptations and counteradaptations, and that biological threats to the integrity and effectiveness of meiosis might in the future result in selection for new macromolecular implementations of (PS). But this is no concession to antireductionism. It is more like an argument that neither (PS) nor (G) reports an explanatory generalization, that they are in fact temporarily true claims about local conditions on the Earth, and so not candidates for postpositivist reduction.

Suppose that (R) can be completed in principle, perhaps because there are only a finite number of ways of realizing a (PS) process but, it is claimed, the disjunction is not a causal or a real property at all. Therefore it cannot figure in an explanation of either (PS) or (G). There are several problems with such an argument. First, the disjuncts in the disjunction of $P_1, \vee P_2 \vee \ldots \vee P_i, \vee \ldots \vee P_m$ do seem to have at least one or perhaps even two relevant properties in common: each was *selected* for implementing (PS) and causally brings about the truth of (G). Second, we need to distinguish predicates in languages from properties in objects. It might well be that in the language employed to express biological theory, the only predicate we employ that is true of every P_i is a disjunctive one, but it does not follow that the property picked out by the disjunctive predicate is a disjunctive property. Philosophy long ago learned to distinguish things from the terms we hit upon to describe them.

How might the antireductionist argue against the causal efficacy of disjunctive properties? One might hold that disjunctive properties will be causally efficacious only when their disjuncts subsume similar sorts of possible causal processes. If we adopt this principle, the question at issue becomes one of whether the disjunction of $P_1, \vee P_2 \vee \ldots \vee P_i, \vee \ldots \vee P_m$ subsumes similar sorts of causal processes. The answer to this question seems to be that the disjunction shares in common the features of having been selected for resulting in the same outcome – PS processes. Thus, the disjunctive predicate names a causal property, a natural kind. If so, antireductionists will be hard pressed to deny the truth and the explanatory power of (R).

Besides its problems in undermining putative macromolecular explanations of (PS), (G), and what (G) explains, antireductionism faces some problems in substantiating its initial claims that (PS) explains (G) and (G) explains individual cases of genetic recombination. The problems of course stem from the fact that neither (PS) nor (G) is a law, and therefore an account is owed of how non-lawlike statements such as these can explain.

We can conclude that so far as postpositivist reductionism and its antireductionist opposition are concerned, neither view is relevant to the real issue about the relation between functional and molecular biology. If there is a real dispute here, it cannot be about the derivability or underivability of laws in functional biology from laws in molecular biology, as there are no laws in either subdiscipline. Nor can the real dispute turn on the relationship between general theories in molecular and functional biology. Once this conclusion is clear, the question of what reductionism was in the postpositivist past can be replaced by the question of what reductionism is now; for the obsolescence of the postpositivist model of reduction hardly makes the question of reductionism or its denial obsolete. Indeed, developments in molecular biology – the rate of sequencing, the success of computational *in-silico* genomics, and especially the elucidation of the molecular mechanism of development – make this question more pressing than ever. But the question has to be reformulated if it is to make contact with real issues in biology.

3.   REDUCTIONISM AFTER POSTPOSITIVISM

The debate between reductionism and antireductionism cannot be one about laws as they have traditionally been understood in philosophy of science. The dispute about whether the rest of biology is reducible to molecular biology will have to be one about the best, most complete, most correct, or most adequate explanation of particular facts about life on Earth, some of these facts obtaining over geological epochs, but all of them ultimately the contingent results of general laws of natural selection operating on the boundary conditions that obtained here 3.5 billion years ago – that is, historical facts. Reductionism needs to claim that the most complete, correct, and adequate explanations of the historical facts uncovered in functional biology are by appeal to other historical facts uncovered in molecular biology, plus perhaps some laws of organic chemistry.

Antireductionism must claim that there are at least some explanations in functional biology that cannot be completed, corrected, or otherwise improved by adducing considerations from molecular biology. The only way to do this is to argue that explanations in functional biology are all ultimately adaptational, as a result of the functional vocabulary in which they are expressed or of their implicit appeal to natural selection in connecting a biological *explanans* to a biological *explanandum*. If the theory of natural selection is not itself reducible to physical science and/or is itself indispensable to explanation in molecular biology, the latter will be no more reducible to physical science than the rest of biology is reducible to it. Examples of such arguments are to be found in Kitcher (1984, 1999), Sober (1993), and elsewhere.

To refute this sort of argument, reductionists need to do two things. First they need to show that evolutionary explanations in functional biology are unavoidably inadequate, and inadequate in ways that can be improved only by evolutionary explanations from molecular biology. And then they need to show that the theory of natural selection is not itself a barrier to the reduction of molecular biology to physical science. These are both daunting tasks.

Let us consider the first challenge, that of showing what makes evolutionary explanations in functional biology inadequate in ways only evolutionary molecular explanations can correct. Suppose we address the question, Why do butterflies have eye-spots on their wings? The explanation is presumably an adaptationalist one that accords a function, in camouflage, for instance, to the eyespot on butterfly wings. Eyespots on butterfly and moth wings have been selected for over a long course of evolutionary history. On some butterflies these spots attract the attention and focus the attacks of predators onto parts of the butterfly less vulnerable to injury. Such spots are more likely to be torn off than more vulnerable parts of the body, and this loss does the moth or butterfly little damage, while allowing it to escape. On other butterflies, and especially moths, wings and eyespots have also been selected for taking the appearance of an owl's head, brows, and eyes. Since the owl is a predator of birds that consume butterflies and moths, this adaptation provides particularly effective camouflage.

The reductionist has no difficulty with this evolutionary explanation, as far as it goes. But, on the reductionist's view, such an explanation does little more than set the research agenda required to cash in the explanation's promissory notes offered the original explanation. The functional explanation leaves unexplained several biologically pressing issues, pressing enough to deny it completeness, correctness, or explanatory adequacy. These are the question of what alternative adaptive strategies were available to various lineages of organisms and which were not, and the further question of how the feedback from adaptedness of functional traits – like the eyespot – to their greater subsequent representation in descendants was actually effected. The most disturbing lacuna in the original explanation is its silence on the causal details of exactly which feedback loops operate from fortuitous adaptedness of traits in one or more distantly past generations to improved adaptation in later generations and how such feedback loops approach the biological fact to be explained as a locally constrained optimal design. These demands all stem from widely shared scientific commitments: the demand for complete causal chains, the denial of action at a distance, and the denial of backward causation. Natural selection at levels higher than the macromolecular is silent on the crucial links in the causal chain that convert the appearance of goal directedness into the reality of efficient causation. Only a macromolecular account of the process can provide them. Such an account would itself also be an adaptational explanation: it would identify strategies available for adaptation by identifying the genes (or other macromolecular replicators) that determine the characteristics of lepidopterans' evolutionary ancestors and that provide the only stock of phenotypes on which selection can operate to move along pathways to alternative predation – avoiding outcomes – leaf color camouflage, spot camouflage, or other forms of Batesian mimicry, repellant taste to predators, Mullerian mimicry of bad tasting species, and so on. The reductionist's more complete explanation would show how the extended phenotypes of these genes competed and how the genes that generated the eyespot eventually become predominant, that is, are selected for. Note that the reductionist's full explanation is still a historical explanation in which further historical facts – about particular genes and biosynthetic pathways – are added and are connected by the same principles of natural selection that are invoked by the original evolutionary explanation. But the links in the causal chain of natural selection are filled in to show how past adaptations were available for and shaped into today's functions (cf. Rosenberg 2000).

Antireductionists will differ from reductionists not on the facts but on whether the initial explanation was incomplete, incorrect, or inadequate. They will agree that the macromolecular genetic and biochemical pathways are causally necessary to the truth of the original evolutionary explanation. But they do not complete an otherwise incomplete explanation. They are merely further facets of the situation that molecular research might illuminate (Kitcher 1999, 199). The original adaptational answer to the question, Why do butterflies have eyespots? does provide a complete explanatory answer to a question.

Who is right here? If explanation follows causation, the reductionism has much to recommend itself as a methodology. However, on an erotetic view of explanations, higher-level and lower-level explanations may be accepted as reflecting answers to differing questions advanced in different contexts of inquiry. The reductionist may admit that there are contexts of inquiry in which the nonmolecular answers to questions satisfy explanatory needs. But the reductionist will insist that in the context of advanced biological inquiry, as opposed, say, to secondary school biology instruction, for example, the nonmolecular question either does not arise or, having arisen in an early stage of inquiry, no longer does. Moreover, the only assurance that anything like the nonmolecular adaptational explanation is on the right track is provided by a molecular explanation that cashes in its promissory notes by establishing the adaptive origins of the functional traits in molecular genetics.

The latter-day reductionist holds that complete explanations can only be provided in biology by adverting to the macromolecular states, processes, events, and patterns on which nonmolecular historical events and patterns supervene. The reductionist does not claim that biological research or the explanations it eventuates in can dispense with functional language or adaptationalism. Much of the vocabulary of molecular biology is thoroughly functional. Nor is reductionism the claim that all research in biology must be "bottom-up" instead of "top-down" research. So far from advocating the absurd notion that molecular biology can give us all of biology, the reductionist's thesis is that we need to identify the patterns at higher levels because they are the explananda for which molecular biology provides the explanantia. What the reductionist asserts is that functional biology's explanantia are always molecular biology's explananda.

There remains a serious lacuna in this argument for reductionism in biology, one potentially large enough to drive a decisive anti-reductionist objection through. Although latter-day reductionism claims to show that nonmolecular explanations must be cashed in for molecular ones, these explanations are still adaptational, still evolutionary – they still invoke the mechanism of blind variation and natural selection operating on macromolecules. And until this principle can show unimpeachable reductionist credentials, it remains open to say that even at the level of the macromolecules, biology is causally autonomous from physical processes. In consequence, an irreducible role for laws governing natural selection in molecular biology will substantiate an autonomy from physical science that no reductionist can accept.

The prospects for reductionism in the end turn on the relationship between the theory of natural selection and physical science. Three alternatives suggest themselves.

(a) The laws governing natural selection are underived nomic generalizations about biological systems, including macromolecular ones, and are emergent from purely physical processes. This alternative would vindicate the irreducibility of all of biology, but it would make biology difficult to reconcile with physicalism. Natural selection would turn out to be a process not fixed by the physical facts.

(b) Biological natural selection is derivable from some laws of physics and/or chemistry. This alternative would vindicate the reductionist vision of a hierarchy of scientific disciplines and theories, with physics at the foundations. Both the temporal asymmetry of natural selection and the multiple realizability of fitness in a vast set of different physical relations between interactors and the environment make this alternative unattainable for reductionism.

(c) Natural selection as a process reflects a hitherto unnoticed and underived law about physical systems (including nonbiological ones), and from it the evolution of biological systems can be derived. This alternative has been little canvassed but would reconcile reductionism's commitment to physicalism and antireductionism's commitment to the autonomy of biology from (the rest of) physical science.

4.   GENIC REDUCTIONISM AND GENOCENTRISM

Independently of the issues of physicalism, the nature of explanation, and macromolecular reductionism as a research program in biology there is another, related issue in the philosophy of biology in which the label 'reductionism' figures. The term is employed to stigmatize a view its opponents call 'genic reductionism.' In this sense, 'reductionism' is not a claim about intertheoretical explanation but a name for the error of ignoring important causal variables in the explanation of some process or phenomenon. Thus for example it would be wrongly reductionistic to trace *the* cause of sickle cell anemia to a mutant hemoglobin gene that results in a serine molecule in the wrong place on the outside of the sufferer's hemoglobin molecules, which initially results in their clumping together, then arterial blockage, and eventually produces anemia. To begin with there also need to be arteries, red blood cells, oxygen, and of course the whole structural and regulatory genetic machinery of hemoglobin synthesis and all the environmental inputs required for the otherwise successful development of the person carrying this gene. The notion that the gene by itself suffices for a trait or even a single protein is a puerile mistake that no biologist could make. But some philosophers and biologists argue that the tendency, widespread among biologists, to accord a special explanatory role in development or somatic regulation to the genes constitutes a more sophisticated but equally egregious version of this reductionistic mistake. "Genic reductionism" is the name given to this reductive error. This sort of erroneous reductionism of course encourages and is in turn encouraged by the prospects for intertheoretical explanatory reduction between nonmolecular and molecular biology.

Of course, intertheoretical reductionism does not imply or require genic reductionism, but it certainly does encourage the view that the gene has a special explanatory role in biology, a view that may cheerfully accept its opponent's label of "genocentrism." According to this view, the special causal and therefore explanatory role of the gene is due to its informational role in storing and transmitting hereditary information, its role in programming the development of the embryo, and its regulatory function in somatic cells. Intertheoretical reduction of nonmolecular to molecular biology encourages these attributions both programmatically, as the DNA structure of the gene is our best current source of information about all biologically active macromolecules, including all enzymes and structural proteins, and systematically, as molecular developmental genetics is beginning to provide the detailed explanation of development that biology has sought but failed to secure since

Aristotle. In the words of one opponent of intertheoretical reduction and genic reduction, the twentieth century was "the century of the gene" (Fox-Keller 2000). According the gene a special causal and explanatory role in heredity, development, and regulation would be a very strong incentive to pursuing intertheoretical explanations of the sort the reductionist advocates. In consequence, there is a lively debate among philosophers of biology and biologists about whether genes have the special causal and explanatory roles genocentrism accords them. The debate is channeled through the following issues particularly about development molecular genetics:

There have been several striking successes in tracing out the details of embryological development among organisms such as *Caenorhabditis eligans* and *Drosophila melanogaster*. This research shows that pretty well all the action is in gene regulation and gene expression (even for epigenetic processes such as methylation). It might be argued against genocentrism that these successes are insufficient to ground the expectation that the rest of the details of *Drosophila* development and behavior are equally intelligible from a purely macromolecular perspective. Why suppose that development among vertebrates should be just like *Drosophila* embryogenesis? Is it reasonable to extrapolate from one case to the claim that a macromolecular program will explain development everywhere and always? The genocentrist will reply that homologies in the regulatory developmental genes of all multicellular creatures (e.g., the *Homeobox* genes) are so great that the burden of proof has been shifted to those who would doubt the extrapolation from the fruit fly to *Homo sapiens*.

A more serious objection to genocentrism, one persistently raised against intertheoretical explanatory reduction as well (Kitcher 1984), is the observation that much of the story of development in these cases is expressed in nonmacromolecular concepts. The description of development advanced in molecular developmental genetics makes repeated reference to cells, their properties and behavior. If the role of whole cells is indispensable to the genetic program for *Drosophila* embryogenesis, we will have to be confident that there is an adequate, purely macromolecular explanation of the role of the cell and its cycle before we can conclude that molecular biology alone, and unaided, provides the explanation for development. And if the explanatory role of the whole cell or its nonmolecular parts is both indispensable and irreducible, then reductionism's explanatory claim must be surrendered. Vindicating genocentrism against this argument requires an analysis that shows that macromolecular explanations adverting to organelles, cells, tissues, and so on, cite their purely physical, topographic, geographic, and spatial properties and so are innocent of seriously presupposing the biological processes they purport to explain, or else it requires that the role of organelles, cells, tissues, and so forth, be themselves fully explained in macromolecular terms. Reductionists are confident both of these tasks are attainable.

A variant on the objection that nonmolecular factors are required even in molecular explanation, expresses the more radical skepticism of a coalition of biologists and others dubious about the genetic explanation of development. Why suppose that the genes have any special role in development? There is a vast range of other conditions – physiological and environmental – causally necessary for fertilization and embryogenesis along with the products of the genes. According to the "casual democracy thesis" advanced for example by Developmental Systems Theorists (Griffiths and Grey 1994, for example), each of these casually necessary factors is on a par with the others; none is even *primus inter pares*; and depending on the explanatory interests and practical focus of a biologist any one of them can take center stage in one or another explanation of development. Accordingly, genocentrism's attribution of a special role in development to the genes is unwarranted, and along with its eclipse we should reject reductionism as well (Lewontin and Levins 1985).

Genocentrists will acknowledge that there are many other factors causally necessary along with the genes for their effects. But they will insist that what makes the genes "special" and vindicates genocentrism is their status as informational molecules and their role in programming the sequence of necessary steps and their materials into building the embryo. On this view the genetic code, which makes sixty-four distinct message units out of the combination of four nucleic acid bases in groups of three bases (the so-called codons), has the literal properties of an information signaling system that (redundantly) names the twenty amino acids and allows for the expression of instructions (in messenger RNA) to the ribosomes to produce all the proteins in the body in the order and amounts required for development and function. The code is on this view an arbitrary convention, in Crick's terms, a "frozen accident" (Maynard Smith 2000). Opponents deny that the genome is literally informational. Some do so by appealing to alternate theories of the origins of the genetic code that purport to show it is nonarbitrary. This debate is vexed by important philosophical problems surrounding the notion of information and the property of intentionality in its "home base" – psychology – and is unlikely to be settled in debate among biologists and philosophers of biology. But the genocentrist is on stronger ground in claiming that the genes are privileged because of their role in literally *programming* the embryo. Here the genocentrist argument is simple and direct. DNA sequences have already been used literally to program computers – albeit not electronic but macromolecular ones – and to enable them to complete calculations that electronic computers are incapable of undertaking, including for example particularly large values of the NP-hard "traveling salesman" problem. If DNA molecules can instantiate a program to solve these sorts of problems, then instantiating a program that solves much simpler problems, such as building an embryo, is if not child's play, at least well within their powers. The sequence of polynucleotide bases in the genome can literally program the embryo.

But to infer from the power of the DNA sequence literally to program development to the conclusion that the genetic program can do so proceeds on the reductive identification of genes with DNA sequences. As we saw in Section 1, this is a highly problematical assumption. In fact, opponents of genocentrism seize upon the difficulty of providing a numerical identity of the gene as a unit of function with the DNA sequence as a structured physical object, to argue against the very coherence of the gene concept and so against the prospects for genocentrism. The notion of the gene is indeed quite unproblematic. Genocentrism and intertheoretical reductive explanation require that there are genes, that they can be distinguished, individuated, counted, and otherwise treated as the relevant units of hereditary transmission and developmental control. But it has been alleged the history of genetics in the late twentieth century has shown that the notion of a gene as one thing indifferently identified by its effects in protein synthesis and by its chemical composition can no longer be countenanced by informed biologists (see Griffiths and Neuman-Held 1999, for example).

Though for a long time productive and fertile in its encouragement of scientific progress, the concept of the gene has been overtaken by events. The complexities in heredity and development that molecular biology has uncovered, on this view, make the gene an obsolete idea. It is an idea that will be eclipsed in the next century by other notions, perhaps ones that will vindicate antireductionist conclusions to the same extent that the twentieth century's fixation on the gene reflected reductionistic ones. And when the gene is superceded, so will genocentrism, its associated reductionistic research program, and its philosophy of biology. Responding to these arguments and proving that the reports of the death of the gene have been greatly exaggerated remain among the chief preoccupations of latter-day reductionism.

8   Mechanisms and Models

> Generally speaking, making models for unknown
> mechanisms is the creative process in science.
>
> Harré 1970, 40

1.   INTRODUCTION: MECHANISMS AND MODELS

Biologists often seek to discover *mechanisms*. Knowledge of biological mechanisms is valuable because descriptions of them often play the roles attributed to general scientific theories. They provide explanations of puzzling phenomena. They enable biologists to make predictions. They aid the design of experiments. They may explain domains of wide scope. They may make possible medical or biotechnological interventions for practical purposes. Especially in molecular biology, theories consist of sets of mechanism schemas, such as those for DNA replication and protein synthesis.

Biologists use many types of *models* to represent and discover mechanisms: diagrammatic models, physical scale models, analogue models, model organisms, in vitro experimental systems, mathematical models, computer graphic and simulation models. Models represent and substitute for the thing modeled, while being easier to understand, manipulate, or study. Choice of an appropriate model depends on the problem to be solved using it. In medicine, animal models are often used when the goal is to understand disease mechanisms in humans. Molecular biologists use bacteria and viruses as models for mechanisms with domains of very wide scope, such as DNA replication.

The topics here are mechanisms in biology and models that aid their discovery. Section 2 provides a characterization of mechanisms, based

139

on cases from molecular and neurobiology. Section 3 introduces the distinctions among mechanism schemas, their instantiations, and incomplete sketches; the term "model" in the sense of a theoretical model may refer to any of the three. Several kinds of models aid the discovery of mechanisms, especially analogue models, model organisms, and in vitro experimental systems; they are the subject of Section 4. Section 5 examines the use of such models in reasoning to discover mechanisms, which is an extended process of generating, testing, and revising mechanism sketches and schemas. Finally, the conclusion points to general philosophical issues and to unanswered questions in this new research program on mechanisms in biology.

2.   CHARACTERIZATION OF BIOLOGICAL MECHANISMS

A mechanism is sought to explain how a *phenomenon* is produced. Mechanisms may be characterized as entities and activities organized such that they are productive of regular changes from start or setup to finish or termination conditions (Machamer, Craver, and Darden 2000, 3). The nature of the *phenomenon* for which a mechanism is sought provides important guidance in discovery. Biologists seek the location of the mechanism and find places for its beginning, ending, topping off, bottoming out, and boundaries, guided in part by the nature of the phenomenon. Many biological mechanisms are *regular* in that they usually work in the same way under the same conditions. The regularity is exhibited in the typical way that the mechanism runs from start to finish, thereby producing and reproducing a given phenomenon. For example, the phenomenon of DNA replication is produced by the mechanism of DNA replication. The mechanism begins with one double helix and ends with two. One double helix unwinds and each half provides a template along which complementary bases are aligned, yielding two identical helices at the end. The description of this mechanism bottoms out at the level of parts of the DNA molecule, including the bases and their charges. The precise hydrogen bonding between bases (usually) produces accurate copying of the order of the bases from the parent strands to the daughter ones. The topping-off point for the description of this mechanism is the entire double helix, a macromolecule within the nuclei of cells.

Mechanisms are composed of both *entities* (with their properties) and *activities*. Activities are the producers of change. Entities are the things that engage in activities. For example, two entities, a DNA base and its complement, engage in the activity of forming hydrogen bonds because of their properties of geometric shape and their arrangements of weak polar charges. Entities and activities are interdependent. For example, polar charges are necessary for hydrogen bond formation. Appropriate shapes are necessary for lock and key docking of enzymes and substrates. This interdependence of entities and activities allows biologists to reason about entities on the basis of what is known or conjectured about the activities, and vice versa. Such reasoning by forward or backward chaining aids discovery of subsequent or prior stages, based on what is known or conjectured about adjacent ones.

For the purposes of a given biologist, research group, or field, there are typically entities and activities that are accepted as relatively fundamental. In other words, descriptions of mechanisms in that field typically *bottom out* and *top off* in particular places. Those places may be more or less arbitrarily chosen. For example, memory mechanisms are investigated at many mechanism levels, from a mouse learning a maze to two neurons (cells) exchanging neurotransmitters (molecules). Alternatively, appropriate bottoming out and topping off may be dictated by the nature of the phenomenon and the kinds of working entities that are active in mechanisms. In molecular biology, mechanisms typically bottom out in descriptions of the activities of molecules (macromolecules, smaller molecules, and ions) and cell organelles (e.g., ribosomes). These entities are the *working entities* of molecular biological mechanisms, such as DNA replication and protein synthesis. Smaller (or larger) entities do not have the requisite sizes, shapes, charges, or other activity-enabling properties to play roles in these molecular mechanisms.

Mechanisms have *productive continuity* between stages: that is, the entities and activities of each stage give rise to the next stage. There are no gaps from the setup to the termination conditions. Mechanisms have a *beginning* and an *end*, again more or less arbitrarily chosen. For instance, a natural beginning point for the mechanism of DNA replication is one double helix, and a natural ending is two double helices. However, in an ongoing series of mechanisms, some of which might be cyclic, the choice of start and stop points may be based merely on the convenience of investigation.

Mechanisms should be distinguished from machines. A machine is a contrivance, with organized parts designed to work together smoothly. Mechanisms are often associated with machines because mechanisms are most conspicuous in human artifacts, such as the mechanical clock. A stopped clock is still a machine, but it is not a mechanism. Mechanisms are active. Human artifacts may exhibit the optimal results of engineers' effective and efficient designs. However, living things are a result of evolutionary tinkering and satisficing. As Michael Ruse noted, the idea that the world is full of designed machines has been replaced by the idea that it contains evolved machines, built in a ramshackle way as evolution fashions their adaptations from available parts. Although organisms may be viewed as ramshackle machines, an organism, as a whole, is not a mechanism. Many mechanisms operate within a living organism. Moreover, looking up instead of down, organisms may be said to play roles in higher-level mechanisms, such as the isolating mechanisms leading to speciation.

3.   DESCRIPTIONS OF MECHANISMS: SCHEMAS, SKETCHES, AND THEORETICAL MODELS

Biological theories represented by sets of mechanism schemas may be contrasted with philosophers' analyses of theories as sets of syntactic formal axioms or as abstract and idealized formal semantic structures. (The sense of formal "model" in this semantic conception of theories will not occupy us here.) Analysis of mechanistic theories in biology does not import a formal structure to understand theories, but instead strives to characterize mechanisms and their representations in a manner faithful to biologists' own usages. Scientists use theories to describe, explain, explore, organize, predict, and control the items in a theory's domain. Descriptions of mechanisms aid all of these tasks (Craver 2002a).

Adequate descriptions of mechanisms include a description of the phenomenon produced by the mechanism, the entities and activities composing the mechanism, their setup conditions, along with their productively continuous spatial and temporal organization. Spatial organization includes localization, structure, orientation, connectivity, and compartmentalization (if any). Temporal organization includes the order in which activities occur; their rate, duration, and frequency; and the overall order of the stages of the mechanism (Darden 2006, chap. 12).

A *mechanism schema* is a truncated abstract description of a mechanism that can be filled with more specific descriptions of component entities and activities. An example is this diagram of the central dogma of molecular biology:

DNA $\rightarrow$ RNA $\rightarrow$ protein

This is a schematic representation (with a high degree of abstraction) of the mechanism of protein synthesis. A less schematic description of a mechanism shows how the mechanism operates to produce the phenomenon in a productively continuous way and satisfies the componency, spatial, temporal, and contextual constraints. The goal in mechanism discovery is to find a description of a mechanism that produces the phenomenon, and for which there is empirical evidence for its features. A mechanism schema can be instantiated to yield such an adequate description.

In contrast, a *mechanism sketch* cannot (yet) be instantiated. Components are (as yet) unknown. Sketches may have black boxes for missing components whose function is not yet known. They may also have gray boxes, whose functional role is known or conjectured; however, what specific entities and activities carry out that function in the mechanism are (as yet) unknown. The goal in mechanism discovery is to transform *black boxes* (components and their functions unknown) to *gray boxes* (component functions specified) to glass boxes (components supported by good evidence). A schema consists of *glass boxes* – one can look inside and see all the parts. Incomplete sketches indicate where fruitful work may be directed to produce new discoveries. The transition from sketch to schema may be a continuous process, as various portions of the mechanism are discovered in a piecemeal way. An instantiated schema shows details of how the mechanism operates in a specific instance to produce the phenomenon. Hence, mechanistic theories explain the phenomena in their domains.

As William Bechtel and Adele Abrahamsen (2005) noted, explaining a phenomenon involves describing the mechanism responsible for it, often by constructing a model that specifies key parts, operations, and organization and that can simulate how their orchestrated functions transform certain parts. The term "model" here refers to a model of a mechanism or what philosophers might call a "mechanistic theoretical model." "Model" in this sense may refer to any of the three terms discussed: a mechanism "schema," an "instantiation" of a mechanism schema, or a "sketch." Sometimes the terms "theory" and "model" in this sense of theoretical mechanistic model are used synonymously, in which case a mechanism schema or a set of mechanism schemas is appropriate. Sometimes a model is said to be an instance of a theory, showing how an abstract theory is to be applied in a particular case, in which case an instantiation of a mechanism schema is appropriate. Sometimes a proposal is called a "model" because components are as yet unspecified, in which case it designates an incomplete mechanism sketch.

The scope of the domain modeled varies. A mechanism schema may represent a single unique case (e.g., a mechanism producing a unique historical event) or a recurring mechanism in only one species (e.g., a disease-producing mechanism in one form of human cancer). More often, mechanism schemas have a "middle range" (Schaffner 1993) of applicability; that is, they are found in some subset of biological cases, such as memory mechanisms in the hippocampus of vertebrates. In a few cases, a schema may be claimed to apply to all known cases, such as all instances of protein synthesis in living things on Earth.

Again, consider the diagram for the mechanism of protein synthesis:

DNA → RNA → protein

This is a simplified and general schema of the protein synthesis mechanism. It is very schematic and abstract; at this degree of abstraction it may be instantiated in a domain of very wide scope. It applies to most instances of protein synthesis in living organisms found on Earth. But compare the following schema:

RNA → DNA → RNA → protein

This diagram is at the same degree of abstraction as the previous one, but it has a domain of much narrower scope, namely, retroviruses. Hence, the degree of abstraction with which the mechanism schema is represented and the scope (that is, the generality) of the domain modeled are distinct. Increasing the degree of abstraction may produce a schema with a higher degree of generality, but not necessarily so, as these examples illustrate.

The amount of detail specified in an abstract mechanism schema is called the "degree" of abstraction. Abstraction hierarchies have an increasing loss of detail as one ascends the hierarchy. Conversely, as one descends the hierarchy, there is increasing "specification" of detail until the schema is "instantiated," resulting in the description of a particular mechanism. Thus, a mechanism schema should not be viewed as a model with merely the two-place relation of a variable and its value; mechanism schema hierarchies may have a range of degrees of abstraction. The term "degree" is used to refer to rungs in abstraction hierarchies while the term "level" refers to rungs in part-whole hierarchies among nested mechanisms (usually represented at roughly the same degree of abstraction).

In biology and philosophy of biology, the term "model" is used in many ways, such as to refer to mechanism schemas, their instantiations, sketches, and hierarchical mechanistic theories. A different but also common usage is employed to refer to something relevantly similar to the mechanism of interest and used in its representation or discovery. To models of this latter type we now turn.

4.   MODELS FOR DISCOVERING MECHANISMS

Many kinds of models aid the discovery of mechanisms. Models have both a "subject" and a "source." The mechanism to be discovered is the "subject" of the model. The "source" of the model may be the subject itself, as in the case of physical scale models and computer simulation models. In contrast, the source of a model may be different from the subject mechanism of interest, as in the case of an analogue model or a model organism used as a substitute for studying mechanisms in humans (Harré 1970).

Diagrams are a type of model used to represent the mechanism of interest; they have the same source and subject. Diagrams are especially propitious for representing many mechanisms. They show overall spatial organization of the parts and depict more or less structural detail of the entities. Activities are more difficult to represent in static drawings. Sometimes arrows illustrate activities, but arrows are also often used to show mere movement or time slices. Cognitive psychologists have studied how humans manipulate visual representations in order to run "mental simulations" of mechanisms. This method enables the person to "see" how some mechanisms work and to use the representation to make predictions (discussed in Bechtel and Abrahamsen 2005). But in more complex cases, humans use aids, such as computer simulations, to represent the complex mechanism and to run a simulation to make a prediction and explore "what-if" scenarios. Jim Griesemer (2004, 438–39) intriguingly noted: "Although interactive graphics extended the tradition of physical modeling, they also constituted a new mode of interaction with numerical data, allowing users to intervene kinesthetically in the simulation process. This is terra incognito for conventional philosophies of scientific knowledge."

A physical scale model of DNA has the same source and subject. In a famous case, the x-ray crystallographer Rosalind Franklin produced x-ray photographs by bombarding crystallized DNA (both the source and the subject). James Watson and Francis Crick used her photographs in choosing the shape and dimensions of their physical scale model of the DNA double helix. Although the x-ray crystallographic data (as well as other data about the chemical composition of DNA) constrained the space of possible models, those constraints were insufficient to determine all the physical properties of DNA (the source and subject) to use in building the scale model. In *The Double Helix*, Watson recounts the moment when he physically manipulated accurately constructed physical models of the DNA bases. That tactile manipulation in two dimensions (based on the assumption that the bases were flat and in a plane) allowed him to discover the geometric fit and hydrogen bonding between complementary bases. This discovery illustrates the role that may be played by physical manipulations of scale models. It was one step among many in this two-year extended discovery episode. The double helix model with its two strands of complementarily bonded bases suggested to its discoverers how it could carry out one of the functions of the genetic material. They immediately proposed the mechanism of DNA replication via the activity of complementary copying to fill the black box of genetic replication in the series of hereditary mechanisms. Thus, the *model of* DNA served as a *model for* investigating the functions of the genetic material. (For more on this distinction of *model of* and *model for*, see Griesemer 2004.)

Two-dimensional diagrams, three-dimensional physical scale models, and computer graphic models may all be generated by using data about the subject being modeled. To see how the structures can possibly function, researchers locate the activity-enabling properties of the entities. Examples include the hydrogen bonds between bases or the active sites of enzymes. Those activity-enabling properties suggest to humans or to computational discovery programs what the activities and stages of the mechanism may be. Drug discovery programs readily exploit such knowledge of active sites on molecules to design new chemicals to play desired roles in disease prevention mechanisms.

Standing in contrast to such models, in which the source and subject are the same, are analogue models, in which the source for the model differs from the subject itself. Distant and near analogies, model organisms, and model experimental systems are examples. An analogue model is more or less similar to the subject of interest. In her classic *Models and Analogies in Science*, Mary Hesse (1966) coined useful terms for the similarities and differences between the analogue and the subject. The components that they both share are the "positive analogy." The components that are dissimilar are the "negative analogy." The components whose relation has yet to be determined at a given stage in the use of the analogy are the "neutral analogy." Scientists have often used analogies in discovering new scientific theories. Examples abound in Keith Holyoak and Paul Thagard's (1995) *Mental Leaps: Analogy in Creative Thought*. The discovery of mechanisms is no exception.

Those working on the use of analogical reasoning to construct scientific hypotheses break it down into stages: problem finding, analogue retrieval, extraction of an abstract causal structure from the analogue, mapping from analogue structure to the subject area, adjustments to fit the subject, and testing of the newly constructed hypothesis. First, one identifies the problem to be solved. For example, one wishes to understand the mechanism of regulation of the genes producing the enzymes for synthesizing the amino acid tryptophan (trp). Then one searches to retrieve an appropriate analogue. For example, one might be familiar with the model for regulating the set of genes for producing the enzymes for digesting the sugar lactose; the lac operon model works via a derepression mechanism. The next stage is extracting an abstract mechanism schema from the detailed analogue. For the lac operon model, one might drop the details specific to the lactose case to construct an abstract derepression mechanism schema. In such an abstract schema, a gene produces a repressor protein molecule that binds to an operator gene on the DNA just upstream for a set of coordinately controlled structural genes. When an external inducer is present (the milk sugar lactose), it serves to bind to the repressor and change its shape; as a result the repressor falls off the DNA. Once the repressor falls off the DNA, the adjacent structural genes become active. This abstract derepression schema may be mapped to the subject area for the tryptophan case. Adjustments must be made because of the differences in the trp case. The trp genes are expressed when certain concentrations of tryptophan are not present. So tryptophan, when bound to the repressor, allows it to bind to the DNA and repress the genes. In the absence of suitable concentrations of tryptophan, the repressor does not bind to the DNA. Thus, one can generate a mechanistic model for the regulation of the tryptophan genes by analogy with the lac operon and appropriate modifications.

Once one has generated a mechanistic hypothesis by analogy, it must be evaluated to see how well it fits the subject area. One uses it to predict the outcome of experiments on the trp system. However, anomalies arose during testing of the trp system. In fact, the trp operon was found to be more complex than the lac operon. The depression mechanism was acting, but something more was also happening. Resolution of the anomalies required the addition of another regulatory mechanism, called "attenuation." This secondary mechanism operated to fine tune the concentration of the enzymes producing tryptophan, depending on the concentration of tryptophan in the bacterial cell. Sometimes when anomalies arise, the unexploited neutral analogy in the original analogue model may be a resource for ideas about how to revise the mechanistic hypothesis. However, in the trp case, new components not found at all in the lac case had to be added to resolve several anomalies (Karp 1989).

After the success of the operon model, it was used as an analogue to construct plausible hypotheses about how other genes were regulated. Some operate by a derepression mechanism, but others do not. The repertoire of types of gene regulation mechanisms continues to grow (Beckwith 1987).

In addition to conceptual analogue models, biologists use physical analogue models, namely, model organisms and in vitro experimental systems. Such model systems may be used, in some cases, to map directly to the subject of interest; in other cases, they may be used in the discovery of a general theory. When the goal is to discover mechanisms in humans, often disease mechanisms, then animal models are sought or constructed and mappings are direct from the animal model to humans. However, in veterinary animal medicine, humans may be the model organisms for mechanisms involving possible pain produced by drugs or procedures. Humans can report their pain sensations, while inferences on the basis of physiological or behavioral cues in animals are much less reliable for judging the presence and severity of pain. Many such considerations guide the choice of model organisms, including the nature of the mechanism sought in the subject of interest, the belief that such a mechanism or a relevantly similar one operates in the model, the ease of manipulation, and the amount of work that has already been done on the model organism that can serve as a basis for further work (Burian 1993).

Schaffner extensively examined the use of model organisms in molecular biology and, more recently, in behavioral genetics (Schaffner 1993, 1998, 2001). As a result, he viewed some biological theories as having the structure of "overlapping temporal models." These are theories of the "middle range": that is, their scope is not universal but, with variations, applicable beyond a single instance to a domain of middle-range scope, such as prokaryotes or vertebrates. Components of such theories are presented as "collections of entities undergoing a process." In the mechanistic perspective proposed here, Schaffner was referring to theories composed of mechanism schemas with varying scope. A model organism or a model experimental system provides the "prototype"; then, how widely the prototypical mechanism (or slight variants of it or its modules) occurs has to be determined empirically.

In biological, as opposed to applied, research, as Schaffner noted, the goal is often to find generalizations. A manipulable model is sought that will provide results that can be generalized. The model organism is the source for the general mechanism schema (the subject), as well as an instance of it. The history of biology is replete with examples not only of excellent model organisms, namely, those with typical mechanisms that were successfully generalized, but also of failed ones, namely, those with odd quirks that led their users astray in the search for general theories. Gregor Mendel's peas (*Pisum sativum*) have what others later discovered to be general hereditary mechanisms, while Mendel's attempts to extend his results to hawkweed (*Hieracium*) failed. Hawkweed, it was later found, can reproduce asexually, and thus was a very poor choice as a model organism for genetic crosses. Hugo de Vries studied the sudden appearance of new true breeding forms of evening primrose (*Oenothera*). He believed that the evening primrose was an excellent model organism for establishing a research program of experimental evolution in his botanical garden. However, the extremely rare chromosomal mechanisms in the evening primrose are not general at all. Considerable empirical work ensued to unravel the quirky mechanisms of hawkweed and evening primrose, more than would likely have been done had they not played the role of anomalous model organisms in the search for general hereditary mechanisms.

In contrast are the triumphal tales. T. H. Morgan's choice of the fruit fly (now named *Drosophila melanogaster*) for his genetic studies yielded understanding of very general hereditary mechanisms. Similarly, Jacques Monod's choice of the bacteria *Escherichia coli* led to the discovery of regulatory genes, a universally found component of gene regulation mechanisms, even though all are not the depression type (Darden and Tabery 2005). However, many anomalies arose for Monod's famous quip: "What's true for *E. coli* is true for the elephant, only more so." Other molecular biologists, desiring to study mechanisms of cellular differentiation not found in bacteria, Sydney Brenner, for example, carefully chose and perfected strains of the nematode worm, *Caenorhabditis elegans* (discussed in Ankeny 2000), and François Jacob (1998) chose the mouse, *Mus musculus*.

Even when it fails, it is a good strategy to generalize from a mechanism discovered in one experimental system to others producing similar phenomena. Evolution often does reuse mechanisms or their components. As Francis Collins, then head of the U.S. Human Genome Project, said:

Because all organisms are related through a common evolutionary tree, the study of one organism can provide valuable information about others. Much of the power of molecular genetics arises form the ability to isolate and understand genes from one species based on knowledge about related genes in another species. Comparisons between genomes that are distantly related provide insight into the universality of biologic mechanisms and identify experimental models for studying complex processes. (Collins et al. 1998, 686–87)

Because of the common evolutionary descent of biological organisms, biology has a stronger basis for appealing to similarity than other fields employing cross-field analogies. Many similarities are a result of evolutionary homology; that is, the similarities result from the subject and source's sharing a common ancestor. Thus, model organisms and model experimental systems may serve as homologues for studying the mechanism of interest. But evolution works both by copying and by editing, that is, both by inheritance and by variation. So, it is an empirical journey to find the appropriate family resemblances, in the literal sense of that term (Schaffner 2001).

As Marcel Weber (2005) pointed out, phylogenetic inferences based on homology provide a sounder basis for generalization than mere induction by simple enumeration. When a mechanism is found in organisms distant on the evolutionary tree, the assumption is made that all the descendants of their common ancestor share the same mechanism. This is, he noted, an argument from parsimony, but one that is plausible because of the unlikelihood that the same mechanism did arise independently in widely separated evolutionary paths. His examples of widely shared homologous mechanisms included the mechanisms of DNA replication and protein synthesis in all eukaryotes.

Even more often than the evolutionary conservation of entire mechanisms, modules of mechanisms are reused in other mechanisms. Model organisms have supplied what Weber called materials for "preparative experimentation." For example, DNA sequences extracted from *Drosophila* were used as probes to fish for homologous DNA sequences in genomic libraries prepared from the DNA of a variety of other organisms. The important homeobox genes discovered in *Drosophila* are an example. Homeobox genes control the development of the front and back parts of the body. DNA sequences almost identical to those from the fruit fly were quickly found in mice and humans (Weber 2005, 162–64).

Now that many whole genomes have been sequenced, the genome databases and the growing protein databases serve as what might be called "canned model organisms." These in silico data allow searches to find "orthologous genes" that can be traced back to a common ancestor. The number of shared genes and proteins with similar activities is surprising. Most organisms share a substantial number of molecular mechanisms or modules of mechanisms that are very ancient. Evolution appears to work by fashioning new architectures from old pieces.

When such orthologous genes are found but their function is unknown in humans, model organisms provide researchers with a unique method for finding the mechanisms in which the genes function. This method is called the "modifier screen" (Hariharan and Haber 2003). Random mutations are induced in the organism known to have a specific mutation in a gene of interest. The added mutations in other genes may modify the usual phenotype, thereby providing clues to the molecular mechanism in which the gene of interest is important. Many different mutants in *Drosophila* can be induced and screened to detect an effect. Genes that undergo a mutation that causes a worsening of the phenotype are called "enhancers," whereas genes that cause a correction of the mutant phenotype are called "suppressors." Additional genetic experiments allow the roles of the enhancers and suppressors in the mechanism of interest to be determined and orthologous genes sought to investigate the scope of the newly discovered mechanism components.

Some philosophers raised concerns about whether the use of simple model organisms might skew results, but Schaffner replied that model organisms "are not only intended to be *representative* prototypes, but also to be 'idealized' in the sense of sharpened and more clearly delineated. The value of sharpened, simplified idealizations is a lesson that the physical sciences can still teach us. ... Once simple prototypes are preliminarily identified ... *then* variations (often in the form of a *spectrum* of mutants) are sought (or re-examined) to elucidate the operation of simple mechanisms" (Schaffner 1998, 280; italics in original).

Thus far in molecular and neurobiology, simple model organisms have proved very useful. The extent to which simple model systems and the search for simple mechanistic accounts must be supplemented in the face of biological complexity remains to be seen (for example of failures of the mechanistic research program in the face of complexity, see Bechtel and Richardson 1993). Nonetheless, many types of models have proved very fruitful in the discovery of biological mechanisms.

5. REASONING STRATEGIES FOR DISCOVERY: GENERATING, TESTING, AND REVISING

Analyzing reasoning in discovery is a much more tractable task when what is to be discovered is a mechanism (rather than a vaguely characterized explanatory theory). Further, discovery is an extended process of generation, evaluation, and revision of mechanistic schemas and sketches. Model organisms and model experimental systems may play many different kinds of roles in all stages of discovery. They may be used for exploratory experimentation, prior to (or in place of) using a conceptual analogy, to discover possible components of the mechanism. In the discovery of the mechanism of protein synthesis, Paul Zamecnik and his colleagues worked to perfect an in vitro experimental system that would incorporate radioactive amino acids into polypeptides (components of proteins). They centrifuged rat livers to extract components, including microscopically visible particles (later called "ribosomes"). They found that they had to put into the in vitro system a particular centrifuge fraction extracted from the rat livers in order to produce incorporation of amino acids into polypeptide chains. This case shows that an in vitro experimental model system can be constructed by physically decomposing an actual organism and then investigating the working parts (Rheinberger 1997). Ideally one can isolate all the mechanism components and determine their roles within the mechanism. But even before a thorough characterization is available, a running mechanism may be constructed in vitro to allow further exploratory experimentation of its parts. Such exploration is one of many ways that generation and testing are closely tied during mechanism discovery (Darden 2006, chap. 3).

More often, model organisms and model experimental systems are used to test plausible mechanistic hypotheses generated via analogy, presumed homology, or other means. Craver (2002b) detailed experimental strategies for testing a hypothesized mechanism. Such experiments have three basic elements: (i) an experimental setup in which the mechanism (or a part of it) is running, (ii) an intervention technique, and (iii) a detection technique. The mechanism sketch or schema being investigated may provide an abstract framework for constructing an experimental protocol: intervene here; detect there. Biologists set up many kinds of experimental models to test mechanistic hypotheses. Intact living organisms have many mechanisms running; the challenge with intact organisms is to find ways of individuating single mechanisms and ruling out confounding factors. In vitro preparations solve some problems encountered with in vivo ones. The challenge is to find the appropriate components and make them work in an in vitro experimental system.

Craver (2002b) discussed several different kinds of intervention strategies that have been used historically to test a mechanistic hypothesis in an experimental model. First are *activation strategies*, in which the mechanism is activated and then some downstream effect is detected. One example of the use of a model organism is to put a rat into a maze and detect activity in its brain cells with a recording device. A common biochemical intervention is to put in a tracer, such as a radioactive element; activate the normal mechanism; and detect the tracer as it runs through the mechanism. Good recording devices and tracers do not significantly alter the running of the activated mechanism; they merely allow observation of its workings.

Second are *modification strategies* that involve not merely activating but modifying the normal working of the mechanism operating in the model system. A way to learn about a mechanism is to break a part of it and diagnose the failure (Glennan 2005). A fruitful way to learn about the action of a gene is to knock it out and note the effects in the organism. As with the notorious ablation experiments in physiology in the nineteenth century, the problem with gene knock-out techniques in intact animals is that such a missing part may have multiple effects that are difficult to disentangle, given the often complex reactions between genotype and phenotype.

Another kind of modification strategy is an *additive strategy*. Some component in the mechanism is augmented or over-stimulated, then effects are detected downstream. Craver's example was of engineered mice with more of a specific kind of neural receptor. Those mice learned faster and retained what they learned longer, thereby providing evidence for the role of such receptors in learning and memory.

Craver (2002b) suggested using all three types of strategies – activation, ablation, and addition. Consistent results strengthen the evidence for the hypothesized mechanism. Each helps to compensate for the weaknesses of the others to yield a robust (Wimsatt 1981) conclusion, namely, a conclusion supported by a variety of types of evidence (Lloyd 1987).

In addition to manipulating an intact, operating mechanism, one may seek evidence for the existence and nature of hypothesized entities, activities, and/or modules separately. For example, an ion channel protein may be isolated and its structure investigated to find its role in a neuronal mechanism. The protein could be genetically altered to have an abnormal additional part to investigate how that affects its functioning. A positive result of such investigation of a hypothesized part of a mechanism is an example of what Elisabeth Lloyd (1987) called "independent support for aspects of the model," to distinguish it from the "outcome of the model" (the latter is often called "testing a prediction of the model as a whole").

Strategies for credentialing experimental evidence, in general, are, of course, important for assessing the evidence obtained from a model organism or model experimental system. These include use of adequate controls, reproducibility of results, appropriate use of randomization, and demonstration of the adequacy of instruments, to name only a few. Finally, evidence from two or more fields further strengthens the claim that the conclusion is robust. Sometimes a single researcher uses techniques from two different fields. Sometimes researchers from different fields provide evidence for different modules of the mechanism, as did the biochemists and molecular biologists for the mechanism of protein synthesis and the working of the genetic code. The study of interfield relations by different research groups and the coordination of their results to provide evidence for a coherent picture of the mechanism is one of the many important social aspects of the collective scientific enterprise.

A description of a particular mechanism may be located in the larger matrix of biological knowledge (Morowitz 1985), which includes hierarchically organized descriptions of mechanisms in which one mechanism serves as a part of a larger one. The matrix also includes longer temporal series of mechanisms that indicate which mechanisms occur before and after a given one. These requirements for an adequate description of a mechanism constrain and guide mechanism discovery, as a description of each is sought, any missing components are filled, and coherence within a wider context is explored.

As we have seen, discovery of mechanisms involves generating of possible and plausible mechanistic hypotheses (e.g., by analogy or homology), testing those hypotheses in model organisms and model experimental systems, and deciding whether a newly proposed mechanism fits coherently into the matrix of other known mechanisms. Another part of the discovery process is error correcting. During testing, a failed prediction yields an empirical anomaly. The mechanistic hypothesis may be in need of revision (Wimsatt 1987). A fruitful and oft employed strategy is to overgeneralize from a successful result, use it analogically in other cases to construct plausible hypotheses, then specialize when anomalies arise. A systematic search for anomalies allows the scope of a mechanism schema to be determined. Further, anomalies guide the generation of hypotheses about alternative, variant mechanisms that do not fit the hypothesized schema.

As in the extended discovery processes of generation and testing, the view that what is to be discovered is a mechanism provides guidance in reasoning to resolve anomalies. Such reasoning in anomaly resolution is, first, a diagnostic reasoning task, and then a redesign task. The location of the failure is sought. Philosophers have been unduly pessimistic about the ability to localize the site of failure in some holistic web of beliefs. In practice, scientists often localize the erroneous part of a mechanism schema and correct it. Diagrams of the mechanism's stages aid localization of the problematic component. Then, depending on the site of localization, a redesign process may be needed to improve the hypothesized mechanism. The hypothesized mechanism or mechanism schema aids both diagnosis to localize the failure and, if required, redesign to supply an improved module.

As a first step in the anomaly resolution process, the anomalous result must be credentialed to ensure that it is not the result of an observational or experimental error. Experiments revealing an anomaly may be reproduced, using careful controls, or investigated, using other credentialing strategies for experimental results (for more on characterizing anomalies, see Elliott 2004).

Once the anomalous result is confirmed, the location of the failure needs to be diagnosed. On the basis of the extent of revision required, an anomaly may be categorized as a monster, special case, or model anomaly. If the anomaly can be localized outside the domain of the mechanism schema, then no revision is required. Another possibility is that the anomaly might result from a disease or other abnormality. Such "monster" anomalies can be barred from requiring a change in the normal mechanism schema. An example of monster barring occurred when lethal gene combinations produced anomalous genetic ratios; normally the combination of two genetic alleles does not lead to the death of the embryo. No revision in claims about normal genetic mechanisms was required with the discovery of lethals; a kind of failure had been found.

Sometimes, the anomaly requires a splitting of the domain in which the mechanism is claimed to operate. If the anomaly only occurs in a small part of the domain, the anomaly is a "special case" anomaly. For the small domain consisting only of retroviruses, a RNA → DNA step was added to the usual mechanism schema for protein synthesis.

In contrast to monster and special case anomalies, model anomalies indicate what is normal for a domain of wide scope. Thus, the anomaly is a model in the sense of an exemplar. There may be no sharp divide between special case and model anomalies as domains are split to accommodate variations in the ways mechanisms operate. The boundary between special case anomalies and model anomalies is not sharp.

Once the anomaly is judged to require revision of a mechanism schema, further guidance results from a diagrammatic representation of a mechanism or other means of locating its modules. In the mid-1950s, the ribosome was hypothesized to play the functional role of the template for transferring the order of the bases in the DNA to the order of the amino acids in a protein.

DNA → template RNA → protein
DNA → ribosomal template → protein

Anomalies began to accumulate for the ribosomal template hypothesis. As Douglas Allchin noted in examining other cases, presence of multiple anomalies localized in the same site of a hypothesis strengthens the confidence that revision is required. Attempting to resolve the anomaly in which the base ratios of DNA and ribosomal RNA did not correspond, Crick (1959) at first systematically generated alternative hypotheses to save the "ribosome as template" hypothesis. This anomaly indicated a problem about the "DNA → RNA" step in the proposed mechanism. He proposed alternatives localized in this module of the mechanism. Each component of this module served as a location for generating "how possibly" redesign hypotheses. This case shows a single researcher systematically generating a set of alternative redesign hypothesis at the site of failure.

Conservatively, the set of alternatives Crick discussed in 1959 did not include the postulation of an as yet undiscovered type of RNA having a base composition like that of DNA. This was the idea of a separate messenger RNA (mRNA), different from the known types of RNA. The discovery of such a messenger RNA was the way the anomaly was soon resolved. Tracer experiments supplied direct evidence for the existence of mRNAs. The functional requirement of a template, at that stage of the mechanism, with appropriate relations to the stages before and after it, acquired a new role filler, namely, messenger RNA (discussed in Darden 2006, chap. 3).

This ribosome anomaly case shows that when what is to be revised is a mechanism schema, that schema furnishes much guidance for anomaly resolution. Diagrams and other representations of the modules of mechanisms guide localization and redesign. When an anomaly is localized to a stage, then redesign may need to be done by adding something before or after the stage or changing hypothesized entities and/or activities within the stage itself. Furthermore, the entities and activities of a stage must give rise to the next, thereby imposing constraints on the components of a subsequent stage, on the basis of what the prior one can produce. Also, the modules of the mechanism not implicated by the anomaly must be shown to continue to function. The desideratum of having a productively continuous mechanism thus aids redesign during anomaly resolution.

In sum: reasoning in the discovery of a mechanism is guided by the description of the phenomenon of interest, aided by the characterization of what a mechanism is, and elaborated by specifying the features that an adequate description of a mechanism should satisfy. Mechanism discovery involves tight relations among generation, evaluation, and revision of mechanism schemas of varying scope. Philosophers should not view discovery as a process of floundering in an unconstrained space of vaguely characterized theories. If the goal is to discover a mechanism, much can now be said about reasoning strategies and experimental models to aid that task.

6. CONCLUSION

Philosophers of biology, working on cases from molecular biology, cell biology, and neurobiology, have characterized mechanisms as used in those fields. Analogue models, model organisms, and model experimental systems aid the discovery of mechanisms. Reasoning in generation, evaluation, and revision converts incomplete mechanism sketches to well-supported mechanism schemas.

This new perspective on mechanisms, arising in the philosophy of biology, allows the reexamination of traditional topics in the philosophy of science. These topics look different when one starts with mechanisms (rather than, e.g., perspectives arising from mathematical physics or formal logic). This research program is just beginning; the citations here point to recent work. Philosophers argue that appeal to mechanisms provides an account of causation (Glennan 1996, Machamer 2004, Tabery 2004, Bogen 2005), discovery (Thagard 2003, Bechtel and Abrahamsen 2005, Glennan 2005, Darden 2006), explanation (Machamer, Darden, and Craver 2000, Glennan 2002, Bechtel and Abrahamsen 2005), functional analysis (Craver 2001), interfield integration and unity (Craver 2005, Darden 2006, chap. 3), and reduction (Craver 2005, Darden 2006, chap. 4). These authors stress the importance of mechanisms in such fields as Mendelian genetics, molecular biology, cell biology, neuroscience, cognitive science, and linguistics. As yet unsolved are the issues of how this view of mechanisms applies to analyzing the mechanism of natural selection (Skipper and Millstein 2005) and to analyzing mathematical models in population genetics and ecology. Mathematical and computer simulation models of mechanisms usually have equations or functions to produce state transitions, while omitting representations of structures and the activities that produce the transitions. Could these impoverished mathematical models be improved by adding the details of the working parts of mechanisms?

This new mechanistic perspective is proving fruitful for reexaming issues in philosophy of science from the point of view of philosophy of biology. It is likely to continue to provide new insights.

9   Teleology

I distinguish between Platonic and Aristotelian teleologies. I detail William Paley's Platonic teleological argument for the existence of God from natural design and offer Darwin's theory of natural selection as an antiteleological response. But, since Aristotle's teleology is distinct from Plato's teleology I ask to what extent is Darwin's theory anti-Aristotelian.

Teleology in biology is making headline news in the United States. Conservative Christians are utilizing a teleological argument for the existence of a supremely intelligent designer to justify legislation calling for the teaching of "intelligent design" (ID) in public schools. Teleological arguments of one form or another have been around since antiquity. The contemporary argument from intelligent design varies little from William Paley's argument written in 1802. Both argue that nature exhibits too much complexity to be explained by "mindless" natural forces alone.

What is so remarkable about complex designs? Compare a watch with a stone. According to Paley, watches are complex and stones are not because a watch's functioning depends on its precise arrangement; a stone's does not. Michael Behe, a contemporary advocate of intelligent design, labels the sort of complexity "irreducible complexity": if you remove any part, the whole structure ceases to function (Behe 1996). The bacterial flagellum, Behe argues, is irreducibly complex. It acts as a tiny propeller spinning at more than 20,000 revolutions per minute. Thirty different proteins are precisely formed and adjusted to produce the motion. If any one of them is removed, the flagellum ceases to propel the bacterium. The example for which Paley is most famous is different – the mammalian eye – but the argument form is the same.

How do we explain the existence of complex designs? Traditionally, there are two main positions, the materialist and the teleologist. A materialist believes, roughly, that all natural phenomena are the product of the causal interactions of matter. There is no room in the materialist's ontology for purpose or goals. Matter is not imbued with it. Material arrangements are not formed by purposive agents. All seemingly goal-directed activity is ultimately the product of matter and cause. Even our own intentions and goals (and consciousness) are explained materially without reference to goal-directed processes. On the materialist view complex items and arrangements are explained by nonpurposive natural forces, matter and cause. Philosophers in the context of the teleology debates sometimes call the materialist a "naturalist". I will use the terms interchangeably.

A teleologist believes otherwise: the explanation for goal-directed arrangements calls for goal-directed causes. There are roughly two sorts of teleologists, one we will call a "Platonist" and the other an "Aristotelian". A Platonist believes that simple matter and cause do not explain the *goodness* or *orderliness* of an arrangement. To illustrate, consider an analogous question, "Why does Socrates sit in prison?" While facts about physiology offer a complete explanation of Socrates' current position in prison, the facts do not provide the real reason for Socrates' predicament. He remains in prison because remaining rather than escaping is what Socrates deems the best course of action. In the *Timaeus*, Plato applies this reasoning to explain cosmological order. He finds the materialist explanation with its reference to the simple motion satisfactory for describing astronomical patterns, but insufficient for explaining the beauty, orderliness, and good arrangement of the cosmos. "The true cause is agency working for the best", concludes Plato (2000, 48a); reference to intentionality is the materialists' missing ingredient.

When it comes to explaining biological arrangement and functioning an Aristotelian eschews Platonic designers for an inherent purposive or goal-directed force that resides in the material properties of living entities. Good arrangements are not the handiwork of a creator; rather they are due to some inner principle of change within living organisms. This force explains, among other things, natural regularities. For example, many carnivores possess sharp teeth in the front and broad molars in the back. This particular dental arrangement allows carnivores to flourish in the wild. Carnivores whose dental arrangements significantly vary tend to perish. The materialist explanation in terms of matter and cause would fail to explain why the carnivorous arrangement is so successful while its variants are not. That is because, according to the Aristotelian, the materialist has no ontological category like Aristotle's "that for the sake of which", or *telos*. Sharp teeth grow in front and broad molars grow in the back for the sake of an organism's flourishing. The goal is inherent in the nature of dental growth.

To sum up, the Platonic and Aristotelian agree that materialist explanations insufficiently explain good arrangements, especially ones that allow the system to flourish or even ones that perform their action in the best possible manner (as Plato believed about the cosmos). They disagree on where to locate the purposive force in nature. A modern day advocate for intelligent design as well as its Victorian cousin (Paley's) is a Platonist. She believes that good natural design must be the product of a designer just as well-designed artifacts must be the product of intelligent and talented designers. The Aristotelian position constitutes an alternative to the Platonic metaphysics, eschewing designers for inherent goal-directed tendencies imbued in the matter of living organisms.

On the face of it, our discussion of the difference between Platonists and Aristotelians changed the essence of what teleological arguments purport to explain. For Paley and contemporary proponents of intelligent design what is to be explained is complexity. Aristotelians and Platonists aim to explain either orderliness in the cosmos or biological arrangements that lead to flourishing of organisms. But, on closer inspection rather than a change of subject our new discussion deepened our previous analysis. Implicitly we defined complexity in functional terms. An item is complex if a change in a part leads to the ceasing of the system's functioning. But what of systems that have a redundancy built in? Although an airplane is composed of many interworking parts, it is not true that the failure of a part will force the plane to cease flying. If the automatic pilot system fails, the pilot can switch to manual; if one engine conks out, the others compensate; if the system responsible for maintaining appropriate oxygen levels in the cabin fails, oxygen masks drop down from above the seat. The redundant systems ensure functioning even when parts fail, but the fact that the parts are not essential for function does not mean that airplane systems are not complex. Quite the reverse, it means that they are extremely complex. I do not mean to engage in a long discussion on exactly what complexity means, but it is worth pointing out that the issue is not exactly straightforward (for opposing accounts see Dembski 1998 and Fitelson, Stephens, and Sober 1999).

1.   PALEY'S INFERENCE TO THE BEST
EXPLANATION

Paley's argument (as with Behe's) starts with an analogy between living organisms and human artifacts. Briefly, if you came across a watch and inquired as to its existence you would not take seriously the conclusion that watches are the product of natural forces. It is highly improbable that natural forces would randomly coalesce matter into a watch. The possible existence of a designer who can manipulate the parts for his own purpose makes the existence of watches much more likely. A Platonic conclusion follows: the existence of a designer best explains watches and living organisms.

On my view many of the more influential teleological arguments in history are instances of "inference to the best explanation" or IBE (see Ariew 2002, Cooper 1987, Johansen 2004). One infers to the truth of a hypothesis from the existence of phenomena that the hypothesis that best explains Paley's inference to the existence of a watchmaker (and later his argument for the existence of a divine creator) works exactly in this way. The existence of watchmakers is supported by the existence of watches since existence of watchmakers best explains how such complex things could come to exist.

An interesting feature of IBE arguments is that with them we can infer the existence of *unobservable* phenomena. As Paley put it, already having inferred the existence of a watchmaker from inspection of a watch found on a dirt path: "Nor would it, I apprehend, weaken the conclusion, that we had never seen a watch made; that we had never known an artist capable of making one; that we were altogether incapable of executing such a piece of workmanship ourselves, or of understanding in what manner it was performed" (Paley 1828, 4). The strength of the inference to a watchmaker depends not on our witnessing watchmakers' making watches but in the relative likelihood that watches would exist *if* skilled watchmakers were to exist. Compare that to the likelihood of the materialist hypothesis that watches would exist if only the random action of natural forces were to exist. Natural processes alone are unlikely to make a watch. A creator with forethought more likely will. While perhaps Paley would have understood the concept of "likelihood" rather colloquially, in the contemporary philosophy of science literature its meaning is made more precise in the context of probability theory (see Sober 2000).

The feature of inferring the existence of unobservable causes distinguishes IBE from garden variety inductive arguments (Ruse 2003, 43). Let us see why. The strength of an inductive argument depends on the size and biasness of the sample. Yet, as Paley suggests (in the preceding quote), we may never have seen watchmakers make watches. If so, our sample size is zero. Likewise for the sample of times in which any of us has seen God creating living things (as Hume pointed out). If teleological arguments were inductively based, then they would be nonstarters. Good thing for Paley that his inference is not inductive. In an IBE the issue is not the features a sample has in common, but rather what explains an observed phenomenon.

Let us take another example to illustrate the difference between IBE and inductive arguments (the example is taken from Sober 2000). Suppose Gregor Mendel's argument for the existence of genes was inductively based. In a sense it was, since he supposedly obtained his definite 3:1 phenotypic ratios for various characters of pea plants on the basis of numerous plantings. But Mendel's theory of genes was not inferred by the repeated plantings; his theory of 3:1 phenotypic ratios was. He then asked a distinctive question: What would explain why the offspring exhibit these ratios from parental crossings from various types? For example, why did he get 3:1 ratios of wrinkled:smooth peas in the offspring crosses? Here, he employed an IBE: suppose the hereditary material acted as a particle that obeys certain rules. That would best explain the ratios (compare to the "blending hypothesis" that Darwin favored). Notice, Mendel had no direct evidence of the existence of these particles (that we now call "genes"). Rather, he inferred them as the best explanation for the repeated 3:1 phenotypic ratios. Some philosophers believe that the use of IBE is one of the hallmarks of good science. If so, perhaps it should be said that the use of IBE in teleological arguments (by both

Paley and IDers) should mark them as scientific hypotheses. I will not push this suggestive line too far, but notice it does give the lie to the typical argument from anti-IDers that the central ID argument is unscientific. IBE arguments are not unscientific.

Paley notes that the strength of the inference to watchmakers is not invalidated if the watch is found to be inaccurate ("Neither ... would it invalidate our conclusion, that the watch sometimes went wrong, or that it seldom went exactly right" [1828, 4]). This is another distinctive feature of teleological explanations, and one not obviously available to materialistic explanations. Recall Aristotle's argument against the materialist: the materialist cannot account for the reason for why certain dental arrangements allow their possessors to flourish while others lead to their demise. The success terms are not part of the materialist vocabulary; the materialist has only matter and cause to explain organic events. And yet the material etiology of the events does not distinguish one that succeeds from one that fails. The teleologist has an extra concept, an evaluative term that explains success and failure. For Aristotle the extra concept is "that for the sake of which". Certain dental arrangements exist *for the sake of* a carnivore's flourishing. Carnivores with defective or alternative arrangements are malfunctioning individuals. The argument is the same for Paley, but the nature of the telos is distinct. All watches have a function, even broken watches. The distinction between broken and running watches is whether or not they fulfill their God-given function. Since the teleological explanation provides an account of the essential difference between functioning and malfunctioning items while the materialist explanation does not (the materialist would have to ascribe the differences to chance), teleological explanations have an advantage over materialist explanations.

1.1   Paley's Self-Replicating Watch

Next, Paley considers what would happen if we found a self-replicating watch. The passage is a lovely early example of science fiction:

Suppose, in the next place, that the person who found the watch should after some time discover that, in addition to all the properties which he had hitherto observed in it, it possessed the unexpected property of producing in the course of its movement another watch like itself – the thing is conceivable; that it contained within it a mechanism, a system of parts – a mold, for instance, or a complex adjustment of lathes, baffles, and other tools – evidently and separately calculated for this purpose; let us inquire what effect ought such a discovery to have upon his former conclusion. (1828, 9)

In addition to serving the function of telling time, this watch has a further extraordinary feature: it produces well-functioning offspring. The "discovery" of a self-replicating watch affects the former con- clusion – the existence of watchmakers – in several important ways. First, it further illustrates that the strength of Paley's inference does not depend on ever having seen any watchmakers make watches. As I argued earlier, this feature distinguishes IBE arguments from inductive ones since it does not depend on sampling from a population of events.

Second, and more important, the discovery of the self-replicating watch *strengthens* the inference to the existence of a designer at the same time that it weakens the inference to the hypothesis that the item is the product of natural forces alone. As Paley puts it, "If that construction *without* this property, or, which is the same thing, before this property had been noticed, proved intention and art to have been employed about it, still more strong would the proof appear when he came to the knowledge of this further prop- erty, the crown and perfection of all the rest" (1828, 9). The prob- ability of natural forces' randomly producing a watch is very very small, but the probability of natural forces' randomly producing something as extraordinary and exquisite as a self-replicating watch is even smaller. The general lesson is, *The more complex the parts, the stronger the evidence of a designer*. It and the next feature play a large role in Paley's ultimate inference, the existence of a God.

The third effect that the self-replicating watch example has on the former conclusion is, in Paley's words, to "increase his admiration of the contrivance, and his conviction of the consummate skill of the contriver". However complex watches are, most of us with average intelligence and skills could imagine learning, after extensive train- ing, how to create watches. But to have the skill of the maker of a self- replicating watch would be extraordinary or even supernatural. The general lesson here is, *The more complex the design, the more intelligent (or skillful) is the designer.*

1.2   Easy Step to Design: Living Organisms

Once the second and third features of the new IBE from self-replicating watches are in place, Paley has only to convince us that living tissues, organs, organisms, and ecosystems are much more complex than self-replicating watches, that their parts are much more attuned to the functions that they serve. In a sense that is the intention of most of Paley's book, *Natural Theology*, from which his famous argument for the existence of an intelligent designer is a relatively small section. The later chapters are more or less a zoological textbook, detailing the wonder of natural adaptations. The most famous passages are found in the section that describes the anatomy and function of the eye. For instance, he expresses amazement at how the anatomy of eyes from animals living in distinct environments differs according to the laws of transmission and refraction of rays of light.

Does Paley's argument succeed in proving that living organisms are created by God? Among his contemporaries he provided a powerful argument against a materialist who seems to have little to account for natural adaptation. Yet, the success of IBE arguments depends on the relative success of the given hypotheses, and, as Darwin would show, there are other hypotheses to consider besides the random action of matter and cause and an intelligent designer. Therein lies a formal limitation of IBE inferences. The strength of an IBE is only as good as the proffered hypotheses. For any given set of hypotheses the interlocutor has always the option to remain agnostic as to the cause of the phenomenon in question. To suggest otherwise, for instance, to argue that since God is a better explanation than matter and cause, God must exist, is to commit "the only game in town" fallacy (Sober 2000). As for proving the existence of an omnipotent and omniscient God, Paley's commentators and critics often pointed out while Paley's inference might have strongly suggested the existence of a supremely intelligent designer (a supernatural designer), it stops short of proving that the designer is God. I suspect that this is the sort of thing that modern IDers like Behe have in mind when they express agnosticism about the nature of the designer (Orr 2005).

1.3   IBE versus Argument by Analogy

Instead of an IBE, some commentators believe that Paley's argument is best read as an instance of an argument by analogy. True, an analogy between functioning organs and artifacts features in a lot of what Paley says. But I think if Paley's argument were essentially an argument by analogy, then it would be all the weaker for it. As we shall see later on, I think the same of those who read Aristotle's many references to "as in art, nature" as confirmation that his teleological arguments are essentially arguments by analogy. There are formal differences between IBE arguments and arguments by analogy.

The strength of an argument by analogy depends on the degree to which the item or phenomenon in question resembles the analog. Winnie-the-Pooh somewhat resembles real-word honey-loving bears, but not enough to warrant the conclusion that real-world bears are made of cotton stuffing. Do watches sufficiently resemble living organisms to warrant the conclusion that the latter are designed because the former are? Probably not if we regard the variety of extant living organisms. Watches are made of metallic cogs and wheels, have glass faces, and fit around one's wrist. Koala bears are endothermic, climb trees, and draw crowds at the San Diego Zoo.

Paley's argument fares much better as an IBE than as an argument by analogy. Sure, the analogy between watches and living organisms plays a central role in the IBE, but as a means to strengthen the IBE inference from complex adaptation to designer, not as a analogy between the features of watches and those of living organisms. In other words, the relevant features in common between watches and living organisms are their complexity and well-suitedness to the function that they serve. If designers are the best explanation for highly functioning artifacts they will be even better explanations for the even more complex and highly functioning forms found in nature.

2.   DARWIN'S TWO ALTERNATIVES TO PALEY'S CREATOR

Earlier I said that the strength of Paley's inference to a designer depends on the relative weakness of alternative hypotheses. Paley's alternative is a straightforward materialist who, like Aristotle's, is forced to explain functional arrangements by citing chance.

Darwin's great reply to Paley is to provide a naturalistic alternative in which chance is not the reason why living organisms are made. In fact, Darwin offered two alternative hypotheses; both serve as an alternative to some aspect of Paley's inference.

Darwin's first hypothesis is that all living organisms are related by common ancestry. Different species evolve from the common lineage as branches grow on a tree. Not all species evolve; some go extinct and others continue their heritage relatively unchanged. The idea that there is a "tree of life" from which all forms evolve is a radical alternative to the Christian idea that each species is the unique creation of an all-good God and thereafter immutable and eternal (for an excellent overview, see Waters 2003).

Darwin's second central idea explains the conditions for evolution of species and extinction, and ultimately, the way new organismal forms emerge from preexisting variation. Darwin argued that all organisms undergo a struggle for existence in part because more individuals are born than could possibly survive given the limited resources of their environment (he got this from reading Malthus). Organisms possess a variety of heritable traits (they pass them to their offspring), and hence organisms are variously equipped to handle the struggle for existence. Those individuals who happen to possess traits that happen to make them better adapted to their local environmental conditions will pass them to their offspring. Those lineages that fail to reproduce their traits will eventually go extinct. Reiterate this "natural selection" over many generations and over many different variants and eventually lineages evolve distinctive features that make them overall better adapted to their conditions than they were before. Darwin did not know how varieties are produced; nor did he understand the details and rules of heritability. But his theory did not need the details; it required only that heritable varieties exist as the "fuel" for natural selection.

Darwin's theory of natural selection is an alternative to both the materialist and the Christian theories for the existence of complex adaptations. As we saw, the materialist is forced to explain complexity and well-adaptedness as a chance side effect of natural processes. Darwin's theory does not resort to chance. He specifies the general biological conditions that determine the existence of adaptations. Darwin's theory is radically distinct from the teleological explanation: adaptations are not created on Darwin's theory, but they emerge out of the confluence of many distinct biological processes, the production of individual differences, the mechanisms of heritability, the basic rules of demography, and the consequence of differential adaptation to local environmental conditions.

It is one thing to offer alternatives to the Christian theories of speciation and biological adaptation and another to claim that the alternatives better explain the natural phenomena we see. First, let us compare Darwin's tree of life hypothesis to the Christian theory of immutable species in the formal context of an IBE argument. We will follow with a comparison of Darwin's theory of natural selection with Paley's hand of God.

How could Darwin demonstrate the superiority of his tree of life hypothesis? Should Darwin infer evolution from the same perfections and intricacies that Paley viewed as evidence for God's handiwork? No. As the contemporary commentator S. J. Gould put it, "Ideal design is a lousy argument for evolution for it mimics the postulated action of an omnipotent creator. Odd arrangements and funny solutions are proof of evolution – paths that a sensible God would never tread but that a natural process, constrained by history, follows perforce" (Gould 1980, 20). Darwin's argument against a creator and for a nonintentional force of nature is found in the awkwardness of developmental patterns, and the seemingly poor designs of nature. Baleen whales develop teeth in neotony only for them to be reabsorbed into the baleen structure that they use to feed on krill. Pandas get at the tender shoots of bamboo through the inefficient process of running the stalks along an inflexible spur of bone that juts out like a thumb. Paley argued that design is evident in mishaps as well, for the purpose is clear even if the system does not achieve it (Paley 1828, 6–7). However, Paley is referring to instances of failed development, that is, deformed individuals. Darwin's mishaps are flaws of type – "design" flaws from a engineer's point of view. One might suppose that the significance of the apparent design flaws is a rhetorical point against the perfection of the creator's powers. We could ask, Why would a perfect designer create such awkward and inefficient designs? But that leads us away from a more significant point and gets us embroiled in the debate about the perfection of God's overall plan. The more significant point is what about these strange biological features reveal to us about their provenance. As Darwin writes, "Rudimentary organs may be compared with the letters in a word, still retained in the spelling, but become useless in the pronunciation, but which serve as a clue in seeking for its derivation" (quoted in Gould 1980, 27). Darwin argued that each lineage is related by common ancestry while the Christian view is that each species represents an independent act of creation. Flawed design in rudimentary organs provides evidence of modification from a common cause rather than independent creation in the same way that the similarity of words for names of numbers among Spanish, French, and Italian suggests that the words share a common ancestral root rather than arising de novo for each language (Sober 2000b, 42). Analogously, the reabsorbtion of whales' teeth in its mother's womb is evidence that whale development is not a separate act of creation but survives as a remnant and modification (by natural selection) of an ancestral developmental pattern. With the use of a strange bone spur to splice bamboo stalks, the panda's eating habits are awkward. Natural selection explains its existence as a modification from the anatomy of panda ancestors.

Once evidence mounts that organs and organisms evolve from preexisting lineages, then the seemingly improbable idea that complex and wondrous adaptations are the result of certain conditions pertaining to blind biological processes rather than the hand of a creator seems much less probable. As Goodrich put it: Darwin's

great merit is to have shown that [evolution] can be seen at work at the present time, can be tested by observation and experiment, and leave no room for any mysterious governing causes in addition; that, in fact, a complete scientific aspect of the process of evolution can be described as an unbroken series of "natural" events, a sequence of cause and effect, a series of steps each one strictly determined by that which came before and determining that which follows after. (Goodrich 1912, 28)

2.1   Darwin and Adam Smith

To see how this works, we will compare Darwin's answer to Paley with Adam Smith's theory of laissez-faire. The analogy brings home the point echoed in Goodrich that Darwin's theory paved the way for naturalistic explanations for adaptations and complexity (this is a similar strategy found in (Schweber [1977] and Gould [2002]). The question for Adam Smith and other economists was, What makes a well-ordered economy? One answer is akin to the teleologist's: congregate economic experts to create and impose principles and rules on the economic activities of the people. Adam Smith's answer is very different; in fact, it is the opposite of the expert theory: you allow individuals to transact without constraint. From the collective of all those unfettered transactions will emerge a well-ordered economy. Accordingly there are no rules or principles; rather order emerges as a side consequence of individual activity whose intent is orthogonal to the collective effect – individuals are acting on their own desire for profit, not for the good of the whole (Gould 2002). Smith invokes the "invisible hand" to describe these distinctive features of this theory: there is no hand of economy creation. Darwin's theory can be viewed in the same light. To achieve good adaptations and harmony in nature you do not need a hand of God; rather these phenomena are the result of the collective activity of individuals that are simply striving to survive and reproduce on their own, without a mind to the collective effect. The individuals vary in their features and in their abilities to survive and reproduce in the local environmental conditions. Natural selection is the invisible hand, or, to mix metaphors, the blind watchmaker. Emergentism is to Paley-style creationism as Copernicus's Sun-centered cosmology is to Ptolemy's Earth-centered universe: it stands the conventional biological wisdom on its head.

An interesting consequence of this analogy between Darwin's theory and Adam Smith's is that, contrary to what is reported in the popular press, the main alternative to ID is not necessarily the specifics of Darwin's theory of natural selection but his broader conception that complex adaptations *emerge* as the cumulative result of blind biological processes (as Goodrich indicates). It might turn out that the specifics of Darwin's conditions for evolution are incomplete or false, but if so, it does not follow that ID is the best explanation for biological complexity. There are other possible naturalistic explanations of complexity, that is, explanations that do not refer to supernatural agency. Hence the debate is not really "Darwin versus God" but "naturalism versus teleology".

3.   PLATONIC VERSUS ARISTOTELIAN TELEOLOGY

Natural selection and laissez-faire are antiteleological theories in that they provide an explanation for the existence of apparently good designs without reference to a creator. Yet, neither is anti-teleological *tout court*. As we saw before, between Plato and Aristotle there are two rather radically distinct theories of the nature of the *telos*. To recap some of these differences: for Plato, the creator governs all motion externally, from on high (as it were), where the action is for the best from a cosmological point of view. For Aristotle, teleology is nonpurposive, nonrational, nonintentional, and immanent, residing in an inner principle of change. The valuation is not about what is best from a designer's point of view, but what is useful to an individual. *Telos* is a built-in feature of matter to account for the final cause of natural arrangements.

So, from an alternative to a Platonic theory of creation it does not follow that Darwin's theory is antiteleological. Perhaps Darwin's theory is Aristotelian. In an exchange with Asa Gray Darwin hints at the possibility that in some sense Darwin's theory is teleological. Asa Gray comments, "Let us recognise Darwin's great service to Natural Science in bringing back to it Teleology: so that instead of Morphology versus Teleology, we shall have Morphology wedded to Teleology" (quoted in Lennox 1993, 409), to which Darwin replies: "What you say about Teleology pleases me especially and I do not think anyone else has ever noticed the point" (Lennox 1993, 409). Let us look more closely at Aristotle's distinctive teleological arguments. In the end I will argue that some of Darwin's theory can be interpreted as providing an alternative to Aristotle's teleology, but some cannot.

Perhaps the reason why Plato's teleology differs so fundamentally from Aristotle's arises from their distinct views about the origins of the cosmos. For Plato, the cosmos began to exist at a particular time. For Aristotle the cosmos is eternal. So, for Aristotle there is no need to explain how the cosmos came to exist because it always existed (Zeyl 2006). Further, Aristotle's teleological arguments tend to concern biological arrangements while Plato's are more "global" in that they concern the existence of the cosmos.

For his biological teleology, Aristotle recognizes two categories, "formal" and "functional". In formal teleology the *telos* is an inherent property of biological development. It explains a developmental event by citing the organism's biological needs. Plants require nourishment, so roots extend downwards "rather than upwards" for the sake of nourishment (*Phys.* 199a29 – all citations are from Aristotle 1984). In functional teleology the *telos* is ascribed to the relation between the placement and functioning of parts for the sake of the whole organism. Sharp teeth are located in the front of the mouth for the sake of tearing (199b24); that is why they are there. In other words sharp teeth persist in nature among carnivores *because* they contribute to the flourishing of carnivores.

3.1   Aristotle's Teleological Argument from Flourishing

Earlier I described Aristotle's argument from flourishing with the example of dental arrangements. Accordingly, the materialist cannot explain why certain dental arrangements (sharp teeth in front, broad molars in the back) regularly lead to the flourishing of carnivores that possess the arrangement while alternative dental arrangements often lead to the carnivore's death. The materialist explanatory toolkit of matter and cause is too limited to explain the difference; hence the materialist is forced to ascribe the difference to chance. But the (immanent) teleologist can refer to the inner principle of change that is characteristic of the species: the winning dental arrangement occurs *for the sake of* the flourishing of the species.

Besides the argument from "flourishing", Aristotle offers at least two more arguments for teleology, the "argument from hypothetical necessity" and the "argument from pattern". Let us look at them both in turn.

3.2   Aristotle's Teleological Argument from Hypothetical Necessity

In *Physics* II.9 Aristotle argues that in addition to the nature and movement of simple bodies (material necessity) and chance, biological explanations require a third ingredient: hypothetical necessity. What is hypothetical necessity? Take eyelids, for example. Eyelids are flaps of skin that protect eyes from easy external penetration. According to Aristotle the eyelid material – the flaps of skin – is necessary for the sake of eye protection. The "necessity" should be read as a constraint on materials given the specific purpose for which the part will be used. Not any material will do for the sake of eye protection, only eyelid material given the specific form of eye protection that humans and other animals require. This is meant to be taken strongly: the actual materials that compose an organ are required for the completion of the process where completion is the goal of development (Cooper 1987, 255). The concept of hypothetical necessity makes clear the relationship between functional teleology and formal teleology. Consider the example: eyelid material is present for the sake of eye protection (that is the function of eyelid material). So, eyelid material has a functional role to play in the growth of eye protection. Further, eye protection is necessary for seeing, and seeing occurs for the sake of the organism's growth. The necessity is granted to matter, eyelids, and is conditional in that it contributes to the goal of natural growth. Eyelid material contributes to natural growth by affording eye protection, which itself is crucial for the function of seeing (Cooper 1987).

Aristotle's argument for the existence of hypothetical necessity is an instance of an IBE. To illustrate, I follow Cooper (1987) and switch examples from eyelids to the development of a newborn from sperm, egg, and the usual background developmental conditions (the example is Cooper's, not Aristotle's). Aristotle would say that the materialist cannot account for the way these materials conspire to produce fetuses nearly every time. In other words, by appeal to simple motion and material cause, materialists cannot fundamentally distinguish between physical forces that are unconstrained to produce a range of different possible outcomes and physical forces that nearly always result in the same product – a newborn. The materialist's only recourse is an appeal to coincidence. Aristotle's reply is that coincidence is insufficient to account for the regularity of the conjugation seen in organic development because chance operates only in unusual circumstances (198b35–199a3). The principle of hypothetical necessity better explains the regularity of development: the materials are there for the sake of producing the conjugation that leads to the development of newborns.

3.3   Aristotle's Teleological Argument from Pattern

In his argument from pattern, Aristotle recognizes that the same teleological scheme applies to explain a particular sort of organization that regularly occurs both within human action and in the nonhuman natural world. The organization he has in mind is exemplified in all of the following: house building, leaves growing to shade fruit, roots descending for nourishment (rather than rising), nest building in birds, and web making in spiders (these are Aristotle's examples). In all of these cases we recognize a certain pattern of arrangement and sequential order. For example, in development of an artifact (such as house building) or in nature (as in roots descending downwards) all the steps of development occur in a sequence that leads up to the final state. These patterns do not happen by accident. Rather they occur in every instance in which the relevant organization is found, for example, in the intentional production of artifacts (house building) or the nondeliberate formation of natural objects (web making, nest making, roots descending, leaves shading fruit). It is in this respect that Aristotle famously remarks that "as in art, so in nature" (Phys. 199a9–10) and "as in nature, so in art" (199a15–16). The same pattern that explains certain organizations found in nature also explains the same organizations found in artifacts (Charles 1995, 115). This "certain organization" is just goal-directed activity. Aristotle infers teleology from patterns of order and arrangement. We will call this the "argument from pattern".

To strengthen this argument, Aristotle presents the first instance in which teleology preserves a distinction between function and accident; however, for Aristotle the term is a "mistake". Mistakes occur when one of the stages required to achieve the goal has failed to complete its role in the production of the goal, for example, when a doctor pours the wrong dosage or when a man miswrites or when monstrosities such as "man-headed ox-progeny" or "olive-headed vine-progeny" develop. The same teleological pattern whereby each stage of development occurs in order for the sake of the goal allows us to explain the difference between what occurs by art or nature, on the one hand, or by mistake, on the other. What occurs by art or nature follows the pattern successfully while mistakes or the creation of monsters features a failed developmental stage. Contrast orderliness among the normal beings with disorder found in monstrosities. For the materialist there is no essential difference between function and malfunction; both are explained in terms of matter and cause. Aristotle argues that teleology better explains the difference: just as purposeful manufacturing produces functional artifacts, disordered manufacturing leads to mistakes.

Perhaps we can view Aristotle's teleology as pointing out a distinctive feature of biological processes that, unlike the cosmological and demographic, seem to beg for a teleological argument. The cosmos featured a regularity or orderliness that required explanation. But biological events feature "means-ends" processes (I borrow here from Amundson 1999, 16, whereby developmental processes and biological arrangements occur for the sake of the good of the organism [recall Socrates' "real reason" for remaining in prison – it is for his own good]).

4.   DARWIN'S ALTERNATIVE TO ARISTOTELIAN TELEOLOGY

Recall the importance of Darwin's answer to Paley's argument to the issue of teleology. Darwin, in gathering evidence for his tree of life hypothesis, debunks Paley's Platonic teleology whereby organic traits are intentional designs of a supreme creator. Evidence from vestigial and "odd arrangements" suggests that organic traits are not derived from a purposeful act of creation but rather organic traits are derived and modified from the traits of their ancestors through natural selection. That is, Darwin replaces the hand of creation with a description of the conditions required for various individuals and their traits to evolve into adaptations. However, by debunking Platonic teleology, it does not follow that Darwin has debunked natural teleology altogether. Platonic teleology is only one sort; Aristotelian teleology is an entirely different sort. So, what is the status of teleology on the Darwinian scheme? In what sense is it anti-Aristotelian?

For Aristotle there are at least two instances in which teleological explanations are called for, to explain functional arrangements, where an item's existence is explained in terms of its usefulness, and to explain regular growth patterns that members of a species or genus share. I would like to argue that Darwin's theory of natural selection offers an alternative to teleological explanations for functional arrangements.

In what way is Darwin's theory of natural selection an alternative to Aristotle's teleological explanation of functional arrangements? For Aristotle, the explanation of functional arrangements depends on the same *telos* that determines growth. So, the existence of a useful arrangement (sharp teeth in front and molars in the back, for instance) is a property of the formal features of growth – they develop for the sake of the usefulness of the arrangement. Recall the notion of hypothetical necessity: the material exists (and hence the development occurs) *for the sake of* preserving the useful arrangement. To put the point differently: the purpose exists first; the process and materials follow thereafter. Darwin's theory differs in at least two ways.

First, Darwin makes a dramatic distinction, antithetical to Aristotle, between the *internal* processes that generate an organism and *external* processes, the environment that determines adaptability (Lewontin 2000, 42). For Darwin, the distinction is absolute: the internal conditions for growth that determine variability and inheritance are not responsive to the environmental forces that ultimately "select" or determine which organismal traits will eventually prevail over time. So, growth explanations are completely distinct from adaptation explanations. In explaining how species come to be so well adapted to their environmental conditions there is no need to mention the factors that determine how variants arise in the first place; variants preexist their selection.

Second (and following from the first point), rather than function's constraining the presence of materials as it does for Aristotle's hypothetical necessity, for Darwin, it is the other way around: materials constrain function. Individual variants are generated by a distinct internal process and preexist their selection. The end result is the existence of "contrivances", organs whose provenance is ancestral yet "modified" by natural selection to the new local environmental conditions. That is why pandas possess such an awkward mechanism for manipulating bamboo shoots. It is a modification of the paw structure of ancestors of pandas. The panda's thumb is a "contraption", modified from the anatomy of what was available to be operated upon by selection.

This last point, I think, begins to explain Asa Gray's remark that Darwin wedded Morphology to teleology. The reference to "Morphology" is to a school of thought that advanced a "unity of plan" theory of organic diversity. Accordingly, members of a taxonomic group are accounted for in terms of resemblances between members of the same and other taxonomic groups. Traits that resemble each other across taxonomic groups are called "homologues" and indicate a "common plan" throughout nature. Morphologists thought that picking out homologous structures constituted picking out essential categories in nature. That is, the existence of homologous structures indicates the fundamental laws of body plans. However, Darwin wondered how to explain the prevalence of variants to the "common plans". To this he invokes natural selection. Natural selection operates over preexisting structures competing for limited resources in a common environment. So, while structures preexist their adaptive uses it is natural selection, a distinctive causal mechanism, that explains morphological change.

Contemporary philosophers of science have taken for granted that Darwin's theory of natural selection "naturalizes" teleological explanation. Yet, as we have seen, there are a variety of teleological explanations in the ancient and early modern literature, only some of which are preserved in Darwin. In particular, the Aristotelian explanation for the existence of traits in terms of their usefulness is preserved in Darwin, but without reference to the *telos* that determines individual growth. Post Darwin when we say that birds have wings *because* wings are *for* flying, we understand the *what-for* explanation in terms of an evolutionary argument whereby possessing wings (which themselves came to be by some mechanism that accounts for individual variation) provides a fitness advantage and hence wings were selected and eventually became prevalent in bird populations (Enç and Adams 1992, Mayr 1988. For contrasting accounts see Cummins 2002, Lewens 2004). So, the form of the teleological explanation remains, but the terms of the explanations are completely naturalized: no reference to the *telos* of growth insofar as growth is a condition but not a mechanism of selection. For this reason, Ernst Mayr argues that a better description of the nature of Darwin's explanation is "teleonomy" rather than "teleology" (Mayr 1988).

Darwin's natural selection also provides us a way of naturalizing the distinction between function and malfunction. "Wings are for flying" explains why birds have wings even if any particular member of a population has a broken wing. For Aristotle, malfunctioning organs are those that fail to achieve their final ends, but for this application of Darwin's theory, malfunctioning organs are those that do not perform the function for which they were selected in the past (Neander 1991).

I have just argued that Darwin's theory of natural selection offers us an alternative explanation to the teleological arguments (both Platonic and Aristotelian) from functional arrangements. Yet, as I have discussed before, Aristotle offers us another reason to infer a *telos*, in the explanation of an individual's growth. Since the root of Darwin's theory is an absolute distinction between the internal conditions of an individual's growth and the external conditions that determine whether or not traits will be selected, it should not be surprising to us that Darwin's theory is silent on the conditions that determine growth. Karl von Baer critiques Darwin on this very point (Lenoir 1982, 270). According to this critique, if "blind necessity" is the only force operating, then the fundamental questions of biology – development, adaptation, and the like – will remain unintelligible. An explanation that strings together mechanical processes lacks the fundamental principle that connects the processes to a particular end (Lenoir 1982, 271). I interpret von Baer's criticism to be close to Aristotle's argument from regularity: the materialist lacks the principle that distinguishes one material process from any other. Consequentially, what distinguishes developmental processes that lead to living newborn from one that fails?

5.   CONCLUSION

At the outset of this essay I claimed that the contemporary argument from intelligent design varies little from William Paley's argument written in 1802. Both argue that nature exhibits too much complexity to be explained by "mindless" natural forces alone. Both conclude the need to postulate the existence of an intelligent designer, a creator with forethought and purpose. But there are differences between Paley's argument and the modern argument from ID. Paley concluded that the Christian God exists, while modern ID supporters claim to be silent about the features of the creator. While Paley's interlocutors were materialists of the sort described by Plato, the target of ID's argument is Charles Darwin. The claim is that Darwin's theory of natural selection cannot account for "irreducibly complex" natural designs such as the flagellum, since its building could have only been done in one step (given its incredible intricacy), not a gradual process of selection. Some critics of ID point out that biology already has examples of natural selection's putting complex designs in place (Ruse 2003, 320, discusses the Krebs cycle). Others point out that elaborate systems might evolve for some reason and then be co-opted for an entirely different function: "Who says those thirty flagellar proteins weren't present in bacteria long before bacteria sported flagelli? They may have been performing other jobs in the cell and only later got drafted into flagellum-building. Indeed, there's now strong evidence that several flagellar proteins once played roles in a type of molecular pump found in the membranes of bacterial cells" (Orr 2005). Still others argue against the criteria of how to detect natural designs (Fitelson, Stephens, and Sober 1998).

The critical part of the present essay has a consistent theme, pointing out argument insufficiencies. Just as I argued that Darwin's and Smith's noncreationist arguments are insufficient to negate an Aristotelian teleology (although Darwin's theory eventually does), I also argued that even if Darwin's version of natural selection does not explain how complex items come to exist in the first place, ID does not necessarily follow. All that is required to defeat ID or creationism is to demonstrate the possibility that complex designs can emerge from naturalistic processes.

ACKNOWLEDGMENTS

In this essay I have relied on an earlier writing, Ariew 2002. I would especially like to thank Daniel Goniprow and Paul DiMartino for comments on an earlier draft.

10 Macroevolution, Minimalism, and the Radiation of the Animals

1. MINIMALIST MODELS OF MACROEVOLUTION

Palaeobiology is our main source of direct evidence about the history of life. But while that history is fascinating in itself, palaeobiology's most distinctive contribution to evolutionary theory is the insight it provides on the importance of scale. Palaeobiologists see the results of evolutionary processes summed over huge sweeps of space and time. As a consequence of that window on the effects of deep time and vast space, we have a chance to see whether the palaeobiological record enables us to identify evolutionary mechanisms that are invisible to contemporary microevolutionary studies with their local spatial, temporal, and taxonomic scales. Palaeobiology, in other words, is the discipline of choice for probing the relationship macroevolutionary patterns and microevolutionary processes.

This chapter will be organised around an important framing idea: that of a "minimalist model" of this relationship. I shall discuss minimalism in detail shortly, but as a rough first approximation, according to minimalism, macroevolutionary patterns are direct reflections of microevolutionary change in local populations; they are reflections of changes of the kind we can observe, measure, and manipulate. For example, Michael Benton (forthcoming) discusses models of global species richness that depend on scaling up in space and time equilibrium models of local ecological communities. To the extent that such scaled-up models are adequate, macroevolutionary patterns are nothing but local changes summed over vast sweeps of space and time. This chapter aims to make explicit the

182

patterns for which this minimalist model is appropriate; to consider the quite different ways in which that model can be enriched; and to highlight some phenomena that suggest that minimalism does indeed sometimes need to be enriched.[1]

The challenge to minimalism is that *scale matters*: there are aspects of the history of life that are not (solely) the result of population-level processes accessible to neobiological investigation. Because scale matters, palaeobiology tells us something we cannot otherwise know about evolutionary processes. But when is a macroevolutionary pattern just a "mere aggregate" of the results of local processes? As I see it, minimal models are simple, perhaps simplified models of the relationship between microevolutionary process and macroevolutionary pattern. But they are simple in four independent aspects.

i. First: they are individualist. In these models of evolutionary change the fitness values that matter are fitnesses of individual organisms. One move beyond minimalism is to develop evolutionary models in which species themselves are selectable individuals in a population of species. In the recent palaeobiological literature, this idea has been centre stage; a good deal has been written on how to characterise and empirically test species selection models (see for example, Vrba 1989, Gould and Lloyd 1999, Sterelny 2003). But while species selection might be important in explaining, for example, patterns of survival in mass extinction events, the case for its importance remains to be made. In contrast, as we shall see in considering the origins and elaboration of multicellularity, some form of group selection is almost certainly important in driving major transitions in evolution.

ii. Second, extinction and speciation probabilities are effects of individual-organism fitness. The speciation and extinction probabilities of a species stand in a simple, direct relationship to selection on individuals in the populations of which the species is composed. For example, if its extinction probability is high, it is high because individual organisms are not well adapted by comparison to their competitors, not because (say) the species lacks genetic variation or because it lacks metapopulation dynamics in which migration buffers individual populations against local extinction. Minimalist models idealise away from the evolutionary consequences of species-level properties.[2]

iii. Third, novelties, key innovations, and the like, arise as ordinary variations in extraordinary circumstances. Some evolutionary innovations (for example, the invention of sex) expand the space of evolutionary possibility open to a lineage. But there is nothing unusual about the genetic, development, or selective origin of those innovations. Novelties arise and are established as ordinary, small variants from ancestral conditions. Possibility-expanding changes are, in Dennett's helpful phrase, "retrospective coronations" (Dennett 1995). Dawkins has argued that the origins of phyla are retrospective coronations: the great branches of animal life began with ordinary speciation events; nothing about those speciations at that time marked those branchings as of especial importance, though it turned out that they were important. Likewise, looking backwards, we can see that (say) the evolution of flowers was a possibility-changing key innovation. It led to an ecological and evolutionary revolution in plant communities. But the importance of this innovation in pollination would not have been identifiable at the time. Key innovations are genuinely important. But their importance can only be recognised retrospectively: a key innovation is one that happens to take place in the right place, time, and taxon. Its origin and establishment in that taxon are not the result of any unusual evolutionary process. We move away from minimalist models if we think that the origin or establishment of novelties (sometimes) requires special explanation.

iv. Fourth, we can idealise away from the changes in the developmental and selective background of genetic variation in phenotypes. We can treat the selective and developmental background of change as a fixed background condition. In microevolutionary studies – for example, models of the response of guppies to sexual and natural selection – we can usually treat the developmental system as a stable background condition of within-population microevolutionary change. We do this when we treat genes as *difference makers*: the substitution of one allele for another makes a selectable phenotypic difference: it (say) makes a male guppy brighter. Genes are difference makers, but only if we hold fixed the causal background in which they act. In such restricted contexts, the concept of a genotype-phenotype map and these associated causal claims makes good sense. Obviously, when our interest turns to macroevolutionary pattern, this assumption becomes much more problematic; models that make these stability assumptions idealise radically, for selective environment and developmental system are labile on deeper time frames. We cannot extrapolate from constraints on short-term evolutionary responses to similar constraints on long-term evolutionary responses.

Understood this way, minimalism is a model, not a doctrine. Everyone would accept that the four elements of minimalism idealise away from some of the complexities of the biological world. No one supposes that selective and developmental environments are invariant. Almost everyone would concede that selection can act on collectives of individuals or that large-effect mutations might occasionally be important. The idea of minimalism is that it is *typically fruitful* to idealise away from these complications. So instead of thinking of minimalism as a doctrine to be defended or undermined, we should instead focus on identifying the range of cases for which minimal models are appropriate, and those cases in which these models need to be supplemented. As we relax the simple picture of the relationship between population level events and species dynamics, we thus get a space of models. Our problem becomes one of identifying the evolutionary phenomena for which models in differing locations in that space are appropriate.

2.   FOUR VIEWS OF LIFE

To show that minimalism is fruitful, we need to show that we can develop insightful minimal models of major macroevolutionary phenomena; that we can explain the tree of life's most striking features. To show the limits of minimalism, we need to identify those palaeobiological phenomena for which minimal models are not adequate. That is beyond the scope of any single chapter, not least because there is vigorous disagreement about the phenomena to be explained. Given the richness of palaeobiological phenomena, and the vigour of palaeobiological debate,[3] my strategy will be illustrative rather than exhaustive. I shall begin with four representative examples of overall views of life's history, and an initial characterisation of their relationship to minimalist models. Two of them seem to fit minimal models; the other two suggest that we need extensions of those models. I then explore in much more detail one crucial and controversial episode in the history of life, the early radiation of the animals.

View 1: The Spread of Variation

In his (1996), Stephen Jay Gould argued that while there really has been a rise in both the maximum and mean complexity of living agents, this fact of history is best understood as an expansion of variation in complexity. If processes of differentiation, speciation, and extinction act independently of complexity, *variation in complexity* will nonetheless increase over time. If we graph change, with complexity on the horizontal axis and time on the vertical, variation will spread both to the left and to the right from the point of life's origin. However, even if differentiation, speciation, and extinction are independent of complexity, the spread need not be equal in both directions, for life's complexity has a minimum bound (set by the biomechanics of metabolism and replication) but no upper bound. Moreover life originated near this minimum bound. So variation will spread to the right, in the direction of greater complexity, but very slightly if at all to the left. Thus maximum and mean (but not necessarily modal) complexity will drift upwards over time. Given a minimum bound and a point of origin close to that bound, the null expectation is an increase in maximum and mean complexity. So even a biased spread of variation need not challenge minimalism. The local evolutionary processes of adaptation of populations to their specific local circumstances, perhaps significantly modified by drift, and by developmental and genetic constraints, would generate spreading variation.[4]

View 2: Escalation and Arms Races

One divide within biology is between those who emphasise the importance of interactions between biological agents and those who emphasise the importance of abiotic factors for the life of organisms. Many evolutionary biologists have emphasized hostile coevolutionary interactions between lineages: arms races between differing competitors for the same resource; between predators and prey, plants and herbivores, hosts and parasites. For example, Geerat Vermeij has defended an "escalation model" of the history of life, taking as his model interactions between bivalves and their predators. As he reads the long history of bivalve evolution, it is one of the gradual improvement of defence. Bivalves have evolved the capacity to dig deeper and faster into the substrate, and they have evolved thicker, spikier shells. Predator efficiency too has ramped up over time. Predators have become better at digging up, drilling into, crushing, or breaking open shells. Vermeij thinks of these bivalve histories as indicative of life's history as a whole: organisms become better adapted over time (Vermeij 1987, 1999).

Thinking of the history of life as an escalation of adaptation driven by arms races is to think of that history in a more selectionist way than does Gould. Escalation scenarios presuppose both that selection tends to drive evolutionary dynamics and that there is a systematic bias in the direction of selection. But this idea is compatible with overall trends in evolution and ecology reflecting locally determined events. The causal engine of escalation is ecological interaction in local communities. Escalation is a minimal model of the history of life, though one whose empirical presuppositions may not be met. External events can interrupt the association between lineages, breaking coevolutionary connections. For example, the changing biogeography and climate of Australia might change the suite of insects to which the eucalypt lineage is exposed, thus aborting arms races between eucalypts and phytophagous insect lineages.

View 3: The Increasing Space of Evolutionary Possibility

When we consider life at a particular time, we should see it as having an upper bound as well as a lower bound (Sterelny 1999d, Knoll and Bambach 2000). For example, the evolution of the prokaryotic cell had significant evolutionary preconditions. Until genes were organised into chromosomes, and the fidelity of gene replication improved, prokaryotic cells were not in the space of evolutionary possibility. There is an upper limit on the complexity of quasi-biotic systems that lack a division of labour between metabolism and replication. The same is true of other grades of complexity. For example, Nicole King has pointed out that only the evolution of signalling and cell adhesion mechanisms in protists made possible the evolution of multicellular life (King 2004). At a time (to use Gould's metaphor) there is a wall to the right as well as on the left. Over time, though, that wall shifts, for the preconditions for a new grade of complexity and of differentiation are assembled. The space of evolutionary possibility for life as a whole expands, because the right wall moves for some lineages. As it expands, some of those possibilities are realised.

The best-known model of this kind is that developed by John Maynard Smith and Ers Szathmary (Maynard Smith and Szathmary 1995, 1999). They see evolutionary history as characterised by a series of major transitions. These include the shift from independently replicating structures to the aggregation of codependent replicators into chromosomes; the shift from RNA as the central replicator to DNA replication; the evolution of the eukaryotic cell; the invention of cellular differentiation and the evolution of plants, animals, and fungi; the evolution of colonial and social organisms from solitary ones; and even the evolution of human language. Many of these transitions have two crucial characteristics: (i) they are revolutions of biological inheritance systems, involving the expansion of the transmission of heritable information across the generations, and (ii) they are revolutions in selection, for they involve the assembly of a new, higher-level agent out of previously independent agents. Minimalist models, trading in the fitness only of individual organisms, do not seem well suited to give an account of such revolutions in selection. Yet, jointly, these revolutions lead to an expansion of the space of biological complexity.

So this view of evolutionary history does raise issues about minimalism, for it suggests that minimalist models cannot give an adequate account of the distinction between two very different kinds of evolutionary change: the *expansion of possibility* versus *the exploration of possibility*. Standard microevolutionary theory seems to be about possibility-exploring change, not possibility-expanding change. This same distinction between possibility-expanding and possibility-exploring innovations seems to be implicit in our next picture of the overall pattern of life's history, too.

View 4: Expanding Ecospace

Andrew Knoll and Richard Bambach (2000) develop a view of evolutionary transitions that focuses on the ecological changes that are consequences of morphological innovations; thus this poses similar challenges in understanding the origin and establishment of novelty. For Knoll and Bambach, the crucial pattern in the history of life is an expansion of the habitats organisms exploit, an expansion of the range of resources organisms exploit within those habitats, and an expansion of the complexity and variety of the ecosystems that are assembled as a result of transitions in morphological complexity (Knoll and Bambach 2000). In developing their model of the expanding ecospace, Knoll and Bambach identity six "megatrends" that map roughly onto the major transitions of Maynard Smith and Szathmary: trends that yoke morphological innovation to ecological revolution. These are (i) the transition from the prebios to life as we know it, (ii) the radiation of the prokaryotes, (iii) the protist radiation, (iv) the evolution of aquatic multicellularity, (v) the multicellular invasion of the land, (vi) the evolution and global dispersal of intelligence.

In their view, these trends all involve ecological revolutions. The radiation of the prokaryotes expanded the range of energy sources into which life could tap. The evolution of the protists, according to Knoll and Bambach, added depth to ecosystems, for it involved the evolution of predation. As prokaryotes mostly extracted energy from abiotic sources, the structure of bacterial ecological communities was fairly simple. Eukaryotes can engulf particles, including living ones, and hence they have added layers to ecological communities, increasing their vertical complexity: grazing, predation, decomposition, are eukaryote specialties.[5] The evolution of marine multicellularity was obviously a very dramatic morphological transition. But, equally, it was an ecological transformation as well. This transition resulted in greatly increased vertical complexity of communities, and greatly increased ecological engineering. For example, with the radiation of marine animals in the Cambrian, the nature of the sea floor changed. Instead of sediments on the floor being stable, they were mixed by animals burrowing through them (this is known as "bioturbation"). Corals and other shelled forms created substrate and habitat for other organisms, as did kelp forests (and much more recently, sea grasses). Rich coevolutionary interactions (both symbiotic and parasitic) between multicelled and single-celled organisms became possible. The invasion of the land, likewise, was an ecological revolution: a whole array of physical environments became habitats. As a result of that invasion, there are new kinds of communities and new kinds of coevolutionary interactions (most strikingly, those between flowering plants and insects). The direction of evolution is marked by the establishment of successively more complex ecologies with more and more vertical layers. Eukaryotes were added to the top of a bacterial foundation, and various multicelled layers were eventually added to persisting though doubtless transformed prokaryote-protist systems.

If Geerat Vermeij is right, these increasingly complex communities are also increasingly dominated by energy-greedy organisms: organisms that harvest the available energy at ever higher rates, and as a consequence have increasing impact on their local environments (Vermeij 1999). In his view, both the radiation of the flowering plants at the expense of the gymnosperms and the radiation of mammals and birds at the expense of amphibians and reptiles exemplify this trend (Vermeij 1999). Ecosystems have become increasingly dominated by these high-activity organisms. At times, these act as defectors in a tragedy of the commons. They harvest more resources than others at the same trophic level, thus sucking resources out of the system at increasing rates, for their numbers expand at others' expense. But because they use so much energy, they exert more power over the local environment, increasing the rate at which energy and nutrients cycle through local ecosystems. Their overall effect is to ramp up the pace of life, thereby further selecting for agents with similarly large energy appetites.

In the next section, I link minimalism to a crucial case: the radiation of the animals. In Section 4 I present – I hope – a near-consensus view of the large-scale history of that radiation. In Sections 5 and 6, I discuss evolutionary explanations of that radiation and the relationship between macro and microevolution. I then very briefly conclude.

3.   MAJOR TRANSITIONS: A CHALLENGE TO MINIMALISM?

In the remainder of this chapter, I shall explore minimalism through consideration of major transitions in evolution and, in particular, the radiation of multicelled animal life. I shall suggest that these transitions take us beyond minimalist models of evolution for three reasons. First, as much of the major transitions literature emphasises, a multilevel perspective is essential to understanding the selective regimes responsible for movements of the right wall. Transitions in complexity have often involved independent agents' coming to share a common evolutionary fate. These transitions involve a trajectory from a population of interacting agents to a more complex collective agent. Perhaps in a few cases (conceivably, the first eukaryote) this transition took just a single step. But in most cases there was (and often still is) a transitional regime in which the fitness of the incipient collective and the fitness of its now semi-independent elements were both important. In understanding such transitional regimes, one crucial problem is to understand the mechanisms that prevent the functional organization of the new collective from being undermined by selection on its components for defection and free-riding. So one theme involves the interaction between levels of selection, and the process through which selection on the collective results in the components of that collective (more or less) ceasing to be Darwinian populations (Buss 1987, Michod 1999).

A second challenge involves the evolution of novelty and the expansion of evolutionary possibility. The possibility space accessible to a lineage depends in part on its current location in morphospace. A lineage in which (say) the arthropod adaptive complex has evolved has evolutionary possibilities open to it that are not open to (say) velvet worms: segmented animals but without the structural complexity and skeletal support of the arthropods. But access to possibility space depends as well on the mechanisms of inheritance and development that characterise a lineage, on the variations from current location that are possible. Developmental mechanisms make some regions of space more accessible from a lineage's current location, and others less accessible (see for example, Arthur 2004, Schlosser and Wagner 2004). As I have noted, minimalist models treat the evolution of novelties as a species of retrospective coronations. Novelties arise and are established in local populations through mechanisms that are indistinguishable from those driving ordinary microevolutionary change. One potential problem for minimalist models of novelty is the role of macromutation (i.e., mutations with dramatic phenotypic effects) in evolution, but in my view, the importance of macromutation for minimalism has been overstated. No one thinks that macromutations are impossible. Even someone as sceptical as Dawkins about the role in evolution of macromutation allows that they have probably played some role in expanding the space of evolutionary possibility, perhaps in the evolution of segmentation (Dawkins 1996). But even if rare macromutational events have played an important role in expanding the space of possibility (as is quite likely), macromutations are certainly not common enough to be the normal explanation of the origin of novelty.[6] Hence in modelling the evolution of novelty and the expansion of possibility, it may well be reasonable to neglect this complication.

In assessing minimalism, it is important to avoid a false dichotomy. On the minimalist model, a possibility-expanding novelty is an ordinary variant crowned retrospectively. One way of being extraordinary is to be the result of a macromutation. But there are other and more important ways, for evolutionary possibility is multifactorial: it depends on a lineage's current position in phenotype space, the array of potential variations in the heritable developmental resources available in that lineage, and in variations in the ways genetic and other developmental resources are used. And while there are good reasons to think that sudden macromutational shifts in morphospace are indeed very rare, those reasons do not generalise to the other factors on which evolutionary possibility depends. For example, in Sterelny (2004) I argued that the formation of symbiotic alliances often involves major shifts in evolutionary possibility. The acquisition of microbial symbionts has given many Metazoans access to lifeways that would otherwise be closed to them. And while the evolutionary origination of symbiotic alliances is not an everyday feature of the biological world, it is not vanishingly rare, either.

Possibility-expanding innovations may be changes in the control of development, and these need not be minimal variations of prior systems of developmental control. The crucial point here is that the mechanisms of developmental plasticity can cushion the immediate phenotypic effect of significant changes in the developmental system. That is why large changes in the genes themselves – chromosome inversions, duplication, and the like – are not always fatal. Significant developmental changes need not result in large (hence almost certainly catastrophic) jumps in morphospace (West-Eberhard 2003). The power of these mechanisms is illustrated by West-Eberhard's extraordinary example of the goat born without front legs. It adopted a two-legged posture and moved in somewhat kangarooish fashion. The adjustment was not just behavioural but morphological. The goat developed with changed hind leg and pelvic structures, a curved spine, strong neck, and associated muscular changes.

Possibility-expanding changes in a lineage may often have their origin in *iceberg mutations*: variations that are not radically different in phenotype from their ancestors and contemporaries, but that generate that phenotype from an importantly changed set of developmental resources. One relatively uncontroversial example of a possibility-increasing change in development is one that increases modularity. If some aspects of phenotype are under modular developmental control, they will be relatively more evolutionarily labile (Lewontin 1985, Kauffman 1993). But there are other examples: for example, Mark Ridley argues that morphological complexity is constrained by the fidelity of inheritance. The error rate characteristic of prokaryote replication would be fatal to multicellular animals with their larger genomes; their evolution depended on the evolution of a more accurate system of error correction (Ridley 2000). In Section 6, we will consider in some detail the claim that the evolution of complex animal bodies depends on the evolution of new mechanisms of gene control.

A third challenge is the integration of internal and external factors in explaining transitions. Selection-driven microevolution is often conceived as a hill-climbing process: a population finds itself suboptimally located in an adaptive landscape and responds to that location by optimisation. As Richard Lewontin has pointed out, even when doing microevolutionary studies we often should not think of environments as fixed and organisms as labile (Lewontin 1985, Odling-Smee et al. 2003). In evolutionary transitions, selective environments and developmental possibilities are labile, so such an idealisation is less likely to be appropriate for possibility-expanding transitions, such as the invention and elaboration of multicellularity.

It is to that transition I now turn. There is no single transition to multicellularity; instead, there is a cluster of transitions that took place at different times and to different degrees. Bonner's 1998 review notes, in addition to the obvious multicellular clades of plants, fungi, and animals, the green, red, and brown algae and a variety of other more exotic cases (Bonner 1998). In these more exotic cases, multicellularity has been established without much differentiation (as in the case of the algae). Indeed, in some of the cases it is not clear whether we should think of these systems as a single multicelled agent or as a social, cooperative population of single-celled agents (as in cellular slime molds and myxobacteria, which aggregate to form stalked structures with spores at the top). I shall focus on just one of these transitions, the Metazoan radiation, for it is an especially important case. So my stalking horse will be the early evolution of complex animal life.

4.   THE METAZOAN REVOLUTION

The "Cambrian Explosion" names the radiation of animal life in the early to mid-Cambrian, from about 543 million years ago (mya) to about 505 mya. The animals of the Cambrian were not the earliest multicelled animals. They were preceded in the fossil record by an enigmatic Ediacaran fauna: an array of discoid and frond-shaped forms whose relations both to one another and to living animals remain controversial (Narbonne 2005). But the Cambrian saw the first appearance of undoubted ancestors of contemporary animals. By then the first bivalves, arthropods, echinoderms, molluscs, and chordates had certainly evolved. There are hints of such animals before the Cambrian, in trace fossils and fossil embryos. But by the mid-Cambrian (about 530 mya), they were richly present. Moreover, this explosion was of enormous consequence. In contrast to (say) the evolution of multicelled red algae and the few multicelled lineages of ciliates and diatoms, the evolution of the metazoa changed the evolutionary and ecological landscape. Arguably, this transition was fast and vast. It resulted in a highly disparate, taxon-rich clade. It was the invention not just of multicellularity but of the control of differentiation and a full division of labour. The metazoa vary widely amongst themselves, yet are characterised by complex though reliable developmental pathways, involving many cell divisions and differentiation into many cell types. All but the earliest (or the most secondarily simplified) Metazoans are vertically complex, with cells organised into tissues, which in turn are organised into organs and organ systems. Many have complex life histories involving radical changes in phenotype over time: many Metazoa, in other words, are adapted not just to a single environment but to different environments at different stages of their life history. In many animal clades there is a fundamental division of labour between reproduction and interaction with the environment, for in these lineages there is an established soma/germline distinction, with the early sequestration in development of those cells that will be the future gametes.

So the Metazoan radiation is not just a transition to multi-cellularity. It is a transition to structurally and behaviourally complex agents. The evolution of such agents depends on the evolution of the reliable developmental control of large numbers of differentiated cells and the complexes they make. Their diversity depends on the proliferation of a vast number of distinct developmental control systems, and the reliable transmission of those control systems to the next generation. Moreover, the invention and radiation of the animals was the invention of fully treelike evolutionary histories, with limited horizontal gene flow and (comparatively) well-defined species taxa. Notoriously, the systematic vocabulary that has been developed to describe animal lineages fits other radiations less well. Moreover, the Metazoa have been profound agents of ecological change. The radiation of the Metazoa established wholly different kinds of ecological communities based on webs of organisms that live by eating other organisms and their products. These communities profoundly changed the physical environment in which organisms lived. David Bottjer, for example, has written of "the Cambrian substrate revolution": Cambrian animals profoundly changed the physical substrate of shallow seafloors. Before this radiation these were stabilised (and hence available as habitat for the Ediacaran biota) by microbial mats. Once burrowing animals radiated, these mats were broken up, and the muds became unstable. The soft seabeds were no longer available to sessile organisms that lacked special stabilising adaptations (holdfasts of some kind) for such seafloors (Bottjer et al. 2000). Likewise, the invention of the turd revolutionised the chemistry of ocean waters: it packed waste chemicals in a form that was heavy enough and compact enough to sink to the seafloor rather than disperse suspended in the water (Logan et al. 1995).

The Metazoan radiation is an especially central case for understanding evolutionary transitions and the limits of minimalist models of evolution. Indeed, if the fossil record is a true record, we go from seas and seafloors with a few simple animals (sponges, jellyfish, perhaps a few wormlike bilaterians) to seas and seafloors teeming with animal life. The Metazoa appear fairly suddenly and richly in the record. Does this rapid diversification in the fossil record reflect evolutionary history, and if so, does it show that unusual evolutionary mechanisms were responsible for this radiation? Famously, in *Wonderful Life*, Stephen Jay Gould argued for affirmative answers to both these questions, arguing that the extraordinary fossils of the Burgess Shale showed that an unprecedentedly disparate animal fauna evolved extraordinarily rapidly in Cambrian seas, and thereafter the mechanisms that generated such great disparity shut down (or, at least, slowed very dramatically) (Gould 1989).

However, our view of the life in the Cambrian has since been transformed. There is reason to suspect that Gould's estimate of its disparity rests on a taxonomic illusion. Gould thought that the Cambrian fauna was more disparate – more morphologically differentiated – than any subsequent fauna. Many of the Burgess Shale fossils did not match the body plans of any of the extant great clades. For example, many Burgess arthropods did not have the characteristics of trilobites; nor did they have the body plan characteristics of spiders and their allies; nor of insects and their allies, nor of crustaceans. But we should not expect early members of a lineage to fit body plan specifications used to identify the living members of a lineage (for a particularly forceful statement of this argument, see Budd and Jensen 2000). Living crustaceans (for example) have limb and segmentation patterns – one that, for example, includes two paired antennae – that (in all probability) were assembled incrementally and have then been inherited by the living crustaceans. Those taxa with that inherited pattern, living and extinct, are the *crown group crustaceans*. On the assumption that this limb/segmentation pattern was assembled gradually, there will be *stem group crustaceans*. These are taxa on the lineage that leads to the living crustaceans, taxa that lived after that lineage diverged from the other arthropods and before the definitive crustacean package was assembled. Stem group taxa are bound to look strange. And there must have been stem group Metazoans aplenty in the Cambrian, for that was the period in which the great Metazoan clades were diverging and acquiring their distinctive morphologies. Gould's extreme Cambrian disparity pulse may be nothing but his encounter with this array of stem group Metazoans.[7] To the extent that the Cambrian challenge to minimalist models was based on the idea that Cambrian animal evolution generated extraordinary amounts of disparity, and then the supply of new disparity dried up, this challenge now looks less pressing.

Even so, if the fossil record is to be trusted, there was a rapid diversification of the Metazoa from the early to the mid-Cambrian. But is that record to be trusted? Over the last ten years or so, molecular methods have served as an independent check on the relationship between the fossil record of early Metazoan evolution and the true pattern of that evolution. A variety of molecular clocks have been used to calibrate the divergence times of the Metazoan phyla. The idea is (a) to compare homologous, slowly evolving genes in different phyla, and calculate the extent to which those gene sequences have diverged one from another; (b) to calibrate the rate of evolution using taxa with a rich, reliable, and well-dated fossil record; (c) to use that rate to calibrate divergence times for other taxa.[8] The earlier results of using these molecular methods were very striking indeed: they pushed the apparent divergence of the basal Metazoans long before the Cambrian. Some of these studies estimated the divergence times between the sponges and other Metazoans (and even the later split between the cnidarians and the lineage leading to the bilaterally symmetrical Metazoa) as over one billion ya (for discussion of these earlier studies, see Bromham et al. 1998, Lee 1999). While the division of a lineage need not imply the simultaneous evolution of their distinct body plans, these dates imply a very long period of cryptic evolution. If they are right, the challenge to evolutionary biology becomes that of explaining why Metazoan animal evolution was invisible for so long, and how and why it suddenly became visible.

However, more recent molecular clock estimates of deep Metazoan splits are more congruent with fossil dates. One deep divide amongst the Metazoa is between the protostome and deuterstome developmental pathways, and Kevin Peterson and his colleagues estimate this divergence in the range 573–656 mya (Peterson et al. 2004). They argue that the very deep divergence times depended on using vertebrate evolution to calibrate the clock; this gave a misleading result because the relevant genes in vertebrates seem to have evolved much more slowly than in other lineages, making the clock run faster than it should.[9] So divergence times are deeper than the fossil record suggests, though not hundreds of millions of years deeper. For example, Peterson and Nicholas Butterfield estimate Metazoan origins at 664 mya (Peterson and Butterfield 2005, 9549). If those dates are right, Metazoans existed for close on 100 million years before they left an unmistakable record of their presence. If the first animals were tiny, soft-bodied elements of the plankton, that invisibility is no surprise. James Valentine has provided a very helpful table of the first appearance in the fossil record of the various animal phyla (Valentine 2004, 186). The Ediacaran is the final era of the Proterozoic eon; it immediately precedes the Cambrian. Only two phyla (Cnidaria and Porifera) have a first appearance in this Ediacaran era, whereas there is a large cluster of first appearances in the Cambrian. However, twelve phyla have no fossil record at all. These are all small, soft-bodied animals. So there is nothing extraordinary in the suggestion that small animals could have been present in the Ediacaran world without leaving a fossil record.[10]

Moreover, while there is no direct fossil evidence of surviving Metazoan lineages until about 570 mya, perhaps there is indirect evidence. Peterson and Butterfield argue that ancient Metazoans have left an indirect ecological signature. They claim that there is a signal of protists becoming vulnerable to Metazoan predation at the base of the Ediacaran period (i.e., at around 635 mya). Sponge-grade organisms, because of the basic design of their bodies, can capture only bacteria and similar size particles. A nervous system and a gut are needed to capture eukaryotes: these innovations were established by the last common ancestor of the cnidaria and the bilaterians. Peterson and Butterfield think there is a palaeobiological signature of that new vulnerability of protists to predation. Before that period, acritarchs (single-celled organisms of unknown affiliation) had hyperstable evolutionary dynamics and little apparent diversity: after the period, they show both diversity and rapid turnover. Peterson and Butterfield think the morphological diversity of the Ediacaran acritarchs is a signature of biological interaction, in particular, a response to predation of protists by early animals.

In summary: our best guess of the pattern of the Metazoan radiation goes something like this. Metazoans probably existed for 80–100 million years or so before the base of the Cambrian. But their ecological and morphological diversity was low. Only a few of the living Metazoan lineages had separated before the Cambrian, and most of these early animals were simple, without much in the way of complex tissues, organs, or musculature. They were, perhaps, mostly very small as well. However, the size and organisational complexity of the Last Common Ancestor of the bilaterian clades – "the Urbilateria" – remains controversial. After the Cambrian, there was a genuinely rapid ecological, phylogenetic, and morphological radiation. A raft of morphological complexes had evolved by the mid-Cambrian; complexes that had existed at best in very rudimentary form at the Ediacaran-Cambrian boundary. The Cambrian Explosion is no illusion. However, it was probably the period in which many of the modern phyla acquired their crown group characteristics rather than the period in which those clades first diverged from their sister taxa. It was, in all probability, an extraordinary period for the evolution of novelty. Puzzles remain. Why did animals remain, collectively, small and unobtrusive for the best part of 100 million years when there was room at the top? After all, the post-Cretaceous radiation of the mammals has taken only two-thirds that time. If unobtrusive microfauna lived for 80 million years or more without much expansion in size or complexity, this suggests the existence of a right wall blocking the evolution of larger and more complex Metazoa. We should look for an external environmental barrier or constraints internal to the Metazoan clade. There remain phenomena to be explained.

5.   MINIMAL MODELS OF THE CAMBRIAN RADIATION

The Metazoan radiation resulted in the evolution of a dazzling array of morphological novelties: it was a period in which the space of evolutionary possibility expanded. I noted in Section 1 that minimalist models of novelty emphasise the role of the external environment in explaining possibility-expanding innovations. Key innovations are ordinary changes in extraordinary times. Theories of the Cambrian explosion that emphasise the role of external environmental triggers fit minimalism. According to this line of thought, armour (say) was indeed a key innovation. But in the generation of relevant variation and its establishment in local populations, the spread of biomineralised structures in (for example) stem group brachiopods looked just like any other small adjustment to local circumstances. The difference between the Cambrian and other periods of animal evolution lay not in the processes through which variations arose and spread in local populations, but in the global extent and importance of driving environmental change. The revolution was a revolution in circumstances, a revolution in what was necessary.

There are quite plausible theories of this kind. One idea is that the Metazoan radiation was initiated by an injection of new resources into the ecosystems of the late Proterozoic world. One such resource is oxygen. Atmospheric oxygen is a biological product: it is not an ancient feature of the world. A threshold level of oxygen is needed to power the aerobic metabolisms of (most) animals. While tiny animals can live in low-oxygen environments, large, active, or well-armoured animals cannot. So one traditional hypothesis was that the radiation was triggered by oxygen's reaching that threshold (Nursall 1959). Knoll discusses this idea sympathetically, arguing that there is indeed geochemical evidence for an increase in atmospheric oxygen in the late Proterozoic, after the youngest of the three severe global ice ages that preceded the Ediacaran biota. The evolution of that biota might well (he argues) have been triggered by that oxygen pulse (Knoll 2003, 217–20).

There are other versions of this resource-driven view of the radiation. Vermeij argues that major pulses of evolutionary innovation are caused externally, by sudden pulses in the availability of resources (Vermeij 1995). In particular, he thinks that the time of Metazoan evolution is characterized by two great innovation revolutions: one from the Cambrian to the mid-Ordovician, the other in the mid-Mesozoic. Undersea volcanism generated a large pulse of extra resources, and those pulses led to biological revolutions.[11] He argues that an external mechanism must play a crucial role in the Metazoan radiation, for many innovations occurred independently in several lineages, for example, the evolution of a skeleton, rapid and controlled locomotion, the capacity to burrow.

The evolutionary dynamics of one lineage are often sensitive to change in others. Perhaps then the Metazoan radiation is an effect of runaway coevolutionary interactions. A key innovation in one lineage triggers coevolutionary responses in others. Those responses themselves may include further key innovation, inducing feedback-driven diversification. There is a raft of mutually compatible suggestions along these lines. The most widely discussed one is based on the invention of macropredation, a breakthrough that triggered a host of defensive counteradaptations and hence further adaptations for predatory lifeways (Vermeij 1987, McMenamin and McMenamin 1990). Building shells and skeletons certainly takes off in the Cambrian: that era saw the evolution of hard structures in bivalves, molluscs, brachiopods, arthropods. There are two more recent ideas: Bilaterians with a true body cavity invented the capacity to tunnel through the substrate both for food and for refuge, destroying one community type (based on sediments stabilized by microbial mats) and establishing others. There was a biological revolution in the nature of the sea floor (Bottjer et al. 2000). Most recently, Andrew Parker has argued that the Cambrian radiation is a central nervous system revolution. The invention of true vision is the invention of a special kind of active agency. Through vision, agents get fast, accurate, and positional information about their local environment, and that sets up selection for rapid, guided response. The result is the evolution of a new kind of agent, one whose behaviour in both foraging and defence is guided by specific and up-to-date information about its local environment. No other sense modality has this combination of range, specificity, and speed of information transmission (Parker 2003).

Such explanations fit with minimalist models of the Cambrian radiation. The selective environment changed in an important, sustained, and global way. Those changes affected different lineages in similar ways: perhaps they eased resource constrains on all of them; perhaps all the Metazoan lineages for the first time were at risk from predation. Thus broadly similar responses evolved in parallel in different Metazoan lineages. As they stand, though, these externalist models are incomplete: they give no account of the origins and establishment of the striking morphological innovations – the key innovations – characteristic of the Cambrian radiation. Minimalist models presuppose rather than deliver a minimalist account of the origins of novelties. Yet the morphological innovations of the Cambrian are truly extraordinary. Moreover, once we recall there are alternatives to minimalism other than models invoking macromutations delivering a whole functioning new system in one go, there is no reason to suppose some minimalist account of novelty *must be true*, even if we do not know what it is. Thus we shall consider in the next section an extended argument for the claim that the bilaterian novelties of the Cambrian depended on fundamental changes in Metazoan developmental systems. We cannot exclude the possibility that key Cambrian innovations really did originate and establish in just the same way that (say) variations in the facets of trilobite compound eyes evolved and established. But in the light of these developmental considerations, it is likely that the Metazoan radiation arose through some form of complex feedback among (a) exogenously caused environmental changes, (b) biotically triggered environmental changes, (c) changes in evolvability. In the next section I discuss the idea that the Cambrian explosion depended on a developmental revolution, and then return to the key problem for this chapter: what does this radiation show about the relationship between microevolutionary process and macroevolutionary pattern?

6.   THE DEVELOPMENTAL REVOLUTION
HYPOTHESIS

Sean Carroll has argued that the diversification pulse of the Cambrian represents a change in developmental program rather than a change in selective regime. He, like others, emphasises the importance of modularity to evolutionary possibility. Innovations that increase the space of evolutionary possibility are made possible (or perhaps, much more probable) by modular construction. If the development of one structure is largely independent of the development of others, those structures can vary independently of one another; the structures can be independently modified. Such developmental compartmentalization decouples phenotypic traits from one another, enabling a lineage to escape from developmental constraints that would otherwise limit the range of variation. Moreover new structures can be made by repetition followed by differentiation, as with arthropod limbs. Carroll argues that there is evidence for an increase in complexity in arthropod limb design (and hence arthropod ecological versatility) since the Cambrian, and he interprets this as an instance of modular development's allowing repetition followed by differentiation (Carroll 2001; see also Lewontin 1985, Kauffman 1993, Wagner and Altenberg 1996). On Carroll's view, these evolutionary changes cannot be understood as the replacement of one allele by another in the context of invariant systems of gene regulation and expression. The Cambrian radiation (and, more generally, the evolution of novelty) requires evolutionary changes in how genes are used as well as changes in the genes themselves. They are essentially evolutionary changes in gene regulation (Carroll 2005).

Carroll has defended one version of the idea that the Cambrian radiation is a radiation in development. But the most articulated developmentalist explanation of the Cambrian radiation is due to Eric Davidson. The Metazoan radiation is really a radiation of one deep branch of the Metazoans. The sponges and jellyfish did not experience an explosive increase in diversity and disparity; that burst took place in the bilateral animals (Knoll and Carroll 1999). Major morphological innovations separate these developmentally simple sponges and jellyfish ("the diploblasts") from the earliest bilaterally symmetrical animals ("the Urbilaterians"). These include a through-gut, a third germlayer, a centralized nerve chord, a body vascular system, primitive organs. Eric Davidson and his colleagues have suggested that these morphological innovations and the diversification that followed from them depended on a developmental revolution. That developmental revolution is the crucial evolutionary change that made the Cambrian Explosion possible (Peterson et al. 2000, Peterson and Davidson 2000, Erwin and Davidson 2002).

Many adult bilaterians develop from larvae that live as very small, but free-living members of the plankton. This life history is known as indirect development, as juveniles are morphologically and ecologically very different from the adults into which they develop. Such larvae consist of only a few thousand cells. They have only ten to twelve cell types. Moreover, they are organizationally simple. They do not have multilayered, organisationally complex internal structures. Finally, these larvae develop in a distinctive way. Their genes are turned on early in development, and thus cell lineages differentiate early, with about ten rounds of cell division. Cells find their final position and role in the organism under local signalling control. In contrast to the standard developmental pattern of adult bilaterians, these larvae do not first differentiate into embryonic regions prior to cell differentiation. The *Hox* genes control this system of developmental regionalisation on the front-to-back axis in adult bilaterians, and those genes are not active in these larvae.

Davidson thinks that the developmental mode exemplified by these planktonic larvae is a relic of the earliest bilaterians, and argues that development with early differentiation and local control suffices only for this grade of morphological complexity. More complex morphologies required a developmental revolution that had two key ingredients. One is regional regulation, and hence delayed cellular differentiation. Crucial genes determining cell type are not switched on until after developmental regions are established. The other is the evolution of a population of "set-aside" cells: cells that retain all their potential for cell division and that are not committed to specific cell fates. These set-aside cell populations exist in the larval form of indirectly developing bilateria, and adult morphological structures are recruited from those set-aside cells. Contemporary adult bilaterians share a developmental recipe that includes a differentiated axis of symmetry from front to back and a system of recursive regionalization. The developing embryo is divided into a set of regions, each of which is under fairly independent developmental regulation. Often these regions are further subdivided until the specific details of adult morphology are constructed.

So how did the developmental revolution take place? Davidson's evolutionary narrative goes something like this: Early-differentiation embryogenesis evolved after the Cnidaria split off from the stem Metazoa. The third tissue layer of the embryo, the endomesoderm, then evolved, and this was crucial to the later evolution of structurally complex organs, for these have their developmental origin in this third layer. After the Ctenophora split off from the stem, bilaterian symmetry and the *Pax-6* genes (involved in vision) evolved. The final innovations, just prior to the crown group radiation of the bilaterian phyla, was the evolution of a full *Hox* cluster with front-to-back differentiation and set-aside cells. *Hox* genes evolved earlier, when the cnidarians split from the basal Metazoans, but they did not originally function to control front-to-back differentiation. The full *Hox* complement with the contemporary *Hox* functions evolved relatively late.

This narrative leads to their portrait of the common ancestor of the extant bilaterian clades: the common ancestor, that is, to the vast majority of multicelled animal designs. Morphologically, the creature was small; perhaps very small, and possibly pelagic. But it was more structurally complex than its Cnidarian sister group. The common ancestor had a mesodermal layer, a central nervous system, and two-ended gut with a mouth and anus. It thus had a front-to-back axis. These are genuine bilaterian homologies not shared by jellyfish. Moreover, some common developmental mechanisms were available to the common ancestor, including the *Hox* system, not just for basic body, front-back, and up-down differentiation, but for structures attached to the body. Thus the morphological similarities between the bilaterian phyla in part depend on crucial and very deep morphological homologies. But they are also the result of parallel evolution working with a homologous developmental toolkit. Many organ systems across the bilaterians as a whole – heads, hearts, sensory systems – are analogs not homologs. But the cell types of which these organs are composed are homologs and explain some of their functional similarities, as do these homologous developmental mechanisms. The common ancestor had the developmental toolkit needed for a complex morphology. But it was not itself complex. While the common ancestor was itself likely to be a relatively small and simple organism, it was preadapted to morphological differentiation in response to the right biotic and abiotic triggers.

As it was initially formulated, the developmental revolution hypothesis entails a puzzle of its own. What selected for these developmental changes? As Graham Budd and Soren Jensen (2000) argue, if developmental innovation *preceded* morphological innovation, the selective advantages of the developmental innovations of set-aside cells, modular developmental regulation, and the *Hox* system are obscure. What was the function of these innovations if they preceded rather than postdated or accompanied growth in size and complexity of bodily organization? Davidson and his colleagues place these developmental changes very deep in the bilaterian stem lineage, long before any trace fossil evidence of bilaterian morphological innovations. They seem to commit themselves to the idea that these animals were minute and nondescript even at the completion of the developmental revolution (Peterson et al. 2000, 12).

There is, however, a natural modification of the idea of the developmental revolution that leads to a more integrated conception of the radiation. Davidson thinks that early-differentiation embryogenesis is not sufficiently powerful to build adult, crown group bilaterian body plans. That could mean that it is simply impossible to build an adult crown group bilaterian by this mode of development. But perhaps the constraint is less absolute than that. Perhaps early-differentiation development can build only a somewhat simple version of an adult bilaterian; for example, one without a complex sensory system or complex locomotor-manipulation system. Alternatively, perhaps the primitive mode of bilaterian development can build complex adults only at the cost of a high rate of developmental error. This more modest view of developmental constraints on complexity leads to a natural ratchet hypothesis: a positive feedback between developmental and morphological change. Early bilaterian morphological innovations (for example, the two-ended gut) would select for improved developmental control, to reduce the rate of disastrous developmental errors. Once these evolve, they permit the evolution of further early bilaterian novelties, which in turn select for further improvements in developmental control, and so on. Early elements in the bilaterian body plan did occur first. But the limits on early-differentiating development in constructing such bodies selected for the key elements of the developmental revolution, both to make development more reliable and to support specific adaptive complexes grafted onto the basic plan.

7.   FINAL THOUGHTS

The key point – the take home message – is that minimal models are indeed minimal, and they can be enriched in a variety of ways. One is by extending temporal scale: on microevolutionary time scales, we can often treat features of both the environment and development as fixed. But these are not fixed on macroevolutionary time scales. In particular, it is unlikely that we can in general model the evolution of novelty in a classically Dawkinsian way, by thinking of alternative alleles as difference makers: one replaces another as each makes a consistent, selectable phenotypic difference, but only relative to a fixed developmental and genetic environment. In the evolution of possibility-expanding innovations, these environments are not fixed. The individualist perspective on selection is sometimes too limiting. It is clear that transitions in individuality require group selection of some kind. But it may well be the case that the structure and organization of species themselves are both important and no simple reflection of within-population change in the constituent demes from which species are formed. It is arguable that species structure plays a role in explaining fine-grained macroevolutionary patterns, for example, the punctuated-equilibrium pattern of typical species life histories (Eldredge 1995, Sterelny 1999b).

Let me finish by connecting these general morals to the Metazoan radiation. First: the origination of the Metazoans involved a transition in individuality, and hence requires a multilevel perspective on evolution. There will have been a period in the early evolution of the Metazoans when the fitness of individual cells within a protoanimal and the fitness of that protoindividual in a population were both important. Second: the radiation of the Metazoans was not an evolutionary radiation taking place within a fairly constant environment. The radiation, whatever its causes, profoundly changed both the selective and the physical environment. Likewise, the Cambrian radiation was evolution in a changing developmental environment. Davidson may not have correctly identified the primitive mode of bilaterian development or the sequence through which the contemporary bilaterian developmental toolkit was built. But it is clear that the radiation involved profound developmental change, for while protist preadaptations for complex development were important, the radiation required the evolution of both cellular differentiation and vertical complexity, as cells are organised into tissues, organs, and organ systems. Very early, simple Metazoans have few cell types: Placazoa have four; sponges have five cell types; cnidarians have ten. Valentine estimates that stem group bilaterians had between twelve and forty cell types, depending on the phylum. Crown group bilaterians have many more (Valentine 2004, 74–75). There is nothing like this in the protist world. Hence the evolution of Metazoans, especially the bilaterians, required a major revolution in developmental control. One cannot model this evolutionary transition as the result of the substitution of variant alleles for their predecessors in a relatively fixed developmental environment. There were crucial changes not just in organisms' genetic complement but in the ways genes are used. Finally, there is the open issue of novelty. Were the crucial novelties of the Cambrian radiation built unobtrusively, with their significance only becoming apparent later? The extent to which we can fit the evolution of novelty into the gradualist framework of minimal models remains open. But if Carroll or Davidson is right, some novelties – phenotypic icebergs – really are different, for their evolution is accompanied by changes in the developmental architecture that make further changes much more likely.

ACKNOWLEDGMENTS

Thanks to Brett Calcott, Patrick Forber, Peter Godfrey-Smith, David Hull, Andy Knoll, Tim Lewens, Dan McShea, Samir Okasha, and Michael Ruse for their helpful comments on an earlier draft of this essay.

NOTES

1. Gould has written here of "extrapolationism" (Gould 2002). I have avoided this term because Gould's work suggests that there is a single contrast between minimalist and extended models of the micro/macro relationship, one that turns on the acceptance or rejection of high-level selection. In contrast, I think there are a number of ways of going beyond minimalism, and some of these have nothing to do with levels of selection.
2. Such properties can be important without species selection's being important, in part because species-level properties may not be heritable.
3. There is considerable debate about the history of life, even at a coarse grain of analysis. It is not surprising that the issue of progress has always been contentious (see Ruse 1996). But even more technical claims about complexity and diversity have generated rich debate: see, for example, McShea 1998a, Knoll and Bambach 2000, Benton 2006.
4. In fitting the *Spread of Variation* view within a minimalist framework, I part company with Dan McShea's important and influential work on these issues. We both think the crucial element of minimalism is local determinism: for minimal models to be adequate, the fate of a lineage depends on the fate of its constituent taxa, and their fate in turn depends on local circumstances. But we have different views on how to identify local determinism. For example, unlike me, McShea counts sensitivity to a left wall of minimal complexity as a *violation* of local determinism: he thinks of it as a feature of the global environment affecting all taxa. See McShea 1996, Alroy 2000, McShea 2000.

5. This may exaggerate the uniqueness of the protist threat to bacteria. There is predation in the bacterial world, and hence probably in ancient bacterial environments. Bacteria predate by lysing and envagination. In the first, a mob of bacteria release an enzyme that causes the membrane of the prey species to breach, spilling the contents of the cell and thereby making the amino acids, nucleic acids, and other building blocks of metabolism available for acquisition. In the second method, a bacterium bumps into, surrounds, takes in, and lyses a smaller microbe (Lyons, personal communication).

6. In saying this, it is important to distinguish between large-effect mutations and macromutations. Large-effect mutations involve significant quantitative changes in existing traits; macromutations involve the single-step creation of new structures. Recent population genetics has embraced the idea that large-effect mutations are important causes of ordinary evolutionary change. Minimalist models are certainly not committed to the view that all structures are built by tiny increments over countless generations. For a good discussion of the resources available to minimalism, see Leroi 2000.

7. For, first, stem-group organisms will not fit taxonomic stereotypes derived from crown group organisms. Second, they will look strange because our sense of a normal-looking organism – for instance, a normal crustacean – is derived from our exposure to a host of crown group crustaceans. Our pattern recognition heuristics are trained on crown group taxa. There remains, as Dan McShea points out to me, the possibility that these early animals are weird, disparity-expanding organisms not just because they fail to fit taxonomic stereotypes but because they have an extraordinary load of unique traits. He may be right; unfortunately, we lack ways of making such intuitions of weirdness rigorous.

8. Simple versions of this approach obviously make risky simplifying assumptions about the constancy of rates of change across times and between lineages. But those concerns can be addressed by using a variety of different genes and different calibration points. For a systematic discussion of the reliability of various clocks and their application to the radiation of the Metazoans, see Bromham 2003.

9. The defenders of the deep dates dispute this, arguing that the (relatively) recent divergence times estimated by Peterson and company depend not on the choice of calibration taxa, but on the treatment of the fossil dates as the maximum age of divergence, as if the first appearance of a fossil in the record was the first appearance of

that organism on Earth (Blair and Hedges 2005). Peterson responds in Peterson and Butterfield 2005.

10. Nothing extraordinary, so long as the preservation potential of soft-bodied organisms has not changed radically over the Ediacaran-Cambrian boundary. This assumption may not be safe: there are Ediacaran fossils of early-stage embryos, and these are of tiny organisms. Moreover, it has been argued that the Ediacaran fossils were formed only because Ediacaran preservation conditions were very different from those of the Cambrian and subsequent eras (Narbonne 2005).

11. There seems to be a serious problem with this idea, for the innovation mechanism rests on the idea that until the resource budget increases, innovations are too expensive. A pulse of resources into the environment eases resource-based constraints on potential innovations. But this assumes that an increase in overall productivity leads to an increase in *per capita* access to resources. But if population growth keeps pace with the growing resource envelope, then the per capita availability of resources may not change. Vermeij notes this problem (1995, 134), but then ends up responding to a different problem, the idea that a sudden resource spurt may be destabilizing, a possibility he argues is confined to relatively undiverse ecosystems.

11   Philosophy and Phylogenetics
Historical and Current Connections

Philosophical arguments have played an influential role in the development of phylogenetic systematics – the field of biology that seeks to reconstruct the genealogical relationships among species, discover the pattern of events that has led to the distribution and diversity of life, and use this knowledge to construct natural classifications of species. Three sets of discussions clearly demonstrate this connection between philosophy and phylogenetics: inference modes and their relevance to competing phylogenetic methods, the nature and treatment of species and higher taxa, and the nature and treatment of phylogenetic evidence (character data). Within each of these areas, systematists have used philosophical arguments to defend particular concepts and methodological approaches, or to propose new ones. And, within each of these areas, philosophers have scrutinized the arguments of systematists and contributed their own.

Vigorous debate amongst systematists regarding these topics is pervasive. A common underlying tension that helps drive such debates revolves around the proper roles of process theories, assumptions, and trained judgment in phylogenetics research. For example, concerns about objectivity and testability have sometimes led systematists to reject methods that depend on evolutionary process theories, but such rejections typically do not 'stick' for very long. Thus, a cyclical pattern is evident – attempts to infuse theoretical dependence into phylogenetics research have repeatedly been countered by charges of non-objectivity and decreased testability, yet attempts to avoid them have repeatedly been countered by charges of operationalism. Two main questions emerge from this: What must be known about evolution in order to analyze phylogeny? What does it mean to be objective as a phylogeneticist?

211

1.   PHYLOGENETIC SYSTEMATICS

Systematics may be the oldest branch of biology, often traced back to Aristotle and the ancient Greeks. Aristotle (384–322 BC) held an essentialistic view of species as eternal and immutable, and characterized features of organisms similarly. This typological view of nature persisted for centuries, and biological classification via logical division (i.e., legs/no legs, blood/no blood) was the dominant approach. Linnaeus' (1707–78) system of classification was fundamentally based on the Aristotelian tradition of logical dichotomization and became formalized under the binomial system of taxonomic nomenclature that is so familiar to all biologists. Additionally, up until the early nineteenth century, a pervasive idea of the natural order of the world was the Great Chain of Being, or Scala Natura (Ladder of Nature), an unbroken sequence from the most primitive organisms to the most advanced (humankind) (Lovejoy 1936). This linear sequence of life was rooted in early ideas about the progressive structure of the world, ever moving towards perfection. However, the observed structure of variation in the biological world eventually rejected hypotheses of progressive ordering.

Darwin's (1859) evolutionary theory laid the groundwork for rejecting an essentialistic notion of species, emphasizing the variability that must exist in order for natural selection and transformation to occur. The emphasis on variability stands in obvious opposition to notions of types. Likewise, it eventually brought an end to the Scala Natura and related ideas about 'natural progression'. Ultimately, these were replaced by 'tree-thinking', with entities related through hierarchies of common ancestry. Darwin also revolutionized the discipline of systematics with the notion that classification should be based on genealogical relationships (Darwin 1859), although not all of his contemporaries agreed with this idea.

During the late nineteenth and early twentieth centuries, biology was strongly influenced by extensive studies of populations and their variability, leading to the Modern Synthesis – a unification of various fields of biology such as palaeontology, systematics, and genetics (e.g., Dobzhansky 1937, Fisher 1930, Huxley 1942, Mayr 1942, Simpson 1953). Building on Darwinian principles and new evolutionary studies, the 'population thinking' of the Modern Synthesis biologists further stressed the uniqueness and variability of organisms and populations. With variation seen as fundamental to biology and the notion of types rejected, a very different worldview emerged – one that is argued to have profoundly affected the discipline of systematics (Mayr 1959).

The last half of the twentieth century witnessed several methodological revolutions in systematics, which are described below. The prominence and reputation of systematics within the broader field of evolutionary biology grew steadily throughout those years. Today, systematics has an intimate connection to many other areas of biology because the results of phylogenetic analysis (phylogenetic trees) allow biologists to test precise hypotheses about evolutionary patterns and processes. Are some groups more diverse than others and, if so, why? Do features of organisms co-evolve? How many times did an ecological association or a structure evolve? Is the evolution of a behavior correlated with the evolution of a morphological feature? How do genetic and developmental regulation vary across groups? Do genetic changes occur more rapidly in some groups than in others? Today, we recognize that answers to all of these questions depend at least partially upon phylogenetic trees.

Modern biology tells us that there is a single evolutionary tree of life for all species – at least 1.7 million species, a staggering number that still does not reflect total historical diversity because of fossil and extant species not yet discovered or described. In its simplest conception, phylogenetic systematics is the organization of this tree of life, or the ordering of biodiversity. The ordering system is a phylogenetic tree, a hierarchical system that groups taxa according to relative recency of common ancestry, based on homologous features derived from comparative studies of phenotypic and genetic data. Thus, the task of the systematist can be seen as the knitting together of species via evidence of common ancestry into a phylogenetic tree. Virtually all contemporary biologists agree that evolution occurs, that the result of it is the vast biodiversity witnessed around us, and that knowledge of historical phylogenetic relationships is necessary for testing evolutionary and ecological hypotheses. However, they still argue about what that means for the practice and methods of systematics.

2.   METHODS FOR MAKING INFERENCES ABOUT PHYLOGENY

Evolutionary taxonomy (e.g., Mayr 1969, Simpson 1961) grew out of the Modern Synthesis, and was heavily rooted in Darwinian evolutionary theory. The methods of evolutionary taxonomy begin with evolutionary first principles such as natural selection, adaptation, and homology. These principles, in conjunction with extensive comparative studies of organisms, are used to assess the relative importance and/or reliability of organismal features (characters) for inferring genealogical relationships and, ultimately, to reconstruct evolutionary relationships among species based on those characters. An emphasis on heterogeneous rates of evolution across groups and on causally important evolutionary innovations leads to the construction of taxonomic groups based on a combination of recency of common ancestry and purported adaptively important similarities. Thus, an evolutionary taxonomic classification may reflect both evolutionary branching patterns and evolutionary disparity between groups. As an example, there is currently considerable support for the idea that birds and crocodilians share a more recent common ancestry than either does with other extant groups (such as turtles, snakes, or 'lizards'). However, evolutionary taxonomists prefer to group crocodilians with turtles, snakes, and 'lizards' in the group 'Reptilia' (to the exclusion of birds). Because birds have many unique characters and are considered to have diverged significantly compared to related groups, they are recognized as a separate taxon despite evidence of a shared evolutionary history with crocodilians. The same kind of argument has been applied to humans in relation to their closest relatives.

Evolutionary taxonomy was criticized for a lack of explicit methodology, subjective judgments about the phylogenetic utility of data, and an eclectic approach that often produced competing classifications for the same group. Evolutionary taxonomists were portrayed as too speculative and intuitive, transcending empirical data to produce authoritarian and untestable views of phylogeny. Most importantly, critics noted the potential for creating artificial (non-monophyletic) groups with these methods since factors other than common ancestry were sometimes used to group taxa. Two very different schools developed in opposition to evolutionary taxonomy – numerical taxonomy and cladistics. However, some fundamental tenets of evolutionary taxonomy remain in systematics today. The architects of evolutionary taxonomy published the first textbooks dedicated to systematic methods, which are widely cited today as landmarks that offered a lexicon and more precisely honed concepts for systematics.

Near the end of the 1950s, some scientists began advocating an approach to systematics that used computer-assisted, quantitative methods. These scientists proposed an explicit and more 'objective' methodology for systematics, leading to the rise of numerical taxonomy or 'phenetics' (Sneath and Sokal 1973, Sokal and Sneath 1963). To a large extent, phenetics may be viewed as a backlash against what were perceived as the subjective and unrepeatable methods of evolutionary taxonomy, combined with the burgeoning application of computer science to various biological disciplines. Pheneticists argued that evolutionary theory should not enter into classification studies; objectivity in systematics was to be found in purportedly 'theory-free', quantitative methods. Indeed, the two principal aims of numerical taxonomy were 'repeatability' and 'objectivity' (Sneath and Sokal 1973, 11). In order to accomplish these goals, pheneticists advocated 1) the use of averaged 'overall similarity' measures for grouping organisms, 2) equal weighting of all characters, 3) the use of large numbers of characters, 4) quantitative character coding, and 5) a 'theory-free' approach to character identification using 'raw similarity' as a guide. A phenetic classification typically depicts groups that are clustered quantitatively on the basis of averaged similarity (or distance) values. Distinctions are not made between homologous versus non-homologous similarity, nor between primitive versus derived similarity.

Phenetics was intended primarily for classification, not genealogy (which was considered unknowable). The approach was meant to produce the most efficient 'information storage and retrieval system', or an all-purpose classification of organisms. It was criticized for many reasons, including the fact that 'overall similarity' is not a biologically meaningful basis for systematics (e.g., Farris 1979, 1983, Mayr 1965). Further, its naiveté vis-à-vis 'theory-free' character identification was described as the "look, see, code, cluster" approach (Hull 1994). Despite the idealistic notion of 'overall similarity', numerical taxonomy also left important legacies to systematics – the numerical coding of characters and the use of computer algorithms to analyze data proved to be lasting changes in systematic methodology. Some would also argue that the antitheory stance of phenetics persists in various forms in the field today.

As did Darwin and others, Willi Hennig (1950) argued that taxonomy should reflect phylogeny, that genealogical relationships among species should be based on 'special similarity' or shared derived characters, and that these relationships should be arranged in a hierarchical manner to reflect the theory of descent with modification. Hennig's phylogenetic systematics emphasized: 1) the use of only shared, derived characters (synapomorphies) as evidence for identifying natural (monophyletic) groups; 2) comprehensive studies of homology determination based on character analysis; and 3) an explicitly genealogical interpretation of relationships among species. In contrast to evolutionary taxonomy, phylogenetic systematics accepts only monophyletic taxonomic groups – for example, those groups composed of the most recent common ancestor of the included species and all of its descendants. In contrast to phenetics, phylogenetic systematics is rooted in the theoretical principle of descent with modification, incorporates biological evaluation of characters, and uses discrete synapomorphies rather than overall similarity values to diagnose groups. The result is a cladogram depicting 'sister-group' relationships, or relative recency of common ancestry among groups.

The important distinctions between monophyletic groups, paraphyletic groups, and polyphyletic groups is one of Hennig's most important legacies. A monophyletic group is diagnosed by synapomorphy and comprises a common ancestor and all of its descendants; a paraphyletic group is diagnosed by symplesiomorphy and comprises a common ancestor and some, but not all, of its descendants; a polyphyletic group excludes the most recent common ancestor of its members because its diagnostic character arose separately in two or more phylogenetically disparate lineages. Only monophyletic groups can be considered 'natural' or 'real' entities according to Hennig because only in those groups is genealogical history captured. In Hennig's system, the important distinction between homologous and non-homologous derived similarity must also be analyzed. Two or more taxa may share a derived similarity (synapomorphy) for either of two reasons: either it was acquired through descent from a common ancestor (homology), or it was acquired convergently (homoplasy). The distinction is revealed through phylogenetic analysis – the analysis of observed features of organisms relative to a hierarchy.

The legacy of Hennig's work in systematics is profound. Indeed, shortly after the translation of Hennig's book into English (Hennig 1966), systematics underwent another revolution with the development of cladistics (e.g., Eldredge and Cracraft 1980, Kluge and Farris 1969, Nelson and Platnick 1981). Expanding on Hennig's views, cladists argued against both evolutionary taxonomy and phenetics. They advocated that phylogenetics ought to be an empirical and testable science (in contrast to the intuitive and/or authoritarian approach of evolutionary taxonomy) and that shared derived features provide the only basis for taxonomy (in contrast to the use of 'raw similarity' in phenetics). From the beginning, cladists have also been closely associated with the idea that the philosophical principle of parsimony should be an integral part of phylogenetic methods – in practice, this principle is used to minimize ad hoc hypotheses of homoplasy in phylogenetic analysis (e.g., Farris 1983). The use of parsimony is usually justified with an appeal to explanatory power – most parsimonious phylogenetic hypotheses are said to explain as much of the available evidence as possible as homology, thereby avoiding ad hoc hypotheses of homoplasy (Farris 1983).

The 'cladistic revolution' in taxonomy is considered a highly significant paradigm change in the field (Hull 1988), initiated by Hennig's strong focus on genealogical relationships between species, and revolutionary in the sense of replacing intensional with extensional thinking in systematics (Dupuis 1984). This may be so, but it is also the case that since the beginning of cladistics, there has existed a tension between those who emphasize genealogical relationships and more or less embrace evolutionary theory and those who emphasize classification and resist the incorporation of evolutionary theory into systematics. The latter group – the 'pattern cladists' – argued that cladistics itself is not about evolution, but only about the pattern of relative relationships amongst taxa as indicated by character distributions (Nelson and Platnick 1981, Patterson 1982, Platnick 1979). Some systematists continue to argue that cladistics is an evolutionary-theory-free classification method.

Pattern cladistics, or 'transformed cladistics', grew out of skepticism regarding the ability of systematists to reconstruct phylogeny, as well as concern about methodological circularity – in other words, if systematists wish to use phylogenetic trees to test hypotheses about evolution, then they should not use evolutionary theory to construct trees. The distinction between observed pattern and explanatory process theory is paramount in these discussions: the explanandum (in this case, the hierarchy of groups within groups) and the explanans (in this case, phylogeny) should not be conflated (Brady 1985). The purported independence of observation and interpretation and the appeal to observation as logically prior to phylogeny seem to be arguments with roots in empiricism and idealistic morphology. In any case, according to pattern cladists, classificatory cladograms – with taxa organized in sets within sets based on the parsimonious distribution of character data – are all that cladistics can claim to achieve. The use of parsimony methods in this context is sometimes justified based on high information content found in parsimonious classifications.

It is, however, difficult to argue for the primacy of 'classification' over 'phylogeny reconstruction' when one examines the utilization of cladograms by biologists. The contemporary literature indicates that systematists are not interested in information storage and retrieval systems, Venn diagrams, or efficient summaries of character distributions. Instead, most systematists today seem to be concerned with phylogeny reconstruction (i.e., inferring historical patterns of common ancestry), and with the use of phylogenetic trees to test broader hypotheses in evolutionary biology – or at least this is how phylogenetic trees are treated once produced, regardless of what is claimed by their authors about their initial ontological status.

Arguments about inference modes have also played an important role in the history of methodological debates in systematics. Farris (1983) proposed a hypothetico-deductive approach to phylogenetics, also suggesting that we should choose those phylogenetic hypotheses with the highest explanatory power. These hypotheses are said to be the most parsimonious ones, which are those that require the fewest hypotheses of homoplasy (convergence or parallelism). The roots of this idea can be found in Hennig's principle that "the presence of apomorphous characters in different species is always reason for suspecting kinship..., and that their origin by convergence should not be assumed a priori" (Hennig 1966, 121). This statement is interpreted by most cladists to mean that homology should be presumed in the absence of evidence to the contrary, or, in other words, that homoplasy should be minimized in phylogenetic analysis.

Early cladists also invoked the falsificationist philosophy of Karl Popper (1959, 1962) as a means to increase the testability of phylogenetic hypotheses, and to support the claim that the least falsified (most corroborated) phylogenetic hypothesis corresponds to the most parsimonious cladogram. Later, cladistics was tied to a Popperian philosophy of science via the 'test of congruence' – the matching versus non-matching of character statements, which play the role of potential falsifiers in this system (e.g., Kluge 1997). According to this, the maximally congruent set of characters gives the most parsimonious tree, which is the hypothesis that is least falsified (and most corroborated) by the data. Some systematists and philosophers disagreed with the idea that cladistics can be construed as a falsificationist endeavor. Many viewed parsimony methods as either inductive inference (relying on the maximal congruence of character statements to obtain the best-supported tree) or abductive inference (inference to the best explanation). The crux of the matter is that all phylogenetic methods permit some level of homoplasy; in other words, phylogenetic hypotheses (particular tree topologies) do not logically forbid any particular character distribution (Sober 1988), making it difficult to conclude that phylogenetic hypotheses can be falsified in a Popperian sense by phylogenetic character data. Nevertheless, the putative hypothetico-deductive nature of cladistics remains an issue of vigorous debate amongst systematists (e.g., de Queiroz and Poe 2001, Hull 1999, Kluge 1997, 2001, Rieppel 2003).

In addition to the arguments described above, a potentially serious 'fly in the ointment' for falsificationism in systematics is the treatment of phylogenetic character data. The stance taken by many contemporary systematists that character data must not be biologically evaluated can cause a serious underdetermination of phylogenetic characters (which are supposed to be potential falsifiers in this system). These systematists eschew investigations of potential character interdependence, developmental or functional correlation of characters, or differential weighting of characters because of concerns about subjectivity, and prefer to use any and all observations as character data using a global congruence test. However, in the absence of any causal grounding for characters, character redefinition and recoding can easily immunize phylogenetic hypotheses against rejection (see Section 4). Thus, the 'character problem' plays an important and neglected role in the debate about the framework of phylogenetic inference.

Felsenstein (1978) identified conditions under which parsimony methods could be statistically inconsistent, laying the groundwork for the rise of maximum-likelihood methods (e.g., Edwards 1972, Fisher 1925) in phylogenetic analysis. Proponents of maximum-likelihood approaches argue that robust hypotheses of phylogenetic relationships are obtainable only on the basis of fairly specific assumptions about the underlying evolutionary process, and with the use of rigorous statistical methods of analysis (Hillis, Huelsenbeck, and Swofford 1994). Unsurprisingly, the rise of maximum-likelihood methods in phylogenetics coincided with the increasing use of nucleotide positions in aligned DNA sequences as character evidence in systematics, and a concomitant interest in developing models of nucleotide evolution. Such models form a major component of maximum-likelihood algorithms for phylogenetic analysis, and are also a major point of criticism by detractors of these methods.

Opponents of maximum-likelihood phylogenetic methods argue that likelihood analyses can be performed only in the context of models that make overly restrictive, simplifying assumptions about evolutionary processes, and that likelihood methods may themselves fail to be statistically consistent under certain conditions (Kluge 2001). Some authors argue against likelihood methods as inductive and 'verificationist' in contrast to the purportedly deductive/falsificationist nature of cladistic parsimony, and have attempted to explicate a relationship between falsificationism and cladistic parsimony using Popper's corroboration formalism (Kluge 1997, 1999), an effort that has stimulated the 'Popper debate' once again (de Queiroz and Poe 2001, Faith and Trueman 2001, Farris, Kluge, and Carpenter 2001, Kluge 2001, Rieppel 2003). Some systematists argue that only cladistic parsimony conforms to Popper's falsificationist philosophy; some argue that likelihood methods of phylogenetic inference are just as consistent with Popper's concept of corroboration as are parsimony methods; some propose a framework for phylogenetics that is purportedly based on Popperian corroboration, yet not on falsificationism; and still others argue once again that Popperianism has nothing to do with phylogenetics. Meanwhile, the field marches on.

Most recently, Bayesian inference methods have been applied to phylogenetics (Huelsenbeck et al. 2001). Unlike cladistic methods (which identify the phylogenetic hypothesis that is most parsimonious given certain assumptions), and unlike maximum-likelihood methods (which identify the phylogenetic hypothesis for which the observed data have the highest probability given a certain model of evolution), Bayesian methods identify the phylogenetic hypothesis with the highest posterior probability. The latter entity is dependent upon the prior probability of the hypothesis and on the probability of the observed data given the hypothesis. As applied to phylogenetic inference, a Bayesian analysis delivers the posterior probability distribution of trees by assigning probabilities to trees conditional on the data. One of the main arguments against Bayesian inference methods in phylogenetics has been the selection of the prior probabilities, which are subjective. Computationally, Bayesian phylogenetic methods are much faster than maximum-likelihood analyses in terms of analyzing large data sets and assessing support for alternative trees, and many systematists prefer them for this reason. However, evaluation and comparison of support values derived from Bayesian versus maximum-likelihood analyses are current topics of debate. Much of the debate over the merits of Bayesian methods mirrors that between cladists and likelihoodists, but there is also an emerging disagreement between likelihoodists and Bayesians, which will be of interest in the coming years. Of course, Bayesian and likelihood methods were debated in statistical fields long before they were applied to phylogenetics, and those debates may be expected to be replayed to some extent in the context of phylogenetic analysis.

The discussions described above illustrate indecision among systematists over the proper methodological framework for phylogenetic inference, as well as some resistance to the use of explicitly statistical approaches. Many systematists strive for a hypothetico-deductive mode of inference in phylogenetic analysis. Some strive for a falsificationist systematics. As attractive as Popper's philosophy of science has been to systematists, the absence of a deductive link between any particular tree topology and any particular character distribution makes it difficult to justify phylogenetic methods on hypothetico-deductive grounds (Sober 1988). Other factors, such as the nature of phylogenetic character statements (see Section 4), may also favor this conclusion.

3.   THE NATURE OF SPECIES AND HIGHER TAXA

Life is wildly diverse, but it is also perceptibly discontinuous; biodiversity comprises more or less discrete entities, which biologists call species. A concept of species is one of the core concepts of systematics and evolutionary biology – that of a fundamental unit of comparison and perhaps a fundamental interactor in the evolutionary process. But what exactly is the nature of these entities that systematists are trying to identify, compare, and classify? This topic has engendered a great deal of conceptual discussion and debate.

Biological species concepts are rooted in the processes thought to create species (such as reproductive and/or geographic isolation) and to maintain species (such as interbreeding and/or cohesiveness). The biological species concept rejects the use of morphological distinctness in recognizing species and instead defines species as groups of populations separated by reproductive gaps: "Species are groups of actually or potentially interbreeding natural populations, which are reproductively isolated from other such groups" (Mayr 1942). This concept was later restated as "A species is a reproductive community of populations (reproductively isolated from others) that occupies a specific niche in nature" (Mayr 1982). Practical problems in applying the biological species concept to all of life exist: asexual, polytypic, and hybridizing entities all occur in nature – are they species? This has led some to suggest that a pluralistic approach to species may be necessary (e.g., Mishler and Donoghue 1994). In addition, documentation of reproductive processes in real populations is difficult at best. Simpson's (1961) evolutionary species concept allows for asexual species: "An evolutionary species is a lineage (an ancestral-descendant sequence of populations) evolving separately from others and with its own unitary evolutionary role and tendencies."

The emphasis on process in the preceding species concepts caused some systematists to note that operationalizing these concepts to recognize species is problematic. Phenetic species concepts define species on the basis of overall phenetic similarity – in other words, species are groups of similar organisms. Pheneticists believed that biological species, just like evolutionary relationships between species, are unknowable in the absence of 'direct proof' and replaced the notion of species as the fundamental unit of classification with 'operational taxonomic units', or OTUs. Phenetic species concepts attempt to avoid theoretical input and to make species identifications stable: "We may regard as a species (a) the smallest (most homogeneous) cluster that can be recognized upon some given criterion as being distinct from other clusters, or (b) a phenetic group of a given diversity somewhat below the subgenus category" (Sneath and Sokal 1973, 365). But phenetic similarity measures are arbitrary, and different ways of measuring similarity will give different 'species'. Moreover, biologists tend to reject typology and recognize that organisms within a species are not always very similar to each other; there are both cryptic and polytypic species. (Since the advent of molecular biology, many cryptic species have been discovered, making species criteria and concepts even more challenging.)

Phylogenetic species concepts identify species as segments of a phylogenetic tree: "A species is the smallest diagnosable cluster of individual organisms within which there is a parental pattern of ancestry and descent" (Cracraft 1983). The emphasis here is on cladogenesis, and on the systematist's ability to diagnose species through phylogenetic analysis. Various permutations of the phylogenetic species concept exist. "We define species as the smallest aggregations of populations (sexual) or lineages (asexual) diagnosable by a unique combination of character states in comparable individuals (semaphoronts)" (Nixon and Wheeler 1990). In general, phylogenetic species concepts tend to focus on diagnosability (Nixon and Wheeler 1990) or monophyly (Donoghue 1985). Operationally, a species is a diagnosable lineage (i.e., where a fixed qualitative difference can be identified). However, if all that is required for species status is a single differentiating feature, then males and females can be separate species, larva and adult can be separate species, and a single mutation can create a new species.

Despite the numerous publications debating the 'species problem', there may be more unity of opinion than appears on the surface (de Queiroz 2005). The major difference between the myriad species concepts is between those that emphasize the primacy of speciation processes (e.g., interbreeding, reproductive or geographic isolation) versus those that emphasize criteria for identifying or delimiting species (e.g., monophyly). According to de Queiroz (2005), if the distinction between species concepts and species criteria is made clear, then there is more underlying commonality among varying species concepts than one might imagine. That commonality is, "Species are segments of population-level evolutionary lineages."

Aside from species concepts, how to think about species is another topic of much current discussion in the field, and that discussion is often philosophically based. Hull (1965, 1976) and Ghiselin (1974) argued that evolutionary theory precludes viewing species as classes or natural kinds[1] because classes and kinds are tied to an essentialism that is inconsistent with an evolutionary worldview. Rather than species representing collections of organisms measured by some degree of similarity, by some defining feature, or by necessary and sufficient conditions, these authors argue that species are diagnosed by their history. For example, despite the absence of limbs, a snake is a tetrapod by virtue of its phylogenetic history (a snake does not have 'no legs,' but 'modified legs'). The related distinction between classification and systematization made by Griffiths (1974) has also been highly influential. Classes or sets impart a membership relation, which makes it difficult to revise them empirically. In contrast, individuals are particulars with spatiotemporal extension; they are not subject to a membership relation but to a part-whole relation. Thus, species are said to be conceptualized as individuals. Whether species and higher taxa[2] can alternatively be viewed as homeostatic property cluster (HPC) natural kinds within a realist perspective (rejecting strictly dichotomous thinking about classes versus individuals) is another issue still being debated (Boyd 1991, 1999, de Queiroz 1992, Ghiselin 1997, Keller, Boyd, and Wheeler 2003, Mayr 1987, Ruse 1987). Such a viewpoint requires divorcing the traditional concept of natural kinds from definitions based on necessary and sufficient conditions in order to accommodate the complexity of the biological world. Indeed, some biologists and philosophers view the strict class/individual distinction as inadequate (e.g., Grene 1990, 2002, Griffiths 1999, Keller et al. 2003, Mayr 1987, Rieppel 2006). These authors suggest that, while variation is prominent, it is not the sole feature of the natural world; the fact that we are able to recognize different species and make scientifically interesting generalizations illustrates something more than strict individualism.

Nevertheless, it is evident that the individuality thesis for species has had a huge impact on the field of phylogenetics, including the current effort by some to overturn the traditional and long-standing rank-based system for governing taxonomic names. For the past 250 years, the Linnaean hierarchy has formed the basis of taxonomy, with ranked taxonomic categories (Kingdom, Class, Order, etc.) to which taxa are assigned during classification. A complex set of rules and conventions governing the naming of taxa is also an integral part of traditional taxonomy. A proposed challenge to the Linnaean system of taxonomy is a phylogenetic system of taxonomy based on the evolutionary principle of descent with modification. Proponents of phylogenetic taxonomy (e.g., de Queiroz 1992, de Queiroz and Gauthier 1990, 1994) argue that species and higher taxa should be ordered into a natural system based on their genealogical relationships rather than the possession of defining characteristics. One of the central issues in phylogenetic taxonomy is the manner in which taxon names are defined. Under the Linnaean system, the name of a family of organisms might be defined as the family that contains certain lower-level taxa; under the phylogenetic system, that family name would be defined as the most recent common ancestor of the lower-level taxa, plus all of its descendants. Thus, the conceptual driving force behind the development of the 'PhyloCode' (Cantino and de Queiroz 2003) is rejection of the essentialism believed to underlie the Linnaean system of classification. Detractors of this 'nomenclatural revolution' argue against phylogenetic nomenclature on various grounds – empirical, philosophical, and practical (e.g., Keller et al. 2003, Nixon and Carpenter 2000, Rieppel 2006). Apparently, whether or not 'PhyloCode' successfully escapes essentialism via the ostensive definition of taxon names remains a matter of debate. In addition, pragmatic issues of nomenclatural stability are of great concern to both sides of the debate. The ultimate acceptance or rejection of phylogenetic nomenclature versus the long-standing rank-based system will be one of the more interesting areas to follow in the coming years for both systematists and philosophers of science.

4.   THE NATURE OF PHYLOGENETIC EVIDENCE

Systematists are in the business of trying to evaluate alternative phylogenetic hypotheses for various groups. They have only the end products of the branching process – organisms and their character-istics – that can be observed today and used as evidence for making inferences about phylogenetic relationships amongst taxa. Features that diagnose groups are proposed to be homologues. Because the relation of homology is an unobservable (i.e., because homology is identified by complex inferences rather than simple observation), character statements that are based on observed similarities and differences in phenotypic or genetic data are used as evidence in phylogenetic analysis. Today, those comparative observations are typically transformed into numerical codes and entered into a data matrix (characters × taxa). Some optimality criterion (e.g., parsi-mony, maximum-likelihood) is then used to analyze that data matrix, usually with the aid of a computer program, and to obtain a phylogenetic hypothesis.

From the very beginning of the history of systematics, there has been great difficulty in determining what the useful phylogenetic characters of organisms might be. The nature of phylogenetic char-acter evidence and the identification of characters continue to gen-erate controversy in the field. Evolutionary theory and comparative studies tell us that organisms are made of parts that are, to some extent, dissociable, recombinable, and changeable over time. These parts are the evidence, or data, of biological systematics. But what exactly constitutes a part? It is clearly inappropriate simply to reduce organisms to aggregates of features, characters, or raw observations because organisms are developmentally and function-ally integrated wholes. However, phylogenetic analysis requires the decomposition of the organismal whole in order to generate char-acter data for phylogenetic analysis. As a result, to propose phylo-genetic characters is far from trivial – among other things, the systematist must decide whether an observed feature is one, two, or many characters, and whether a specific character is a reliable indicator of homology or possibly a misleading convergence. Most systematists agree that the characters capable of indicating phylo-genetic affinity are not just any features, but evolutionary homo-logues. And, at least since Darwin, the definition of homology for most biologists is a correspondence of parts due to common descent. From this viewpoint, it would seem that insight into underlying causality in character evolution would be helpful to systematists in their work of identifying and coding characters. However, history shows that this is not always the case, and for familiar reasons.

The evolutionary taxonomists' approach to homology and characters was rooted in extensive organismal studies, and character weighting was based on presumed phylogenetic reliability. Issues such as potential non-independence of characters due to evolutionary processes of constraint, selection, adaptation, and correlation were considered very important. Such evaluation is admittedly imprecise, requiring judgments about the relative phylogenetic utility of organismal features, a comprehensive understanding of the characters and organisms under study (the 'expert problem'), and consideration of evolutionary processes acting upon character evolution (a consideration that many systematists see as too assumption-laden).

Hennigian phylogenetic systematics also emphasized initial character analysis as a necessary guide to homology. Hennig (1966) used a variety of criteria – detailed comparative morphological studies, topology, connectivity, ontogeny, functional anatomy, geological precedence in the fossil record, and ecology – to identify, analyze, and polarize characters. Evaluation of character quality and utility was based on both theoretical justifications and empirical investigations. Although one may disagree with the use of any or all of these guidelines for character delineation, it is instructive to note that homology was something to be comprehensively investigated prior to tree construction for Hennig, not solely the result of phylogenetic analysis. Character quality and utility were evaluated using theoretical justifications, empirical investigations, and estimations about the likelihood of convergence versus homology (see also Hennig and Schlee 1978).

Pheneticists considered such judgments about characters arbitrary and subjective. Sokal and Sneath (1963, 87) emphasized that approaches to character data need not be based on biological evaluation, but should be objective, explicit, quantitative, and repeatable: "One way to deal with problems of homology is to ignore details of structure." (It is important to note that in this conception of 'objectivity' both theory dependence and qualitative descriptions of character states diminish 'objectivity'.) Fundamentally, the phenetic approach to character data reduces characters to raw observations, and this uncritical empiricism is one factor that ultimately led to the method's demise. However, the overall philosophy does not seem to have been completely overcome in modern systematics, at least for morphological characters.

Some contemporary systematists paradoxically acknowledge that no theory-free observation is possible, yet they reject theoretical and empirical evaluations of characters in favor of a putatively rigorous method of testing – congruence of characters relative to a hierarchy. A related argument emphasizes our ignorance with respect to all of the causal correlates of phylogenetically informative characters and seeks as unbiased an approach to character delineation as is humanly possible. Both approaches maintain that biological evaluation of characters is irrelevant and impossible, and that any observation can be a character, and both ultimately defer to congruence under parsimony as the sole method of testing homology. Citing the principle of 'total evidence', they advocate that phylogenetic studies should include all previously published character data in a global congruence test, this being the most objective and rigorous way to test characters and homology. This stance has generated a new debate about the 'character problem' amongst systematists (e.g., Kearney and Rieppel 2006, Kluge 2003, Rieppel and Kearney 2002).

The heart of the debate seems to be that some systematists give the phylogenetic tree logical priority over critical comparative studies of character data – from such a viewpoint, it is only the tree, not empirical character evaluation, that can inform us about homology and what a legitimate character might be (Härlin 1999). Other systematists acknowledge the limits and difficulty of character evaluation but are uncomfortable with the contention that knowledge of homology and phylogeny can be derived from the simple coherence of theory-free observation reports. As Ruse (1988, 60) notes: "As soon as one starts breaking organisms into parts, one must bring in theory ... Take two bears, one white and one brown. Do they differ in one feature, or does one take each hair separately ... The point is whether someone who explicitly eschews the theory has the right to combine all the hairs into one feature."

It is instructive to note in today's context that numerical taxonomists previously stressed the 'empirical approach' in taxonomy, with an emphasis on 'firm observation' rather than phylogenetic or evolutionary assumptions. Today, most systematists would agree that no such theory-free 'observation language' exists, yet many still admit (at least potentially) any observation report into the total evidence under evaluation and disallow empirical rejection of the same. One concern about this approach is the threat of instrumentalism – that character statements may become mere instruments used to achieve a hypothesis of phylogeny, rather than being grounded empirically and causally in the organisms under study. A related concern is that the stance against evaluation of characters, or against any criteria for homology hypotheses, can cause a serious underdetermination of phylogenetic hypotheses (Richards 2002, 2003). Through definition and redefinition, virtually any character statement (certainly of morphological characters) can be made to cohere with any set of other such statements, and through splitting or lumping of the number of character statements, the same can be achieved. This is particularly true if 'anything' can be a character on the sole condition of its coherence with other characters relative to a hierarchy. Thus, while coherence of character statements relative to a hierarchy may be a necessary condition of phylogeny reconstruction, it seems unlikely to be a sufficient condition.

The claim that severity of test increases exclusively with an increasing *number* of characters used in phylogenetic analysis, no matter the nature of those characters, also seems questionable. This might be true if each character corresponded to some bit of information that could be empirically grasped by every working systematist and were fully independent from all other bits of information. This, however, is not the case, for biological as well as epistemological reasons. In contrast, to bring the insights of developmental biology, functional anatomy, and other evolutionary considerations to bear on character delineation and interdependence applies theory to the problem of character delineation. Criteria such as topological correspondence and connectivity have more or less successfully been used to help make the common historical origin of homologues empirically accessible, even in face of the fact that topological relationships can themselves evolve. It is assumed that this is so, not because of any arbitrary notion of similarity, nor because of a merely conventional use of topology and connectivity in the search for homology, but because these guides are at least approximately aligned with causal evolutionary and developmental processes. Such criteria are arguably what allow transcendence of 'primitive' similarity (i.e., the outermost ear ossicle of a mammal and the lower jaw of a shark are not phenotypically similar but they share similar topological relations; such guidelines have arguably led to the successful discovery of homology whereas 'primitive' similarity could not). But systematists also recognize that such criteria are not foolproof, and thus character congruence is an important part of evaluating homology hypotheses.

5.   DISCUSSION AND CONCLUSIONS

Although conceptual and methodological dialogues in systematics seem to replay an eternal debate in different forms, the field has also transcended these debates to a great extent – real progress has been made in understanding the tree of life for many groups, and systematics continues to become more and more integrated with other areas of evolutionary biology. It is now recognized as the foundation for research in evolutionary biology, ecology, behavior, and biogeography. In addition, the field continues to be influenced by numerous developments, from new discoveries about evolutionary mechanisms of inheritance and development, to the widespread use of computers that can analyze large amounts of data, to novel methods for extracting and sequencing DNA, and others.

Yet, contained within the debates described above is evidence of a persistent struggle with notions of objectivity, theory dependence, and testability. This was expressed in the methodological debate between pheneticists and evolutionary taxonomists, and in the different methodological viewpoints of phylogenetic cladists versus pattern cladists. Today, a similar tension exists between likelihoodists who seek to incorporate information about the evolutionary process into systematics through model-based analyses, and other systematists who reject the use of these models as too theoretically assumptive. Within debates about species, some suggest that species are the smallest phylogenetically diagnosable units, whereas others suggest that something more may be necessary. Different approaches to character data also reflect this theme. Pheneticists advocated analyzing as many traits as possible 'objectively' into quantitative unit characters, in contrast to the biologically steeped approach of evolutionary systematics. Early cladists rejected the tenets of numerical taxonomy, yet phenetic tendencies in character delineation persist.

Concerns about 'objectivity' and its connection to 'testability' have led systematists to critique and sometimes reject methods that are dependent upon theories or judgment. However, attempts to avoid theory and trained judgment in phylogenetics often reach dead ends, which may illustrate that such avoidance does not work. The character debate is an excellent example of this – reliance on atheoretical observations as characters yields the predicament of myriad, user-defined ways to delineate characters, and an approach that fails to transcend subjectivity. Indeed, in the absence of causal grounding, observations simply become more definitional and phylogenetic hypotheses less testable. In contrast, it may be argued more successfully that linking observations to causal mechanisms may increase objectivity.

Many systematists and philosophers of biology have noted that the influence of evolutionary theory has not yet been fully integrated in systematics. One explanation offered for its incomplete integration is that systematists still fail to grasp the distinction between classification and systematization – that is, the distinction between ordering things into classes on the basis of properties and ordering things into systems on the basis of a natural process through which their parts are related (e.g., de Queiroz 1988). Perhaps there is another reason, one that may be resolved by further discussions between philosophers and systematists: incorporating theoretical and causal considerations into phylogenetics research without sacrificing objectivity or testability has proved to be difficult. Fertile ground for future discussion between systematists and philosophers lies in the critical examination of what it means to be objective *and* scientific within an evolutionary worldview.

ACKNOWLEDGMENTS

Thanks to David Hull and Michael Ruse for inviting me to contribute to this volume, and to R. Boyd, K. de Queiroz, D. Hull, A. Larson, B. Patterson, R. Richards, and O. Rieppel for reading the chapter and offering helpful criticism. This work was supported, in part, by grants from the National Science Foundation (DEB-0235628 and EF-0334961).

NOTES

1.   'Class' has a special use in this debate, meaning something close to 'set defined by necessary and sufficient ahistorical membership conditions.' On alternative conceptions (e.g., Boyd, 1991, 1999), species and higher taxa could be historically defined kinds that lack necessary and sufficient defining conditions, rather than individuals.
2.   It is not clear that the homeostatic clustering of characters honors the prevailing conception of monophyly, making the HPC conception for higher taxa potentially more complex than that for the species level. At the species level, both the HPC conception and the species-as-individuals approach may be able to explain the historicity of species.

12 Human Evolution

The Three Grand Challenges of Human Biology

Man is but a reed, the weakest in nature, but he is a thinking reed.

Blaise Pascal, *Pensées*, number 347

A SUMMARY OF THE ARGUMENT

Human biology faces three great research frontiers: ontogenetic decoding, the brain-mind puzzle, and the ape-to-human transformation. By ontogenetic decoding, or the egg-to-adult transformation, I refer to the problem of how the unidimensional genetic information encoded in the DNA of a single cell becomes transformed into a four-dimensional being, the individual that grows, matures, and dies. Cancer, disease, and aging are epiphenomena of ontogenetic decoding. By the brain-mind puzzle I refer to the interdependent questions of (1) how the physicochemical signals that reach our sense organs become transformed into perceptions, feelings, ideas, critical arguments, aesthetic emotions, and ethical values; and (2) how, out of this diversity of experiences, there emerges a unitary reality, the mind or self. Free will and language, social and political institutions, technology and art, are all epiphenomena of the human mind. By the ape-to-human transformation I refer to the mystery of how a particular ape lineage became a hominid lineage, from which emerged, over only a few million years, humans able to think and love, to develop complex societies and subject to ethical, aesthetic and other values. The human genome differs little from the chimp genome.

233

The egg-to-adult transformation is essentially similar, and similarly mysterious, in humans and other mammals. The brain-to-mind transformation and the ape-to-human transformation are distinctively human; they define the *humanum*, that which makes us specifically human. No other issues in human evolution are of greater consequence for understanding ourselves and our place in nature.

Erect posture and large brain are two of the most significant anatomical traits that distinguish us from nonhuman primates. But humans are also different from chimpanzees and other animals, and no less importantly, in their behavior, both as individuals and socially. Distinctive human behavioral attributes include tool making and technology; abstract thinking, categorizing, and reasoning; symbolic (creative) language; self-awareness and death awareness; science, literature, and art; legal codes, ethics, and religion; complex social organization and political institutions. These traits may all be said to be components of human culture, a distinctively human mode of adaptation to the environment that is far more versatile and successful than the biological mode.

Cultural adaptation is more effective than biological adaptation because (1) its innovations are directed, rather than random mutations; (2) it can be transmitted "horizontally," rather than only "vertically," to descendants; and (3) because cultural heredity is Lamarckian, rather than Mendelian, acquired characteristics can be inherited.

LIFE TO HUMAN

The oldest known fossil remains of living organisms are dated somewhat earlier than 3,500 million years ago, just a few hundred million years after the Earth had cooled. The organisms were microscopic, individual cells, but having already considerable complexity of organization and elaborate biochemical machinery to carry on the functions of life. We do not know when life started, but it likely was at least one hundred million years earlier.

There are several hypotheses about how life first started, but none of these hypotheses is sufficiently well supported by evidence and, thus, none of them is accepted by all scientists. But the fact that it took "only" one or a few hundred million years from the formation of the Earth to the appearance of the first single-cell organisms, suggests that life in some form is likely to appear in any planet that has water and a few other elements (notably, in our planet, carbon, nitrogen, phosphorus, and sulfur). The temperature must also be "right", within a certain range, as it is the case for planet Earth, because of the 150 million kilometers that separate it from the Sun, so that water can exist in liquid phase (rather than only as either ice or vapor, if the temperature is too low or too high).

There are three large groups of organisms on Earth: eucaryotes, bacteria, and archaea. The eucaryotes include animals, plants, and fungi. Eucaryotes are organisms that have their genetic material enclosed in a special capsule, or organelle, called the nucleus. Humans are eucaryotes. Animals, plants, and fungi are the only organisms that we can directly experience with our senses, and thus they were the only organisms whose existence was known to humans up to three centuries ago. Yet they account for only a fraction of the total diversity of the eucaryotes. The other eucaryotes are all microscopic. Some cause well-known diseases, such as *Plasmodium*, which causes malaria, or *Entamoeba*, which causes severe intestinal maladies.

A second group of organisms are the bacteria. Humans have known of the existence of bacteria for more than a century. We associate them with diseases, but bacteria perform many useful functions, including the incorporation of nitrogen from the atmosphere, nitrogen that animals and plants need but are not able to get directly from the atmosphere (where it is very abundant, about 75 percent of the total; the rest is mostly oxygen). Also, bacteria are responsible for the decomposition of dead matter, a process that is essential in the maintenance of the cycle of life and death, because it makes again available, for new organisms, valuable components that had been incorporated into the now dead organisms. The genetic diversity and number of species of bacteria are at least as large as in the eucaryotes. There are many more kinds of bacteria than there are kinds of animals, plants, and fungi combined. And they are so abundant that their total weight (their "biomass") is at least as great as (and probably much greater than) that of all plants, fungi, and animals combined, even though individually they are so much smaller. This is a humbling thought. We see ourselves, the human species, as the summit of life and we are the most numerous of all large animals; and we see animals and plants as the dominant forms of life on Earth. However, modern biology teaches as that, numerically as well as in biomass, the nearly two million known species of animals (including humans) amount only to a very small fraction of life on Earth. From the perspective of numbers and biomass, the bacteria alone count much more than we do.

There is another group, the archaea, likely to be about as large as the eucaryotes or the bacteria. The existence of the archaea is a very recent discovery of molecular (modern) biology. Because these organisms do not directly interact much with us, biologists were not aware of their existence. Three decades ago scientists only knew a few species, such as those that exist in the hot springs of Yellowstone National Park in the United States and in other volcanic hot springs, where they thrive at temperatures approaching the boiling point of water. Biologists thought that these were some unusual forms of bacteria. Now we know them to belong to a very diverse and numerous group of organisms, abundant in the top water layers of the seas and oceans. A bucket of sea water studied with the modern techniques of molecular biology may yield tens or hundreds of new archaea species.

The number of living species on Earth is estimated to be between 10 and 30 million, but some biologists think that there may be as many as 100 million species, if bacteria and archeaea are included. Animals represent a small fraction of all species now living. More than 99 percent of all animal species that lived in the past have become extinct without issue. This is most likely true for all other kinds of organisms as well. Thus, the total number of species that have existed since the beginning of the Earth is more than one billion. We humans are but one of them.

Humans are animals, but a very distinct and unique kind of animal. Our anatomical differences include bipedal gait and an enormous brain. But we are notably different also, and more importantly, in our individual and social behaviors, and in the products of those behaviors. With the advent of humankind, biological evolution transcended itself and ushered in cultural evolution, a more rapid and effective mode of evolution than the biological mode. Products of cultural evolution include science and technology; complex social and political institutions; religious and ethical traditions; language, literature, and art; radio and electronic communication.

HUMAN ORIGINS

Our closest biological relatives are the chimpanzees, who are more closely related to us than they are to the gorillas, and much more than to the orangutans. (The chimpanzees include two species closely related to one another, but both equally related to humans, *Pan troglodytes*, or common chimpanzee, and *Pan paniscus*, or bonobo.) The hominid lineage diverged from the chimpanzee lineage 7–8 million years ago (mya) and it evolved exclusively in the African continent until the emergence of *Homo erectus*, somewhat before 1.8 mya (Cela-Conde and Ayala 2001). The first known hominids are the recently discovered *Sahelanthropus tchadensis* (dated 6–7 mya; Brunet et al. 2002; Vignaud et al. 2002), *Orrorin tugenensis* (dated 5.8–6.1 mya; Senut et al. 2001), and *Ardipithecus ramidus* (dated 5.2–5.8 mya; Haile-Selassie 2001). They were bipedal when on the ground, but retained tree-climbing abilities. It is not certain that they all are in the direct line of descent to modern humans, *Homo sapiens*; rather, some may represent side branches of the hominid lineage, after its divergence from the chimpanzee lineage. *Australopithecus anamensis*, dated 3.9–4.2 mya, was habitually bipedal and has been placed in the line of descent to *Australopithecus afarensis*, *Homo habilis*, *H. erectus*, and *H. sapiens*. Other hominids, not in the direct line of descent to modern humans, are *Australopithecus africanus*, *Paranthropus aethiopicus*, *P. boisei*, and *P. robustus*, who lived in Africa at various times between 3 and 1 mya, a period when three or four hominid species lived contemporaneously in the African continent (see Cela-Conde and Ayala 2001 for an extensive review of hominid evolution).

The first intercontinental wanderer among our ancestors was *H. erectus*. Shortly after its emergence in tropical or subtropical eastern Africa, *H. erectus* dispersed to other continents of the Old World. Fossil remains of *H. erectus* are known from Africa, Indonesia (Java), China, the Middle East, and Europe. *H. erectus* fossils from Java have been dated 1.81 ± 0.04 and 1.66 ± 0.04 mya, and from Georgia between 1.6 and 1.8 mya. Anatomically distinctive *H. erectus* fossils have been found in Spain and in Italy, deposited about 800,000 years ago, the oldest known in Western Europe.

Fossil remains of Neanderthal hominids (*Homo neanderthalensis*), with brains as large as those of *H. sapiens*, appeared in Europe around 200,000 years ago (200 kya) and persisted until 40 kya. The Neanderthals were thought to be ancestral to anatomically modern humans, but now we know that modern humans appeared at least 100 kya, much before the disappearance of the Neanderthals.

Moreover, in caves in the Middle East, fossils of modern humans have been found dated nearly 100 kya, as well as Neanderthals dated at 60 and 70 kya, followed again by modern humans dated at 40 kya. It is unclear whether the two forms repeatedly replaced one another by migration from other regions, or whether they coexisted in the same areas. Recent genetic evidence indicates that interbreeding between *H. sapiens* and *H. neanderthalensis* never occurred.

The origin of anatomically modern humans is controversial. Some anthropologists argue that the transition from *H. erectus* to archaic *H. sapiens* and later to anatomically modern humans occurred consonantly in various parts of the Old World. Proponents of this "multiregional model" call attention to fossil regional continuity in the transition from *H. erectus* to archaic and then modern *H. sapiens*. They postulate that genetic exchange occurred from time to time between geographically separate populations, so that the species evolved as a single gene pool, even though geographic differentiation occurred and persisted, just as geographically differentiated populations exist in other animal species and in modern humans. This claim of interbreeding between *H. erectus* populations depends on the postulate of persistent migrations and interbreeding between distant populations, even from different continents, of which no direct evidence exists, although it is not theoretically unlikely to have occurred. However, it is difficult to conciliate the multiregional model with fossil evidence of the contemporary coexistence of different species (*H. erectus* and *H. sapiens*) or forms (archaic and modern *H. sapiens*) in China, Indonesia, and other regions.

Other scientists argue instead that modern humans first arose in Africa between 150 kya and 100 kya, and from there spread throughout the world, replacing elsewhere the preexisting populations of *H. erectus* or archaic *H. sapiens*. This is called the "Out of Africa" hypothesis, which is now favored by most evolutionists. Genetic and molecular evidence shows greater difference between African and non-African populations than between all non-African human populations. This pattern of differentiation endorses the hypothesis that the origin of anatomically modern humans was in Africa, whence modern humans expanded to the rest of the world, starting about 100 kya. It is not possible, however, to exclude completely a partial participation of archaic *H. sapiens* from the Old

World in the origin of modern humans. Some observations evince the persistence of older anatomical traits in modern human populations of Central Europe and traces of ancient mitochondrial DNA have been found in Australian populations (Wolpoff et al. 2001, Adcock et al. 2001). In any case, genetic analysis supports the occurrence of at least two, not just one, major migrations out of Africa, well after the original range expansion of *H. erectus* (Templeton 2002).

I wrote earlier that *Homo sapiens*, our species, is only one of more than one thousand million species that have lived on Earth since the beginning. From that perspective, humans are but a speck on our planet. This is also the case from the perspective of time. The hominids diverged from the apes about 7–8 mya, and modern humans come into existence about 100 kya. Yet, life has existed on Earth for more than 3,500 my.

It is difficult to think in millions of years. So let me transform the time line of evolution into a one-year scale, so that life arises in our planet on January 1, at zero hours, and so that it is now midnight on December 31. In this one-year scale, for the first eight months there is only microscopic life; the first animals appear around September 1, they are marine animals. The land is colonized around December 1; the primates originate on December 26; the hominids separate from the chimpanzees on December 31, at noon; and modern humans arise on that last day of the year at twenty-three hours forty-five minutes. We have been around for a total of fifteen minutes. That also is a humbling thought. But I hasten to add that even though we are "but a reed," as Pascal famously put it, we are a *thinking* reed, and to this I shall presently return.

THE HUMAN GENOME SEQUENCE

Biological heredity is based on the transmission of genetic information from parents to offspring, in humans very much the same as in other animals. The genetic information is encoded in the linear sequence of the DNA's four nucleotide components (the "letters" of the genetic alphabet, represented by A, C, G, T) in a similar fashion to encoding of semantic information in the sequence of letters of a written text. The DNA is compactly packaged in the chromosomes inside the nucleus of each cell. Humans have two sets of twenty-three chromosomes, having received one set from each parent. The total number of DNA letters in each set of chromosomes is about three thousand million. The Human Genome Project, which was undertaken in 1989, has deciphered the sequence (except for a number of small segments) of the three thousand million letters in the human genome (that is, in one set of chromosomes; the human genome sequence varies among individuals).

I estimate that the King James Bible contains fewer than three million letters, punctuation marks, and spaces. Writing down the DNA sequence of one human genome demands one thousand volumes of the size of the Bible. The human genome sequence is, of course, not printed in books, but stored in electronic form, in computers where fragments of information can be retrieved by investigators. But if a printout is wanted, one thousand volumes will be needed just for one human genome.

The two genomes (chromosome sets) of each individual are different from one another, and from the genomes of any other human being (with the trivial exception of identical twins, who share the same two sets, since identical twins develop from one single fertilized human egg). Therefore, printing the complete genome information for just one individual would demand two thousand volumes, one thousand for each of the two chromosome sets. Surely, again, there are more economic ways of presenting the information in the second set than listing the complete letter sequence; for example, by indicating the position of each variant letter in the second set relative to the first set. The number of variant letters between one individual's two sets is about ten million, about one in three hundred.

The Human Genome Project of the United States was initiated in 1989, funded through two agencies, the National Institutes of Health (NIH) and the Department of Energy (DOE). (A private enterprise, Celera Genomics, started in the United States somewhat later but joined the government-sponsored project in achieving, largely independently, similar results.) The goal set was to obtain the complete sequence of one human genome in fifteen years at an approximate cost of three thousand million dollars, coincidentally about one dollar per DNA letter. A draft of the genome sequence was completed ahead of schedule in 2001. In 2003 the Human Genome Project was finished. The sequence has become known with as much precision as wanted.

Proponents of the project had used inflated rhetoric to extol its anticipated achievements. The project was called the "Holy Grail" of biology, which would meet the biblical "Know thyself" injunction. The Nobelist Walter Gilbert said about a computer disk encoding an individual's DNA sequence information, "this is you".[1] (The Nobelist and first director of the project, James Watson, asserted that "our fate is in our genes".)[2] Daniel Koshland, editor at the time of *Science*, proclaimed that with knowledge of the genome sequence, "we may be able to prevent the damage" caused by violent behavior.[3] Has the Human Genome Project accomplished any of these lofty objectives? Has knowledge of the human genome sequence accomplished the anticipated promise of curing human diseases?

THREE FRONTIERS OF HUMAN BIOLOGY: BEYOND THE HUMAN GENOME

Human biology faces three great research frontiers: ontogenetic decoding, the brain-mind puzzle, and the ape-to-human transformation. This transformation involved the emergence of cultural heredity and cultural evolution, a new and much more effective mode of adaptation to the environment than the biological mode. The conundrum is how this was accomplished through the change of less than 2 percent of the genome.

One can refer to these three issues as the egg-to-adult transformation, the brain-to-mind transformation, and the ape-to-human transformation.

Knowing the DNA sequence of human beings is of great use as a database to biologists and health scientists. But such knowledge about the human genome does not by itself contribute much to the solution of any of the three conundrums I have identified here, or to the solution of any other fundamental biological problem.[4]

ONTOGENETIC DECODING

The instructions that guide the ontogenetic process, or the egg-to-adult transformation, are carried in the hereditary material. The theory of biological heredity was formulated by the Augustinian monk Gregor Mendel in 1866, but it became generally known by biologists only in 1900: genetic information is contained in discrete factors, or genes, that exist in pairs, one received from each parent. The next step toward understanding the nature of genes was completed during the first quarter of the twentieth century. It was established that genes are parts of the chromosomes, filamentous bodies present in the nucleus of the cell, and that they are linearly arranged along the chromosomes. It took another quarter-century to determine the chemical composition of genes – deoxyribonucleic acid (DNA). DNA consists of four kinds of chemical components (nucleotides) organized in long, double-helical structures. As pointed out earlier, the genetic information is contained in the linear sequence of the nucleotides, very much in the same way as the semantic information of an English sentence is conveyed by the particular sequence of the twenty-six letters of the alphabet.

The first important step toward understanding how the genetic information is decoded occurred in 1941 when George W. Beadle and Edward L. Tatum demonstrated that genes determine the synthesis of enzymes; enzymes are the catalysts that control all chemical reactions in living beings. It became known later that a series of three consecutive nucleotides in a gene codes for one amino acid (amino acids are the components that make up enzymes and other proteins). This relationship accounts for the precise linear correspondence between a particular sequence of coding nucleotides and the sequence of the amino acids that make up the encoded enzyme.

But chemical reactions in organisms must occur in an orderly manner; organisms must have ways of switching any gene on and off. The first control system was discovered in 1961 by François Jacob and Jacques Monod for a gene that determines the synthesis of an enzyme that digests sugar in the bacterium *Escherichia coli*. The gene is turned on and off by a system of several switches consisting of short DNA sequences adjacent to the coding part of the gene. (The coding sequence of a gene is the part that determines the sequence of amino acids in the encoded enzyme.) The switches acting on a given gene are activated or deactivated by feedback loops that involve molecules synthesized by other genes. A variety of gene control mechanisms were soon discovered, in bacteria and other microorganisms. Two elements are typically present: feedback loops and short DNA sequences acting as switches. The feedback loops ensure that the presence of a substance in the cell induces the synthesis of the enzyme required to digest it, and that an excess of the enzyme in the cell represses its own synthesis. (For example, the gene encoding a sugar-digesting enzyme in *E. coli* is turned on or off by the presence or absence of the sugar to be digested.)

The investigation of gene control mechanisms in mammals (and other complex organisms) became possible in the mid-1970s with the development of recombinant DNA techniques. This technology made it feasible to isolate single genes (and other DNA sequences) and to multiply them, or "clone" them, in order to obtain the quantities necessary for ascertaining their nucleotide sequence. One unanticipated discovery was that most genes occur in pieces: the coding sequence of a gene is divided into several fragments separated one from the next by noncoding DNA segments. In addition to the alternating succession of coding and noncoding segments, mammalian genes contain short control sequences, like those in bacteria but typically more numerous and complex, that act as control switches and signal where the coding sequence begins.

Much remains to be discovered about the control mechanisms of mammalian genes. The daunting speed at which molecular biology is advancing makes it reasonable to anticipate that the main prototypes of mammalian gene control systems will be unraveled within a decade or two. But understanding the control mechanisms of individual genes is but the first major step toward solving the mystery of ontogenetic decoding. The second major step will be solving the puzzle of differentiation.

A human being consists of 1 trillion cells of some two hundred different kinds, all derived by sequential division from the fertilized egg, a single cell 0.1 millimeter in diameter. The first few cell divisions yield a spherical mass of amorphous cells. Successive divisions are accompanied by the appearance of folds and ridges in the mass of cells and, later on, of the variety of tissues, organs, and limbs characteristic of a human individual. The full complement of genes duplicates with each cell division, so that two complete genomes are present in every cell. Moreover, experiments with other animals (and some with humans) indicate that all the genes in any cell have the potential of becoming activated.[5] Yet different sets of genes are active in different cells. This must be so in order for cells to differentiate: a nerve cell, a muscle cell, and a skin cell are vastly different in size, configuration, and function. The differential activity of genes must continue after differentiation, because different cells fulfill different functions, which are controlled by different genes.

The information that controls cell and organ differentiation is, of course, ultimately contained in the DNA sequence, but probably only in very short segments of it. What sort of sequences are these controlling elements, where are they located, and how are they decoded? In mammals, insects, and other complex organisms, there are control circuits that operate at higher levels than the control mechanisms that activate and deactivate individual genes. These higher-level circuits (such as the so-called *homeobox* genes) act on sets rather than individual genes. The details of how these sets are controlled, how many control systems there are, and how they interact, as well as many other related questions, are what need to be resolved to elucidate the egg-to-adult transformation. The DNA sequence of some controlling elements has been ascertained, but this is a minor effort that is only helped a little by plowing the way through the entire three thousand million nucleotide pairs that constitute the human genome. Experiments with stem cells are likely to provide important knowledge as scientists ascertain how they become brain cells in one case, muscle cells in another, and so on.

The benefits that the elucidation of ontogenetic decoding will give to humankind are enormous. This knowledge will make possible the understanding of the modes of action of complex genetic diseases, including cancer, and therefore their cure. It will also confer an understanding of the process of aging, the unforgiving disease that kills all those who have won the battle against other infirmities.

Cancer is an anomaly of ontogenetic decoding: cells proliferate although the welfare of the organism demands otherwise. Individual genes (oncogenes) have been identified that are involved in the causation of particular forms of cancer. But whether or not a cell will turn out cancerous depends on the interaction of the oncogenes with other genes and with the internal and external environment of the cell. Aging is also a failure of the process of ontogenetic decoding: cells fail to carry out the functions imprinted in their genetic codescript or are no longer able to proliferate and replace dead cells.

In 1985, health care expenditures in the United States totaled $425 billion; in 2004 they surpassed $1 trillion. Most of these expenditures go for supportive therapy and technological fixes that seek to compensate for the debilitating effects of diseases that we do not know how to prevent or truly cure. By contrast, those diseases whose causation is understood – tuberculosis, syphilis, smallpox, and viral childhood diseases, for example – can now be treated with relatively little cost and the best of results.[6] A mere 3 percent of the nation's total health care expenditures is devoted to basic research. Doubling or tripling this percentage would result in only a modest rise in total expenditures, but would yield large savings in the near future, as cancer, degenerative diseases, and other debilitating infirmities become preventable or curable, and thus no longer require the expensive and ultimately ineffectual therapy now in practice.

THE BRAIN-MIND PUZZLE

The brain is the most complex and most distinctive human organ. It consists of 30 billion nerve cells, or neurons, each connected to many others through two kinds of cell extensions, known as the axon and the dendrites. From the evolutionary point of view, the animal brain is a powerful biological adaptation; it allows the organism to obtain and process information about environmental conditions and then to adapt to them. This ability has been carried to the limit in humans, in which the extravagant hypertrophy of the brain makes possible abstract thinking, language, and technology. By these means, humankind has ushered in a new mode of adaptation far more powerful than the biological mode: adaptation by culture.

The most rudimentary ability to gather and process information about the environment is found in certain single-celled micro-organisms. The protozoan *Paramecium* swims apparently at random, ingesting the bacteria it encounters, but when it meets unsuitable acidity or salinity, it checks its advance and starts in a new direction. The single-celled alga *Euglena* not only avoids unsuitable environments but seeks suitable ones by orienting itself according to the direction of light, which it perceives through a light-sensitive spot in the cell. Plants have not progressed much further. Except for those with tendrils that twist around any solid object and the few carnivorous plants that react to touch, they mostly react only to gradients of light, gravity, and moisture.

In animals the ability to secure and process environmental information is mediated by the nervous system. The simplest nervous systems are found in corals and jellyfishes; they lack coordination between different parts of their bodies, so any one part is able to react only when it is directly stimulated. Sea urchins and starfish possess a nerve ring and radial nerve cords that coordinate stimuli from different parts; hence, they respond with direct and unified actions of the whole body. They have no brain, however, and seem unable to learn from experience. Planarian flatworms have about the most rudimentary brain known; their central nervous system and brain process and coordinate information gathered by the sensory cells. These animals are capable of simple learning and hence of variable responses to repeatedly encountered stimuli. Insects and their relatives have much more advanced brains; they obtain precise chemical, acoustic, visual, and tactile signals from the environment and process them, making possible complex behaviors, particularly in their search for food and their selection of mates.

Vertebrates – animals with backbones – are able to obtain and process much more complicated signals and to respond to the environment more variably than insects or any other invertebrates. The vertebrate brain contains an enormous number of associative neurons arranged in complex patterns. In vertebrates the ability to react to environmental information is correlated with an increase in the relative size of the cerebral hemispheres and of the neopallium, an organ involved in associating and coordinating signals from all receptors and brain centers. In mammals, the neopallium has expanded and become the cerebral cortex. Humans have a very large brain relative to their body size, and a cerebral cortex that is disproportionately large and complex even for their brain size. Abstract thinking, symbolic language, complex social organization, values, and ethics are manifestations of the wondrous capacity of the human brain to gather information about the external world and to integrate that information and react flexibly to what is perceived.

With the advanced development of the human brain, biological evolution has transcended itself, opening up a new mode of evolution: adaptation by technological manipulation of the environment. Organisms adapt to the environment by means of natural selection, by changing their genetic constitution over the generations to suit the demands of the environment. Humans, and humans alone, have developed the capacity to adapt to hostile environments by modifying the environments according to the needs of their genes. The discovery of fire and the fabrication of clothing and shelter have allowed humans to spread from the warm tropical and subtropical regions of the Old World, to which we are biologically adapted, to almost the whole Earth; it was not necessary for the wandering humans that they wait until genes would evolve providing anatomical protection by means of fur or hair. Nor are humans biding their time in expectation of wings or gills; we have conquered the air and seas with artfully designed contrivances, airplanes and ships. It is the human brain (the human mind) that has made humankind the most successful living species, by most meaningful standards.

There are not enough bits of information in the complete DNA sequence of a human genome to specify the trillions of connections among the 30 billion neurons of the human brain. Accordingly, the genetic instructions must be organized in control circuits operating at different hierarchical levels, as described earlier, so that an instruction at one level is carried through many channels at a lower level in the hierarchy of control circuits. The development of the human brain is indeed one particularly intriguing component of the egg-to-adult transformation. But we must focus now on the issue at hand, namely, how this awesome organ, the human brain, works.

Within the last two decades, neurobiology has developed into one of the most exciting biological disciplines. An increased commitment of financial and human resources has yielded an unprecedented rate of discovery. Much has been learned about how light, sound, temperature, resistance, and chemical impressions received in our sense organs trigger the release of chemical transmitters and electric potential differences that carry the signals through the nerves to the brain and elsewhere in the body. Much has also been learned about how neural channels for information transmission become reinforced by use or may be replaced after damage, about which neurons or groups of neurons are committed to processing information derived from a particular organ or environmental location, and about many other matters. But, for all this progress, neurobiology remains an infant discipline, at a stage of theoretical development comparable perhaps to that of genetics at the beginning of the twentieth century. Those things that count most remain shrouded in mystery: how physical phenomena become mental experiences (the feelings and sensations, called "qualia" by philosophers, that contribute the elements of consciousness), and how out of the diversity of these experiences emerges the mind, a reality with unitary properties, such as free will and the awareness of self, that persist through an individual's life.

I do not believe that these mysteries are unfathomable; rather, they are puzzles that the human mind can solve with the methods of science and illuminate with philosophical analysis and reflection. And I will place my bets that, over the next half-century or so, many of these puzzles will be solved. We shall then be well on our way toward answering the injunction "Know thyself."

THE APE-TO-HUMAN TRANSFORMATION

Knowing the human DNA sequence is a first step, but no more than one step, towards understanding the genetic makeup of a human being. Think of the one thousand Bible-sized volumes. We now know the orderly sequence of the three thousand million letters, but this sequence does not provide an understanding of human beings any more than we would understand the contents of one thousand Bible-sized volumes written in an extraterrestrial language, of which we only know the alphabet, just because we would have deciphered their letter sequence.

Human beings are not gene machines. The expression of genes in mammals takes place in interaction with the environment, in patterns that are complex and all but impossible to predict in the details – and it is in the details that the self resides. In humans, the "environment" takes a new dimension, which becomes the dominant one. Humans manipulate the natural environment so that it fits the needs of their biological makeup, for example, using clothing and housing to live in cold climates. Moreover, the products of human technology, art, science, political institutions, and the like, become a dominant feature of the human environment. As I have mentioned earlier, a distinctive characteristic of human evolution is adaptation by means of "culture," which may be understood as the set of non–strictly biological human activities and creations.

Two conspicuous features of human anatomy are erect posture and large brain. We are the only vertebrate species with a bipedal gait and erect posture. Birds are bipedal, but their backbone stands horizontal rather than vertical (penguins are a minor exception); kangaroos are mostly bipedal, but without proper erect posture or bipedal gait. Brain size is generally proportional to body size; relative to body mass, humans have the largest (and most complex) brain. The chimpanzee's brain weighs less than a pound; a gorilla's slightly more. The human male adult brain has a volume of 1,400 cubic centimeters (cc), about three pounds in weight.

In earlier decades, evolutionists raised the question whether bipedal gait or large brain occured first, or whether they evolved consonantly. The issue is now resolved. Our hominid ancestors had, since at least four million years ago, a bipedal gait, but their brain was still small, no more than 450 cc, a pound in weight, until about two million years ago. Brain size started to increase notably with our *Homo habilis* ancestors, who had a brain about 650 cc and also became tool-makers (hence the name *habilis*), and who lived for a few hundred thousand years, starting about two and a half million years ago. Their immediate descendants were *Homo erectus*, with adult brains reaching up to 1,200 cc in size. (I use the name *Homo erectus*, as it is often used, in a broad sense that encompasses a fairly diverse group of ancestors and their relatives, which current paleoanthropologists classify in several species, including *Homo ergaster*, *Homo antecessor*, and *Homo heidelbergensis*.) Our species, *Homo sapiens*, has a brain of 1,300–1,400 cc, about three times as large as that of the early hominids. Our brain is not only much larger than that of chimpanzees or gorillas, but also much more complex. The cerebral cortex, where the higher cognitive functions are processed, is in humans disproportionally much greater than the rest of the brain when compared to that of apes.

BIOLOGICAL EVOLUTION VERSUS CULTURAL EVOLUTION

Culture, as I define it here, has an individual and a social component. It includes ideas, habits, dispositions, preferences, values, and beliefs of each individual. It also includes the public results of human intellectual activity; technology; humanistic and scientific knowledge; literature, music, and art; codes of law and social and political institutions; ethical codes and religious systems. The individual and social components of culture correspond to the World

2 and World 3 of the eminent philosopher Karl Popper. The difference between the two becomes apparent when we consider that the extinction of humankind on Earth would eliminate World 2, while World 3 could survive in part or on the whole and could be assimilated by humans or humanoids from a different planet. The advent of culture brought with it cultural evolution, a superorganic mode of evolution superimposed on the organic mode, which has, in the last few millennia, become the dominant mode of human evolution.

There are in humankind two kinds of heredity – the biological and the cultural – which may also be called organic and superorganic, or endosomatic and exosomatic systems of heredity. Biological inheritance in humans is very much like that in any other sexually reproducing organism; it is based on the transmission of genetic information encoded in DNA from one generation to the next by means of the sex cells.

Cultural inheritance, in contrast, is based on transmission of information by a teaching-learning process, which is in principle independent of biological parentage. Culture is transmitted by instruction and learning, by example and imitation, through books, newspapers and radio, television and motion pictures, through works of art, and by any other means of communication. Culture is acquired by every person from parents, relatives, and neighbors and from the whole human environment (Dobzhansky 1962, Ehrlich 2000, Cavalli-Sforza and Feldman 1981, Boyd and Richerson 1985, Richerson and Boyd 2005).

Cultural inheritance makes possible for humans what no other organism can accomplish – the cumulative transmission of experience from generation to generation. Animals can learn from experience, but they do not transmit their experiences, their "discoveries" (at least not to any large extent) to the following generations. Animals have individual memory, but they do not have a "social memory." Humans, on the other hand, have developed a culture because they can transmit cumulatively their experiences from generation to generation. Some cultural transmission has been identified in chimpanzees and orangutan populations, but the "cultures" developed by these apes amount to trivial rudiments when compared to human cultures (Whiten et al. 1999, Whiten 2005).

Cultural inheritance makes possible cultural evolution, a new mode of adaptation to the environment that is not available to nonhuman organisms – adaptation by means of culture. Organisms in general adapt to the environment by means of natural selection, by changing over generations their genetic constitution to suit the demands of the environment. But humans, and humans alone, can also adapt by changing the environment to suit the needs of their genes. (Some animals build nests and modify their environment also in other ways, but the manipulation of the environment by any nonhuman species is trivial compared to humankind's, even in the case of the apes.)

For the last few millennia, humans have been adapting the environments to their genes more often than their genes to the environments. In order to extend its geographical habitat, or to survive in a changing environment, a population of organisms must become adapted, through slow accumulation of genetic variants sorted out by natural selection, to the new climatic conditions, different sources of food, different competitors, and so on. The discovery of fire and the use of shelter and clothing allowed humans to spread from the warm tropical and subtropical regions of the Old World to the whole Earth, except for the frozen wastes of Antarctica, without the anatomical development of fur or hair. Humans did not wait for genetic mutants promoting wing development; they have conquered the air in a somewhat more efficient and versatile way by building flying machines. People travel the rivers and the seas without gills or fins. The exploration of outer space has started without waiting for mutations providing humans with the ability to breathe under low oxygen pressures or to function in the absence of gravity; astronauts carry their own oxygen and specially equipped pressure suits. From their obscure beginnings in Africa, humans have become the most widespread and abundant species of mammal on Earth. It was the appearance of culture as a superorganic form of adaptation that made humankind the most successful animal species.

Whenever a need arises, humans can directly pursue the appropriate cultural "mutations," that is, design changes to meet the challenge. These changes are the discoveries and inventions that pervade human life. The invention and use of fire, the construction of bridges and skyscrapers, the telephone and the Internet, are examples of technological cultural mutations; science, art, political institutions, codes of ethics and religious systems also are cultural mutations. On the contrary, biological adaptation depends on the accidental availability of a favorable mutation, or of a combination of several mutations, at the time and place where the need arises.

Cultural heredity and biological heredity drastically differ in their mode of transmission, with important consequences in the speed with which a favorable adaptation spreads. Biological heredity is transmitted only vertically, from parents to their offspring, while cultural heredity spreads "horizontally" as well as vertically, as noted earlier. A favorable genetic mutation newly arisen in an individual can be transmitted to a sizable part of the human species only through innumerable generations. However, a new scientific discovery or technical innovation can be transmitted to the whole of humankind, potentially at least, in less than one generation. Witness the worldwide spread of cellular phones or the Internet in less than a decade or of the personal computer in less than a quarter-century.

Biological heredity is Mendelian because only the genes received from one's own parents are transmitted to the progeny. (The presence in an individual of newly acquired gene variations by spontaneous mutation does not materially challenge this statement.) But acquired characteristics, that is, the inventions, technological developments, and any kind of learning or experience acquired throughout an individual's life, can all be transmitted to other humans, whether or not they are direct descendants of the individual. Cultural heredity is Lamarckian in this sense, because "acquired characteristics," and not only inherited ones, can be transmitted to others.

The draft DNA sequence of the chimpanzee genome was published on 1 September 2005.[7] In the genome regions shared by humans and chimpanzees, the two species are 99 percent identical. The differences appear to be very small or quite large, depending on how one chooses to look at them: 1 percent of the total seems very little, but it amounts to a difference of 30 million DNA letters out of the three billion in each genome. Of the enzymes and other proteins encoded by the genes, 29 percent are identical in both species. Out of the one hundred to several hundred amino acids that make up each protein, the 71 percent of nonidentical proteins differ by only two amino acids, on the average. The two genomes are about 96 percent identical if one takes into account DNA stretches found in one species but not the other. That is, a large amount of genetic material, about 3 percent or some 90 million DNA letters, has been inserted or deleted since humans and chimps initiated their separate evolutionary ways, 7 or 8 million years ago. Most of this DNA does not seem to contain genes coding for proteins.

Comparison of the two genomes provides insights into the rate of evolution of particular genes in the two species. One significant finding is that genes active in the brain have changed more in the human lineage than in the chimp lineage. Also significant is that the fastest evolving human genes are those coding for "transcription factors." These are "switch" proteins, which control the expression of other genes, that is, when they are turned on and off. On the whole, 585 genes have been identified as evolving faster in humans, including genes involved in resistance to malaria and tuberculosis. (It might be mentioned that malaria is a much more severe disease for humans than for chimps.) Genes located in the Y chromosome (the chromosome that determines maleness; females have two X chromosomes; males have one X and Y chromosome, the Y being much smaller than the X) have been much better protected by natural selection in the human than in the chimpanzee lineage, where several genes have incorporated disabling mutations that make the genes nonfunctional. There are several regions of the human genome that seem to contain beneficial genes that have rapidly evolved within the past 250,000 years. One region contains the *FOXP2* gene, which had earlier been discovered to be involved in the evolution of speech.

Extended comparisons of the human and chimp genomes and experimental exploration of the functions associated with significant genes will surely advance considerably our understanding, over the next decade or two, of what it is that accounts for the *humanum*, what makes us distinctively human. Surely also, full understanding will only result from the joint solution of the three conundrums that I have identified. The distinctive features that make us human begin early in development, well before birth, as the linear information encoded in the genome gradually becomes expressed into a four-dimensional individual. In an important sense, the most distinctive human features are those expressed in the brain, those that account for the human mind and for human identity. It is human intelligence that makes possible human culture.

NOTES

1.   Cited by D. Nelkin and M. S. Lindee, *The DNA Mystique. The Gene as a Cultural Icon*, W. H. Freeman, New York, 1995, p. 7.
2.   Quoted in Leon Jaroff, "The Gene Hunt," *Time*, 20 March 1989, pp. 62–67.
3.   D. Roshland, "Elephants, Monstrosities and the Law," *Science* 25 (4 February 1992), p. 777.
4.   I am not challenging here that the Human Genome Project has many public health applications or that the deciphering of the genomes of other species is of great consequence in health care, agriculture, animal husbandry, and industry. The question is how much it can contribute to solve the three fundamental problems faced by human biology that I am expounding.
5.   The sheep "Dolly" was conceived using genes extracted from a cell in an adult sheep.
6.   This statement is overly optimistic, and it may be outright erroneous if the phrase "understood causation" is not precisely construed. Malaria and AIDS are two diseases whose causation is understood at a number of levels, yet we fail to treat them "with relatively little cost and the best results." In any case, one can anticipate that increased knowledge of the etiology of these diseases may lead to successful development of effective vaccines or drugs.
7.   *Nature* 437 (1 September 2005); see also *Science* 309 (2 September 2005).

13 Varieties of Evolutionary Psychology

INTRODUCTION

What is evolutionary psychology? The answer to this question is complicated by the fact that the term "evolutionary psychology" is commonly used in two distinct senses. In one sense, evolutionary psychology is simply the study of human behavior and psychology from an evolutionary perspective. In this sense, evolutionary psychology is a *field of inquiry*, a loose confederation of research programs that differ significantly in theoretical and methodological commitments. These diverse research programs attempt to explain a wide variety of phenomena, ranging from foraging and birth spacing in traditional hunter-gatherer societies to encephalization (the progressive increase in brain size relative to body size in the human lineage) and the evolution of altruism and language. What unites these research programs is not a shared commitment to specific theories regarding the evolution of human behavior and psychology, but only a commitment to articulating questions about human behavior and psychology, and articulating answers to those questions, with conceptual and theoretical tools drawn from evolutionary theory.

In this broad sense, evolutionary psychology dates back to Darwin's *The Descent of Man* (published in 1871) and *The Expression of the Emotions in Man and Animals* (published in 1872). But, despite Darwin's early efforts, there was relatively little concerted study of human behavior and psychology from an evolutionary perspective until the latter half of the twentieth century, when several research programs emerged and attracted significant numbers of researchers (Laland and Brown 2002). The earliest of these

255

research programs was *human ethology*, exemplified by Konrad Lorenz's 1963 book *On Aggression*. The field really took off, however, with the emergence of *human sociobiology* in the 1970s, and in the ensuing decade additional research programs known as *evolutionary anthropology* and *human behavioral ecology* emerged. These research programs differ in the methods by which they apply evolutionary theory to the study of human behavior and psychology, and they differ in their theoretical accounts of how evolution has affected the human mind. Nonetheless, in the broad sense of the term, "evolutionary psychology" encompasses all of these research programs.

In a narrower sense, the term "evolutionary psychology" often designates just a *specific research program* within the field of evolutionary psychology, the foremost theoreticians of which are the anthropologists John Tooby and Donald Symons and the psychologists Leda Cosmides and David Buss. This group of researchers is united in the belief that adoption of an evolutionary perspective on human psychology immediately entails a number of very specific theoretical and methodological doctrines, and often the term "evolutionary psychology" specifically refers to this set of doctrines. So as to clearly distinguish the field of inquiry from the specific research program, I will refer to the field of inquiry as "evolutionary psychology" (in lowercase) and the research program as "Evolutionary Psychology" (capitalized).

Since its emergence in the late 1980s, Evolutionary Psychology has become the single most dominant research program in the field of evolutionary psychology, having garnered the lion's share of attention both within academia and throughout the popular media. But there is more to evolutionary psychology than Evolutionary Psychology. In particular, while Evolutionary Psychology has occupied the limelight, human behavioral ecology has quietly become a vibrant research program with impressive credentials. Indeed, it is the strongest rival to Evolutionary Psychology within the field of evolutionary psychology. In this chapter, I strive to give some sense of the diversity of research in evolutionary psychology by comparing and contrasting the theoretical and methodological principles of Evolutionary Psychology and human behavioral ecology. Both of these research programs, however, grew out of human sociobiology, so that is where we will begin.

HUMAN SOCIOBIOLOGY

Although many researchers have contributed to the program of human sociobiology, without doubt its leading theoretician has been Edward O. Wilson. In the mid-1970s, Wilson published several works that showcased numerous applications of evolutionary theory to the explanation of animal behavior and that articulated a theoretical framework within which to view them. Wilson called this framework *sociobiology*, which he defined simply as "the extension of population biology and evolutionary theory to social organization" (1978, x). Wilson further argued that the very principles that successfully explain the social organization of bee hives and dominance hierarchies in spider monkeys could be extended to human social behavior as well. This extension of sociobiology to human behavior became known as *human sociobiology*, and Wilson conceived it as the study of the biological basis of human social behavior.

The core idea of Wilson's sociobiology was that behavior has evolved under natural and sexual selection just as aspects of organic form have. Evolution by natural or sexual selection occurs when organisms in a population exhibit phenotypic variation, that variation is heritable, and organisms with one of the phenotypic variants are, on average, better adapted to their environment than organisms with the alternative phenotypes. When these conditions are met, selection causes the better-adapted phenotype to increase in frequency in the population, and the population as a whole becomes better adapted to its environment. Over very long stretches of time, selection has this effect on many different phenotypes, and populations thereby become well adapted to the environments they inhabit. The simple idea at the foundation of Wilson's program was that these explanatory principles are applicable to *behavioral*, not just morphological and physiological, phenotypes. For example, females of many species choose a mate on the basis of the quality of male courtship displays. If males' courtship displays vary in quality, and that variation is heritable, then sexual selection will cause the superior display to increase in frequency, and males will become behaviorally adapted to female preference. In this way, selection can shape the way that organisms behave just as it shapes their bodies.

This simple idea has two important corollaries. First, it entails that behaviors that have been shaped by selection are *adaptations*.

Thus, just as organisms in a population possess anatomical adaptations, they possess behavioral adaptations as well. Accordingly, part of Wilson's program was an effort to provide adaptationist explanations of how certain forms of behavior evolved. Second, since selection has shaped behavior to the environment in which it occurs, and since an organism's total environment includes its social environment, Wilson's simple idea entails that some behaviors are *adaptations to social life*. Accordingly, Wilson's program was principally concerned with explaining how individuals in a population are behaviorally adapted to social life with one another – explaining behavioral adaptations for dominance hierarchies, for manifesting and dealing with aggression, and for mating. Indeed, Wilson took the central theoretical problem of his program to be explaining the evolution of *altruism* – explaining why so many organisms have evolved to perform acts that benefit other organisms at a cost to themselves.

To illustrate these aspects of Wilson's program of human sociobiology, consider sex differences in human mating behavior. Both sexes need to reproduce in order to be successful in the evolutionary long haul, but reproduction entails very different costs for the two sexes. In order to produce a single child, a woman must invest one of her very limited number of eggs, physiological resources for a nine-month gestation, and the metabolic costs of lactation (often lasting two or three years). Moreover, during pregnancy and lactation, a woman is unable to reproduce with males other than – and possibly *better than* – the father of her child. In contrast, in order to produce a single child, a male need only invest the energy expended in copulation and the contents of a single ejaculate. After a fruitful copulation, a man can reproduce with other women, whereas a woman is committed to the costly act of childbearing. This is a radical asymmetry in the *minimum obligatory parental investment* required of the sexes in order to produce a single offspring: Women are obligated to a far higher investment in offspring than are men. Given this asymmetry, selection should have made women very choosy when selecting a mate, since they have to invest a great deal in a single offspring and, hence, have a great deal to lose by choosing a poor sire. In contrast, since men incur such a minimal obligation in order to produce an offspring, and since they can (theoretically) impregnate innumerable women during the time it takes a woman to bear one man's child, selection should have made men indiscriminately promiscuous. As Wilson says, selection should have created "males to be aggressive, hasty, fickle, and undiscriminating," and "females to be coy" (1978, 125). In other words, in humans, male promiscuity and female coyness are behavioral adaptations.

EVOLUTIONARY PSYCHOLOGY

The starting point of Evolutionary Psychology is a corrective to the core idea of Wilson's sociobiology. Evolutionary Psychologists argue that treating behavioral phenotypes as just like morphological and physiological phenotypes obscures a fundamental difference between them, for behaviors are *events*, which are the output of an information-processing brain reacting to informational input about the current conditions in both the environment and the brain itself. The only way that selection can affect behavior, then, is by altering the information-processing structure of the brain (Tooby and Cosmides 1992). So, when a behavior has evolved under selection, there is an important sense in which it is not the behavior itself that has been selected for, but rather the *psychological mechanism* (cognitive or motivational) that is causally responsible for producing that behavior under appropriate conditions. Since behavioral evolution involves selection for the psychological mechanisms that cause behavior, the adaptations that emerge in the process of behavioral evolution are the psychological mechanisms that cause behavior. Consequently, Evolutionary Psychologists conclude, sociobiology was mistaken in seeking adaptation at the level of behavior; adaptation must be sought at the level of the psychological mechanism (Tooby and Cosmides 1992). The goal of Evolutionary Psychology is thus to discover and describe the information-processing structure of our psychological adaptations (Buss 1995).

From this starting point, Evolutionary Psychologists derive a number of theoretical and methodological doctrines. First, they argue, our psychological adaptations are undoubtedly complex, and the construction of complex adaptations typically requires hundreds of thousands of years of cumulative selection. Our ancestors spent the Pleistocene – the epoch spanning 1.8 million to 10,000 years ago – living in small hunter-gatherer groups, but only the past 10,000 years living as agriculturists and the past few hundred years living in industrial societies. Consequently, it is highly improbable that humans have evolved adaptations to post-Pleistocene environments. Rather, Evolutionary Psychologists argue, our psychological adaptations must have been designed during the Pleistocene to solve the adaptive problems faced by our hunter-gatherer ancestors (Symons 1992). As Cosmides and Tooby colorfully put it, "Our modern skulls house a Stone Age mind" (1997, 85).

Adaptive problems are commonly characterized as problems whose solutions enhance the ability to survive or reproduce. And the adaptive problems faced by our Pleistocene ancestors ranged from acquiring mates and forming social alliances to avoiding predators and inedible flora. These problems are very diverse in character, and each requires a unique behavioral solution; a successful behavioral solution to one problem would not have transferred to another. Thus, Evolutionary Psychologists argue, each adaptive problem would have selected for its own dedicated problem-solving psychological mechanism (Symons 1992). Moreover, since our Pleistocene ancestors faced such an enormous variety of adaptive problems, Cosmides and Tooby conclude that "the brain must be composed of a large collection of circuits, with different circuits specialized for solving different problems. One can think of each specialized circuit as a minicomputer that is dedicated to solving one problem. Such dedicated minicomputers are sometimes called *modules*" (1997, 81). Indeed, Cosmides and Tooby estimate that the human mind contains hundreds or thousands of such modules, and this view has accordingly been dubbed the *massive modularity thesis*.

According to Evolutionary Psychologists, evolved modules have the following properties (Cosmides and Tooby 1997, Tooby and Cosmides 1992). First, they are *domain specific*, specialized to deal only with a restricted task domain. As such, their information-processing procedures are activated by, and sensitive to, only information about a particular aspect of the world, in much the way the ear is responsive only to specific vibratory frequencies. Second, they are equipped with substantial innate knowledge about their proprietary problem domains and with a set of innate procedures specialized in employing that knowledge to solve problems in their domains. And, third, they develop reliably, and without formal instruction in their problem domains, in every "normal" member of our species.

Since evolved modules are complex adaptations, and since "selection usually tends to make complex adaptations universal or nearly universal in a species," Evolutionary Psychologists argue that "humans must share a complex, species-typical and species-specific architecture of adaptations" (Tooby and Cosmides 1992, 38). Indeed, Evolutionary Psychologists believe that evolved psychological modules constitute a "universal and uniform human nature" (Tooby and Cosmides 1992, 79). Accordingly, Evolutionary Psychologists interpret differences between individuals within the same culture, and differences between individuals in different cultures, as "the product of a common, underlying evolved psychology operating under different environmental circumstances" (Tooby and Cosmides 1992, 45).

However, because our network of modules – our universal human nature – evolved to solve the adaptive problems faced by our Pleistocene ancestors, and because the environments we now inhabit differ enormously from those inhabited by our Pleistocene ancestors, Evolutionary Psychologists argue that our evolved modules often fail to produce adaptive behavior among modern humans. For example, fear evolved as an emotional alarm that signals a threat to survival. But, since human fears evolved during the Pleistocene, humans tend to fear snakes but not cars and guns, despite the fact that more people are killed by cars and guns than by snakes. In addition, people in modern industrialized societies could maximize their reproductive success by donating their sperm or eggs to cryobanks, but very few people pursue this reproductive option. The reason is that this option was not available in the Pleistocene, and we have minds designed to maximize reproductive success only under Pleistocene-like conditions, in which such success was achieved only through the pursuit of copulation. Because of this mismatch between human nature and contemporary human environments, Symons argues that the study of whether contemporary human behavior is adaptive will "rarely shed light on human nature or the selective forces that shaped that nature" (1992, 146). Thus, Evolutionary Psychologists claim, in order to discover the evolved design of the mind, we must "reverse engineer" the mind from the vantage of our evolutionary past.

The method by which Evolutionary Psychologists propose to reverse engineer the evolved structure of the mind is *evolutionary functional analysis* (Tooby and Cosmides 1992, Buss 1995). Evolutionary functional analysis begins with the specification of an adaptive problem that Pleistocene humans presumably faced. That adaptive problem is then analyzed into a number of subproblems whose solutions collectively constitute a solution to the adaptive problem. (For example, Pleistocene era males faced the problem of intrasexual competition for reproductive access to females, and solving this problem presumably required solving the subproblems of acquiring the resources desired by females, successfully courting females, and retaining mates, among other things.) The next step is to determine what forms of behavior would have constituted adaptive solutions under Pleistocene conditions to these subproblems. A module is then postulated, which is assumed to have evolved to generate solutions to all of these subproblems. The final step is to determine the information-processing procedures by which the module generates its behavioral solution(s) from its inputs. Evolutionary Psychologists then conduct standard psychological experiments in order to determine whether people behave in ways predicted by the modular hypothesis generated in these last two steps.

Evolutionary Psychologists claim to have made many discoveries regarding the evolved nature of the mind by employing this method. Consider just one example by way of illustration. Throughout our evolution as Pleistocene hunter-gatherers, men invested resources (food, protection, and paternal care) in the offspring of their mates. But because ovulation is concealed and fertilization occurs internally in our species, a Pleistocene human male could never be 100 percent certain when he was likely to impregnate his mate or, if his mate was pregnant, whether it was he who had impregnated her. This posed the following problem for an ancestral male: If his mate was surreptitiously unfaithful, a man could waste his resources on a child that was not his own. Pleistocene human women, in contrast, were always 100 percent certain that offspring born to them were their own. An ancestral woman's problem was that infidelity by her mate could lead to his falling in love with another woman, abandoning her, and withdrawing the resources on which she depended to rear her children successfully. Evolutionary Psychologists argue that jealousy evolved as an emotional alarm to protect against these respective potential losses due to a mate's infidelity. However, since the threats posed by infidelity were different for the sexes,

Evolutionary Psychologists argue, males and females must have evolved different psychological mechanisms: "The inputs that activate jealousy for men will focus heavily on the sex act per se, whereas for women they will focus on cues to the loss of the men's commitment and investment" (Buss 1995, 14). Evolutionary Psychologists have conducted numerous studies to test this prediction, and they claim that it is confirmed by the evidence (Buss 1995, 14–15). Thus, they conclude, men and women have evolved distinct psychological adaptations for monitoring, and emotionally responding to, cues of potential infidelity.

HUMAN BEHAVIORAL ECOLOGY

Whereas Evolutionary Psychology is an attempt to blend evolutionary theory with cognitive psychology, human behavioral ecology derives from the branch of biology known as *behavioral ecology*. Behavioral ecology is the study of how animal behavior is adaptively responsive to conditions in animals' physical and social environments. The fundamental premise of behavioral ecology is that "animals are maximizers of one sort or another – efficient predators or foragers, or elusive prey. The usual ground for believing this is the presumption that natural selection has made them so" (Grafen 1991, 5). Behavioral ecologists view animals as behaving so as to maximize their shares of a variety of "currencies" that are correlated with survival and reproductive success. These evolutionarily significant "currencies" include caloric intake, offspring survivability, clutch size, territory, number of copulations, quality of mate, number of sperm in an inseminate, number of inseminates "harvested" per fertile period, and number of mates per fertile period. Behavioral ecologists presuppose that animals tend to adopt behavioral strategies that enable them to maximize these "currencies" in the particular environmental conditions in which they find themselves. And this presupposition, in turn, entails that animals are capable of behaving adaptively across a very wide range of ecological conditions, flexibly altering their behavior in response to current conditions in order to maximize their chances of survival and reproductive success.

Behavioral ecologists study animal behavior with optimality models and evolutionary game theoretic models. Such models begin with the specification of some *currency* whose maximization is to be studied. For example, a model may study "clutch size" (that is, number of offspring born and cared for at the same time) in some species of bird. The models then identify a number of alternative *strategies* that animals may pursue by way of attempting to maximize that currency. If the currency is clutch size, the alternative strategies would be various clutch sizes: one strategy would be to have two chicks, another would be to have three, and so on. The models then identify the *costs and benefits* associated with each of the available strategies. In the case of clutch size, the benefits of the alternative strategies are easily measured in terms of number of offspring reared to reproductive viability. Accordingly, benefits appear to increase with increasing clutch sizes. However, offspring need to be fed and cared for, and those activities exact a high cost in parental energy; indeed, the greater the number of fledglings, the more food that needs to be captured and returned to the nest. Moreover, if clutch size becomes too large, parents cannot adequately provide for all the chicks in the brood, and fledgling mortality increases. So there are also costs associated with each strategy (each clutch size). Behavioral ecologists calculate the costs and benefits of each strategy in order to determine which of the available strategies maximizes the average ratio of benefits to costs – that is, in order to determine which is the optimal strategy. For example, behavioral ecologists may predict that, for a particular species of bird, five fledglings is the optimal clutch size. They then predict that the studied animals will pursue that optimal strategy, and they test their prediction against the actual behavior in a population of the studied species.

There are two points to note about modeling in behavioral ecology. First, the particular costs and benefits associated with a particular behavioral strategy depend heavily on the specific features of the environment in which that strategy is pursued. In an environment in which food is scarce and difficult to obtain the optimal clutch size will be smaller than in an environment in which food is abundant and easily obtainable. Thus, predictions regarding the optimal strategy in a population are always relative to the particular environment inhabited by the population. Second, although sometimes an animal's optimal strategy is independent of the strategies of other population members, at other times it is not. For example, for many animals there is an optimal amount of time spent foraging for food, which maximizes the energy intake per unit of foraging time, and this optimum is independent of the amount of time other population members spend foraging. However, when population members directly compete with one another for resources (including members of the opposite sex, who are reproductive resources), the optimal strategy for any particular population member will depend on the strategies of other population members. If most males competing for territory only engage in threatening displays and retreat when attacked, a tactic of extreme aggression may be greatly beneficial. But, if most males are extremely aggressive, then aggression could entail the costs of injury or death. So the costs and benefits of a behavioral strategy in a competition depend on the strategies adopted by other population members. In such cases, a population may be characterized by an *evolutionarily stable ratio* of alternative behavioral strategies.

When behavioral ecologists find that animals are, in fact, pursuing the strategy predicted by an optimality or evolutionary game theoretic model, they are confident that their model has correctly identified the selective forces in the environment to which animal behavior is responsive and the cost-benefit structures of the available alternative strategies in that environment. However, if animal behavior fails to conform to the predicted optimal strategy, behavioral ecologists assume that the model needs to be revised. Models can be revised by altering the set of strategies assumed to be available to the population or by changing the costs and benefits associated with the strategies in the set. But when a model does not accurately predict behavioral strategies, behavioral ecologists typically assume that the model has failed to include some variables to which animals are responding in "choosing" a behavioral strategy. In particular, behavioral ecologists typically assume that the studied animals are not pursuing the predicted strategy because of a *trade-off* among competing life demands. The assumption is that the need to maximize another currency places constraints on the ways in which population members can pursue maximization of the currency in the model.

In fact, for the typical animal, life is little more than a series of trade-offs (Laland and Brown 2002, 117–18). In very general terms, animals face a trade-off between somatic effort (effort expended toward bodily growth and maintenance) and reproductive effort. Within the category of reproductive effort, there is a trade-off between mating effort (effort expended to increase the number of offspring) and parenting effort (effort expended to care for already produced offspring). And, within the category of parenting effort, parents of two or more offspring face a trade-off between caring for one offspring and caring for another. Accordingly, when animal behavior fails to conform to the predictions of a model, behavioral ecologists typically assume that the animals are trying to simultaneously maximize several currencies and that efforts to maximize one currency place constraints on efforts to maximize another. "Unlike a robot designed to excel at sweeping or stamping," behavioral ecologists believe, "natural selection is unlikely to design organisms to maximize outputs of any particular task; rather, selection should favor organisms that optimize these abilities (trade off amounts and efficiencies in each), thus maximizing their chances of surviving and reproducing" (Smith, Borgerhoff Mulder, and Hill 2001, 130). Thus, the presupposition underlying modeling in behavioral ecology is that selection has designed animals to achieve an *optimal allocation of effort* among competing life demands. In the ideal limit, then, behavioral ecology aims to provide a set of interconnected models showing how animal behavior strikes the optimal compromise in pursuing all evolutionarily significant currencies.

Human behavioral ecology is simply the application of these ideas to humans, and it thereby involves several theoretical commitments regarding human behavior. First, human behavioral ecology assumes that human decision making is flexibly responsive to current environmental conditions, resulting in the choice of behavioral strategies that will optimize the allocation of effort among competing life demands and maximize lifetime reproductive output relative to the constraints imposed by the environment (Borgerhoff Mulder 1991, 70). As a result, second, human behavioral ecology sees behavioral differences between individuals as adaptive responses to differing environmental conditions. Human behavioral ecology thus seeks "to determine how ecological and social factors affect behavioural variability within and between populations" (Borgerhoff Mulder 1991, 69). Accordingly, human behavioral ecologists often interpret human behavior as the result of *conditional strategies*, behavioral strategies of the form "In environmental conditions A, do x; in conditions B, do y; in conditions C, do z" (Smith et al. 2001, 128). Third, human behavioral ecologists assume that human behavior is adaptive across a very wide range of environmental conditions, including many environmental conditions to which our species was never exposed during its evolutionary history. Thus, whereas Evolutionary Psychology expects human behavior to be frequently maladaptive in contemporary environments (because evolution in our psychological adaptations is lagging behind the rapid changes in post-Pleistocene human environments), human behavioral ecology expects human behavior "to be well-adapted to most features of contemporary environments, and to exhibit relatively little adaptive lag" (Smith 2000, 30).

In addition, human behavioral ecologists believe that adaptive behavioral responses can be produced and reproduced by a variety of different mechanisms. The same adaptive behavior could be achieved by one individual through the output of an innate module, but by another individual as the result of domain-general learning. Moreover, the same adaptive behavior could be genetically transmitted across generations through genes for modules or learning biases, through direct teaching by others, or through indirect cultural transfer of learnable information. Since adaptive behavior can be achieved through a variety of different mechanisms, human behavioral ecologists adopt a methodological strategy known as the *phenotypic gambit*: They ignore details about underlying mechanisms (which are typically not known anyway) in the belief that these details will not matter with respect to understanding human behavior. That is, human behavioral ecologists believe that a focus on evolutionarily significant ecological conditions, and the adaptive demands these place on humans, will enable them to understand why humans behave as they do even in the absence of knowledge of the mechanisms responsible for producing that behavior (which, in any case, may vary from one individual to another). Thus, human behavioral ecologists are "generally agnostic about mechanisms (including the question of cognitive modularity)" (Smith 2000, 30).

To illustrate these principles of human behavioral ecology, consider the phenomenon of *polyandry*, a marital system in which one woman has more than one husband. Nearly all systems of marriage in ethnographically recorded human societies are either monogamous or polygynous (in which one man has more than one wife). But, of 849 recorded societies, polyandry is practiced in four, all of which are located in the Himalayan highlands (Borgerhoff Mulder 1991, 82). At first glance, polyandry appears to defy evolutionary logic, for a woman's lifetime reproductive output is limited by the number of pregnancies she can carry to term, whereas a man's lifetime reproductive output is limited only by the number of women he can impregnate. At the theoretical limit, a woman can achieve her maximal reproductive output with a single mate, whereas a man can achieve his maximal reproductive output only with multiple mates. Thus, polyandry appears to entail no reproductive benefits for women, while involving a vastly suboptimal reproductive arrangement for men. From an evolutionary cost-benefit standpoint, polygyny would appear to provide the greatest benefits for men, while nonetheless allowing women to achieve their maximal lifetime reproductive output. So why would men ever agree to enter a polyandrous marriage?

Human behavioral ecologists study polyandrous populations with an eye to understanding the ecological factors that may make polyandry an adaptive choice, and they have identified several ecological factors that may affect the cost-benefit calculations in the decision making of those who enter polyandrous marriages. Human behavioral ecologists have discovered that polyandrous marriages are typically fraternal – that is, marriages in which one woman is married to two or more brothers. This helps, in part, to offset the costs of polyandry to the cohusbands, since their resources are pooled to rear only offspring to which they are all genetically related. Moreover, human behavioral ecologists have also discovered that polyandry typically occurs among brothers who have inherited farmland that is too small to be divided into parcels that could each sustain a family. In addition, farming the inherited land is highly labor intensive, so that no one of the brothers could successfully cultivate it in order to support a family. Finally, where polyandrous marriages occur, there are not alternative sources of income available to the brothers; cultivating the family farm is the only available means of subsistence. Thus, human behavioral ecologists have concluded, polyandry pays brothers under such circumstances, since they do better by maintaining joint possession of the farm, working it together, marrying one woman, and rearing their joint offspring than they would do by trying to go their own ways (Borgerhoff Mulder 1991, 84). As further confirmation of this hypothesis, human behavioral ecologists have discovered that when alternative sources of income sufficient to raise a family became available, younger brothers typically leave their polyandrous marriages in order to start a family of their own (Laland and Brown 2002, 123–24). Human behavioral ecologists therefore believe that polyandry is an adaptive marriage system in the ecological conditions of those who choose it.

COMPLEMENTARY RESEARCH PROGRAMS OR COMPETING PARADIGMS?

There are several apparent differences between Evolutionary Psychology and human behavioral ecology (summarized in Table 13.1). First, whereas Evolutionary Psychology strives to discover psychological adaptations to Pleistocene environments, human behavioral ecology studies human behavior and how it is adaptively responsive to ecological conditions. Second, Evolutionary Psychology expects human behavior to be frequently maladaptive in contemporary environments, because of adaptive lag in psychological evolution,

Table 13.1. *Comparison of Evolutionary Psychology and Human Behavioral Ecology*

	Evolutionary psychology	Human behavioral ecology
What is evolutionary theory employed to explain?	Psychological adaptations	Adaptive behavioral strategies
Is contemporary human behavior generally adaptive?	No	Yes
From what vantage point are evolutionary principles applied?	Our Pleistocene past	The present
Committed to massive modularity?	Yes	No
Committed to a universal human nature?	Yes	No

while human behavioral ecology expects human behavior to be fairly well adapted to contemporary environments. Accordingly, third, Evolutionary Psychology believes that the evolved nature of the human mind must be "reverse engineered" from the vantage of our species's Pleistocene past, whereas human behavioral ecology believes that evolutionary principles can be applied in studying human behavior in contemporary environments. Fourth, whereas Evolutionary Psychology postulates that human behavior is caused by hundreds or thousands of modules, which are special-purpose minicomputers adapted to specific adaptive problems faced by our Pleistocene ancestors, human behavioral ecology is agnostic about the nature and number of psychological mechanisms that are causally responsible for adaptive human behavior. Finally, Evolutionary Psychology strives to discover a universal human nature underlying behavioral differences between cultures and between individuals in the same culture, whereas human behavioral ecology studies how environmental differences between individuals affect behavioral differences between them.

Some have argued that these differences between Evolutionary Psychology and human behavioral ecology are more a matter of explanatory emphasis than substantive scientific disagreement (Smith 2000, 33–36; Laland and Brown 2002, chap. 8). According to this ecumenical view, Evolutionary Psychology is simply the investigation of the psychological mechanisms about which human behavioral ecology remains agnostic in its focus on behavior. Whereas human behavioral ecology studies the behavioral "outside" of the human organism, Evolutionary Psychology studies the behaviorally generative psychological "inside." So, while human behavioral ecology aspires to explain how our behavior is adaptively responsive to our ecological conditions, Evolutionary Psychology aspires to explain how our psychological adaptations cause that behavior. Similarly, the argument goes, human behavioral ecology seeks to explain how variation in ecological conditions affects behavioral variation both within and between populations, whereas Evolutionary Psychology seeks to explain how this behavioral variation is caused by a universal human nature responding differentially to differing environmental conditions. Thus, it is possible to see Evolutionary Psychology and human behavioral ecology as offering complementary, rather than competing, explanations.

But the ecumenical position greatly exaggerates the extent to which the two research programs are compatible. For, while human behavioral ecology is compatible with *some* evolutionary account of the mechanisms underlying the behavior it studies, it is not compatible with Evolutionary Psychology's account of those mechanisms.

To see why, reconsider Evolutionary Psychology's massive modularity hypothesis. According to this hypothesis, each adaptive problem our lineage faced in its Pleistocene past was solved by a dedicated module; adaptive behavior was achieved in each problem domain by an "expert system," which was designed to achieve adaptive performance in its problem domain, but was ineffective outside its area of expertise. Indeed, Evolutionary Psychologists claim, humans often fail to behave adaptively in contemporary environments because modules, with their "tunnel cognition," are incapable of functioning effectively when not encountering precisely the conditions for which they were designed.

This contrasts sharply with human behavioral ecology's presupposition that humans can flexibly alter their behavioral strategies so as to strike optimal trade-offs among numerous adaptive problems. According to human behavioral ecologists, "effective adaptive design requires integrative mechanisms for measuring tradeoffs (which themselves vary in complex and contingent ways), and adjusting behavior according to the weighted effect of different activities" on reproductive success (Smith et al. 2001, 130–31). But, since the problem of striking the optimal trade-off *across* adaptive problem domains is not a problem *in* any of those domains, no domain-specific module could weigh the costs and benefits of alternative trade-off strategies and adjust behavior accordingly. Such strategic trade-offs could be struck only by some *domain-general* psychological mechanism. Moreover, that domain-general mechanism could not simply be a mechanism that turned on and off the various modules that are relevant to one's circumstances, leaving the modules to solve their own problems in their own ways. It would have to be a domain-general mechanism that could adjust behavior *within* each adaptive problem domain in a way that allowed for the optimal allocation of effort and efficiencies among competing demands. But any psychological mechanism capable of adjusting behavior within problem domains so as to strike optimal trade-offs would not need to be supplemented with mechanisms that are specialized for functions within each problem domain. Thus, although human behavioral ecology's presupposition that humans can flexibly alter behavior so as to optimize trade-offs is compatible with *some* account of the psychological mechanisms that make such adaptive trade-offs possible, it is not compatible with Evolutionary Psychology's massive modularity hypothesis. If human behavioral ecology is right about the flexible adaptiveness of human behavior, Evolutionary Psychology is wrong about the psychological mechanisms underlying that behavior.

This substantive difference between the two research programs is related to another difference, which Evolutionary Psychologists have taken to be substantive, but is only partly so. Because human behavioral ecology seeks to explain adaptive human behavior, while remaining agnostic about the psychological adaptations underlying that behavior, Evolutionary Psychologists have often claimed that it is not a genuinely evolutionary theory of human behavior. According to Evolutionary Psychologists, the theory of evolution by natural selection is a theory of *adaptation*, so "nothing in the theory of evolution by natural selection justifies an adaptation-agnostic science of adaptiveness" (Symons 1992, 150). Since Evolutionary Psychology's goal is to discover human psychological adaptations, it claims to be the only truly evolutionary theory of human behavior and psychology.

But human behavioral ecology is not agnostic as to *whether* adaptations underlie the adaptive behavior it studies. Indeed, human behavioral ecologists assume that "human decisions are guided by complex processes of observation, evaluation, recalled experience, experimentation and strategizing which ... have themselves been shaped by past selection pressures" (Borgerhoff Mulder 1991, 70). In this respect, human behavioral ecologists are no more agnostic about adaptations than Evolutionary Psychologists. Human behavioral ecology merely refuses to commit itself to hypotheses regarding the precise causal mechanisms comprising the adaptations that underlie adaptive human behavior. There are two reasons for this restraint, and these reasons substantively differentiate human behavioral ecology from Evolutionary Psychology.

First, human behavioral ecologists believe that the best way to discover human psychological adaptations is to study the ways in which humans respond adaptively to their ecological conditions rather than to attempt to "reverse engineer" them from the vantage of our Pleistocene past (Smith et al. 2001, 131–32). Thus, we will achieve knowledge of the adaptations underlying human behavior only after we understand the actual decisions humans make regarding survival and reproduction. Second, the belief that humans can flexibly alter their behavior, so as to behave adaptively even in evolutionarily novel environments, presupposes that human psychological adaptations are mechanisms of *adaptive plasticity* (in which a single genotype produces more than one phenotype by responding appropriately to environmental conditions). Since mechanisms of adaptive plasticity are not yet well understood, human behavioral ecologists currently treat them as "black boxes" in their studies of human behavior. But whatever the details about the causal workings inside such black boxes, mechanisms of adaptive plasticity contrast sharply with Evolutionary Psychology's modules, which are functionally specialized to produce particular forms of behavior and which develop reliably across a broad range of environmental conditions. Consequently, while human behavioral ecology does not presuppose hypotheses about specific psychological adaptations, it differs from Evolutionary Psychology regarding the *kinds* of adaptation that underlie human behavior.

Thus, although human behavioral ecology can be fruitfully supplemented with explanations of the nature and evolution of the psychological mechanisms underlying adaptive human behavior, its theoretical commitments preclude being conjoined with the particular explanations that Evolutionary Psychology offers. Despite some superficial complementarities, human behavioral ecology and Evolutionary Psychology are actually competing paradigms rather than complementary research programs. As the theoretical and empirical fortunes of one program wane, those of the other will wax.

CONCLUSION

The widely popularized research program of Evolutionary Psychology is not the only game on the field of evolutionary psychology. Indeed, human behavioral ecology is a vibrant alternative paradigm for understanding human behavior from an evolutionary perspective. And, since much recent research has detailed numerous problems with the theory and methodology of Evolutionary Psychology (see, for example, Buller 2005), human behavioral ecology is the paradigm that holds the greatest promise for the future of evolutionary psychology.

14 Neurobiology

Most of the issues found in traditional philosophy of science are recapitulated in the philosophy of neurobiology. In particular, philosophers of neurobiology worry about what counts as appropriate empirical justification for a theoretical claim, how to determine which level of organization is the correct one for a scientific explanation, what explanations should look like, whether all explanations will or should reduce to some primitives, and how what we learn about the mind/brain should affect larger social, economic, and political decisions.

In addition, philosophers of neurobiology concern themselves with some traditional aspects of philosophy of mind, including worrying how it is a brain can represent, if it does, and how and whether this representation ties to other notions of representation in cognitive science and beyond. It is difficult to focus on only one of these concerns to the exclusion of the rest. Most likely, as we come to understand some particular aspect of the practice of neurobiology, we will also understand others as well. In what follows, I discuss these areas of concern as they *differ* from traditional arguments. This discussion therefore should be laid on top of and be seen to complement the very rich literature in traditional philosophy of science and philosophy of mind.

1. THEORIES IN NEUROBIOLOGY

Brains are complicated and messy affairs; theories about brains share these same traits. The difficulty is that in order to make a simple generalization about how some aspect of the brain functions, scientists have to retreat to such a broad level of abstraction that their assertions become almost empirically meaningless. In order to

275

make their claims testable in a laboratory, neurobiologists have to confine their ideas to particular animals, to particular experimental tasks, or to both. As a result, scientists end up with neurobiological "theories" that contain two distinct parts: a broad statement of theoretical principle and a set of detailed descriptions of how that principle plays out across different animal models and experimental tasks. Though the detailed descriptions fall under the general principle, they are not immediately derivable from it. Moreover, the detailed descriptions can be incompatible with one another, though each will maintain a family resemblance with the others. (See Hardcastle 1995, Schaffner 1993, Suppe 1989 for similar approaches to understanding theories in the biological sciences.)

At a gross level, mammalian brains are remarkably similar to one another. Indeed, the central nervous system (CNS) in invertebrates is not all that different from the mammalian CNS either. There are innumerable homologous areas, cell types, neurotransmitters, peptides, chemical interactions, and so forth. However, once we scratch the surface of different animal brains, we do find important differences.

For example, consider the semicircular canal. All mammals have roughly the same five end organs in their ears to support their auditory and vestibular systems, and they all work to keep their lateral semicircular canals in their ears parallel to the horizontal plane relative to the Earth, for keeping it in that position allows them to get the best possible information about head position in space. (The lateral canal is maximally excitatory to a yaw (left-to-right) head motion; keeping the canal in line with the horizontal plane allows the organ to detect this motion with the greatest accuracy.) But rodents ambulate with their necks extended, which keeps their heads in an extreme dorsal position, while humans incline their heads about twenty degrees when walking naturally. In general, scientists can correlate the differences in the shape of the semicircular canals in the ear with skull shape and the position that an animal's head is normally in. (It is an unanswered but intriguing question whether scientists find the canal structures they do because different heads evolved to be oriented in different directions or whether animals naturally hold their heads in different positions because their semicircular canals evolved differently.)

For another example, consider the retina. There are striking differences between herbivores and predators in brain structure, for creatures who munch on grasses and trees require much less precise environmental information than those who hunt moving targets in order to survive. As a result, rodents have no foveae. To maintain visual fixation on a point, they move their necks, using what is known as the vestibular-colic response. The vestibular system in their ears tells them how their head is oriented and they use that information to reorient their heads in order to keep whatever object currently fascinates them in their line of sight.

In contrast, primates have foveae and they move their eyeballs to keep their target within the foveal area, using the vestibular-ocular response. This is a much more precise orienting mechanism, which allows them to move their eyes to compensate for changes in head position such that they can keep objects foveated for as long as they wish. For some indication of how important computing horizontal eye motion is to our brains, consider that the abducens (or sixth) nerve in humans, which controls horizontal eye abduction, feeds into one of the biggest motor nuclei in the brain stem. This ocular nucleus, which controls only one very tiny muscle, is only slightly smaller than the nucleus that controls all of our twenty or so facial muscles.

In more striking contrast still, bats do not maintain ocular position in the same fashion as the rest of the mammals. Because they fly and so have greater freedom to move in three-dimensional space, maintaining body position relative to horizontal is not an easy option. As a result, they use other sense organs, primarily hearing (the other half of the eighth nerve), to determine how their eyes should be oriented. Consequently, they need not rely on vestibular-ocular responses as we do, even though their bodies are equipped with such reflex machinery.

All of these anatomical and physiological differences are important when neurobiologists want to investigate something like the way the brain learns to compensate for damage to the vestibular pathways. What may seem as small and insignificant differences from a broad mammalian perspective become hugely important as scientists seek to understand the particular mechanisms of brain plasticity. Can they use animals with no foveae and a vestibular-colic response to learn about how foveated mammals recover their vestibular-ocular response? More generally, they need to know how well particular animal models translate across the animal kingdom. Should they be allowed to generalize from experiments on a single species (or set of species) to the way nature functions?

In all vertebrates, a unilateral labyrinthectomy (UL), or a lesion of the labyrinthine structure in one ear, gives rise to two types of ocular motor disorders. There are static deficits, such as a bias toward looking toward the lesioned side when the head is not moving, and dynamic deficits, such as abnormal vestibular-ocular reflexes (VORs), which occur in response to head movements. In only two or three days after the UL procedure, the brain starts to compensate for its loss and the static deficits disappear. Since labyrinthine structures do not regenerate, and peripheral neurons continue to fire abnormally, whatever the brain is doing to recover has to be a central effect (Shaefer and Meyer 1974). Single neuron recordings from a variety of animals indicate that the vestibular nuclei (VN) on the same side of the brain as the lesion start to show normal resting rate activity as the brain learns to compensate for its injury. Scientists do believe that whatever the mechanism is, it is also likely to be a general procedure the brain uses for recovery, for they find similar resting rate recoveries of the sort they see with the ipsilateral vestibular nuclei after denervation in the lateral cuneate nucleus, the trigeminal nucleus, and the dorsal horn, among other areas. Exactly how an argument to defend these convictions is supposed to run, though, is unclear, since it is fairly easy to find significant differences in the ways organisms recover and compensate for vestibular damage across the animal kingdom. Frogs, for example, appear to rely on input from the intact labyrinth to regulate the resting activity of the vestibular nuclei. Mammals, however, do not. The recovery of their vestibular nuclei occurs independently of transcommissural inputs (Flohr et al. 1981). In addition, static symptoms follow different time courses in different animals. In rats, spontaneous nystagmus disappears within hours after UL, while in the rabbit and guinea pig, it persists for several weeks (Baarsma and Collewjin 1975, Sirkin, Precht, and Courjon 1984). In humans, it may continue in one form or another for several years (Fisch 1973).

There is a fundamental tension in neurobiology between the big picture story and what is found in particular instances. All sciences strip away features of the real world when they devise their generalizations. Physicists neglect friction; economists neglect altruism; chemists neglect impurities, and so on. However, what neurobiologists are doing is not analogous to what physicists, economists, and chemists are doing. In each of the other cases, the scientists are simplifying the number of parameters they must consider in order to make useful and usable generalizations. In contrast, if neurobiologists were to ignore the differences they find across species, then they would have no data left to build a theory with. There is not anything left over, as it were, once neurobiologists neglect the anatomical and physiological differences found in the brain across the animal kingdom. There is much left over when physicists neglect friction; most of classical mechanics is left, in fact. In distinction to the other sciences, in neurobiology we find a tension between the general rules one hopes to find that describe all brains and the particular cases neurobiologists happen to study.

What should the scope and degree of generalization for neuro-biological theories be? It appears scientists are confronted with an unpleasant choice. Either they settle for large-scale abstract generalizations, which gloss over what may be important differences, or they focus on the differences themselves, at the expense of what may be useful generalizations. However, despite appearances, they do not have an either-or proposition that they have to resolve before they can move ahead, for a proper neurobiological theory contains both general (and fairly vague) abstractions and detailed comments on specific anatomies and physiologies. The paradigm theories for physics are simple elegant equations with universal scope. Theories in neurobiology read more like a list of general principles plus detailed commentaries. One feels the tug of the dilemma posed above only if one is operating with a restricted notion of what a scientific theory is. Some theories are pithy and succinct; some are not. Neurobiological theories are of the latter sort.

In neurobiology, scientists start with a theoretical description at the most general level; it is what we might call the "theoretical framework" – the most general component in a neurobiological theory. Once they adopt the framework, they can make more precise hypotheses as a way of filling out their theoretical proposal. These claims can be local to particular phyla or species; hence, they are not intended to be a more detailed specification of the general framework. Instead, they can be thought of as instances or examples of how the framework might be cashed out in particular cases.

However, it is not the case that all "fillings out" fail to generalize. For example, the dynamic symptoms of UL recover by using a different mechanism (probably). One hypothesis is that brains use a form of sensory substitution to compensate for the vestibular-ocular reflex (Berthoz 1988, Miles and Lisberger 1981). In this case, the brain uses internally generated signals from the visual or somatosensory systems to compensate for the vestibular loss. It may substitute computations from the saccadic or a visual pursuit system, both of which (probably) reconstruct head velocity internally, for vestibular throughputs. Data drawn from experiments on frogs, cats, and humans indicate that they all apparently use the same mechanism, though it remains to be seen whether this proposal will be applicable to all creatures and whether it can be generalized much beyond vestibular reflexes.

There are different degrees of abstraction one might use once some theoretical framework is adopted. Some discussions are going to be restricted to a single species, or maybe even one developmental stage within a species; others will include several unrelated species or phyla. Both are legitimate ways of cashing out the framework in particular instances, and neither is to be preferred to the other. The data will dictate the scope of subhypotheses, and scope can vary dramatically.

And this is how theories in neurobiology are built and structured. Detailed conclusions regarding a single animal model give rise to general theoretical principles. These principles inspire new experiments done with other animal models, which in turn give us new (and probably incompatible) details but also new general principles. These new principles then connect to other detailed studies using different protocols on still other animals, and so it goes.

At the end of the day, we have a set of related theoretical principles that jointly compose a general theoretical framework. And these principles are held together by the detailed data from a wide variety of animal studies. Neurobiology continually moves between two different ways of understanding the nervous system, first in broad and sweeping strokes and second by submergence in the minutiae. General theoretical principles arise out of and then feed back into particular animal experiments done on different animal models. Because physiology differs across species, specific experimental protocols are appropriate only for specific models. Sometimes the data arising out of the different animal models and different experimental procedures overlap, but largely they do not. Hence, sometimes the detailed conclusions are consistent, but sometimes – a lot of the time – they are not. Neurobiologists weave a story through their animal models and experimental protocols united by a common guiding theoretical thread. They both find commonalties and define differences. And this entire exercise, taken together, fashions the theoretical structure of neurobiology.

2. THEORY-LADEN OBSERVATIONS AND SINGLE-CELL RECORDINGS

It is almost a truism in philosophy of science that there is no unproblematic distinction between observation and theory. That is, any scientific observations we make are filtered through and by a prior theoretical framework. Raw data become observations as we interpret the ways they either fit or belie our hypotheses (Woodward 1989). In short: what counts as an observation and how that observation functions in the business of science are heavily mediated by theory. In neurobiology in particular, it is easy to change the fundamental nature of our observations using accepted methodological techniques for manipulating raw data.

Good data allow scientists to discriminate among competing claims about phenomena (Suppe 1989). The particular practices of the scientific subfield tell us how to judge whether data are good. Sometimes these practices involve explicit calculations and formal derivations; sometimes they involve matters of personal judgment and skill. The cases in neurobiology involve both. In particular, it is a matter of personal judgment in the world of single-cell recording when to employ certain computational procedures. Different sorting techniques give rise to different data, so which techniques to employ is an important question. But that is also a question for which no easy or accepted answer exists.

It has only been during the last decade or so that neurobiologists have been able to record from the extracellular space of a large number of neurons from awake and behaving animals. When they record with an electrode near a single cell, they do pick up the cells' action potentials, which are commonly believed to be the means through which neurons communicate. But they also record things that look like action potentials, but are instead voltages generated by axonal bundles or the field potentials from parallel sets of dendrites. Moreover – and especially if the microelectrode has a relatively low impedence – extracellular electrodes pick up signals from several neurons at the same time, recording from all the cells in a nearby area.

The problem is how to differentiate the contributions of the different cells and cell parts from a single lump recording. In many cases scientists only care about one particular action potential; the rest, from their perspective, is background noise. The challenge is how to separate what they want from all the electrical signals they do not want. The challenge is how to move from the recordings of the electrode's output to genuine, reliable, and informative data.

This challenge is compounded by the noisy nature of the recordings themselves. Some of the noise is mechanical and arises from the amplifiers themselves, but some is biological and comes from the neurons. Brain cells jitter around constantly (cf. Connors and Gutnick 1990). Neurons are not quiet until they fire off a spike, as some might think. Instead, they are always producing some activity or other. All in all, scientists have to cull their data from quite a din.

Finally, because the components in a recording are not constant, it is difficult to get a theoretical hook into the waveform. Spike shapes can change over time; electrodes can drift during recording session, changing position relative to the cells, which would also alter the spike amplitudes; and the electrical properties of electrodes vary with changes in tip condition or background impedance. Gathering data from single unit activity presents neurobiologists with a serious technical challenge.

In order to get usable data – to get genuine observations – out of what the electrode transmits, scientists must isolate each neuron's contributions to the recorded waveform. They first need to ascertain exactly how many neurons the recorded waveform reflects. How can they do this if they have a mess of overlapping action potentials and field potentials from a variety of cells at different and unknown distances from the electrode? This question becomes particularly vexing if other neurons in the same area have spikes of the same or a similar shape and amplitude.

There are several decomposition algorithms; however, each is imperfect (see Lewicki 1998). Each represents a different way to move from raw output to interpreted and interpretable data, giving scientists different ways of refining the waveforms they have recorded so that they can later interpret them. Each is what philosophers are thinking about when they talk about the theory-ladenness of data. Scientists have to choose what to do with their measurements in order to get something that can be scientifically useful. And the way they choose is determined by previously accepted theories.

But even with all these advanced sorting techniques, it is still hard to predict the number of neurons eliciting the data. Ideally, scientists would like to claim that one neuron generates each cluster of spikes we have identified, but if the cells are firing in complex bursts, or if there is nonstationary noise, or if the spike trains overlap one another, they cannot get accurate classifications at all. It is simply an unsolved problem how to decompose coincident action potentials with variable spike shapes. The best scientists can do at this point is guess. Their guesses are informed by their years of experience, but they are guesses nonetheless.

Guessing is not quite what philosophers of science have in mind when they talk about the theory-ladenness of observation. Their vision of creating data is one of more "scientific method." That is, to pull data out of the dial movements or changes in color or squiggles on the page, philosophers generally hold that there is some explicit background theory, devised in some other scientific inquiry, that scientists learn and then use to interpret what they are seeing or measuring as something useful for their studies. But there is a theoretical gap, as it were, in the move from raw recordings to genuine data, a gap scientists cannot fill with any sort of decision-making algorithm. The best scientists can do at this point is simply leap across the gap, on blind faith, with an eye to where they want to go.

Neurophysiology travels in a cognitive circle; scientists use what they know to cull data that support what they believe to be the case. Nevertheless, progress is not stymied. Knowledge accrues in small increments, with each set of single-cell recordings altering the face of what is known a wee bit at a time. Because neurobiological sorting techniques rely so heavily on previously accepted neurobiological hypotheses, there will likely never be an abrupt or dramatic conceptual revolution. But what is known can evolve slowly but surely until the final resting position is quite far removed from the place where the investigation began.

3.   LOCALIZATION AND REDUCTION

When scientists do single-unit recordings from a set of neurons they assume that they are examining a discrete system. They have been wildly successful using this strategy, identifying at least thirty-six different topographical visual processing areas in cortex (De Gelder 2000), differentiating the "what" from the "where" object processing streams (DeYoe and Van Essen 1988; Mishkin, Ungerleider, and Macko 1983), and distinguishing motion detection from contour calculations (Barinaga 1995), to name but a few examples. Maps of brain function are getting more and more complicated as more and more is learned about the processing capacities of individual cells. And all these projects are founded on the belief that brains have discrete processing streams that feed into one another.

Yet the most neurons scientists have ever been able to record from simultaneously are a few hundred; the most cells they can ever see summed local field potential activity over are a few thousand. But brain areas have hundreds of thousands of neurons, several orders of magnitude more than can be accessed at any given time. And these neurons are of different types, with different response properties and different interconnections with other cells, including other similar neurons, neurons with significantly different response properties, and cells of completely different types. Any conclusions scientists draw about the behavior of whatever cells they are recording from are going to be limited to very basic stimulus-response and correlation analyses of whatever neuronal subtype they are currently examining. Hence, the functionality they ascribe on the basis of these relatively meager sorts of experiments might be much more restricted than what the cells are actually doing.

They insert an electrode in or near a cell and then record what it does as they stimulate the animal in some fashion. They record from a cell in a vestibular nucleus and then move the animal's head about to see whether doing so changes the activity of the neuron. If it does, then they move it more or they move it differently and see how that changes the neuronal output. If it does not, then they either try another nearby cell or try some other stimulus. But what they cannot do is record from all the neurons in some isolated area, even if the area is very small. And what they cannot do is test any given cell for all the known functional contributions of brain cells in general. So, what they conclude about any cell will only reflect the cells they have actually recorded from using stimuli they have actually used. This research strategy systematically underestimates when neurons actually respond and under what conditions.

Unit studies attempt to combine scores, hundreds, or even thousands of single-unit recordings to try to analyze the population. Theoretically, scientists could perhaps, in principle, delineate a nervous system region stereotaxically if it had reproducible correlations between afferent and efferent connections such that they could ultimately articulate the neurobiological function of the defined region. However, the likelihood of success for this type of study decreases as the complexity of the organism increases. Scientists can draw functional conclusions regarding the activities of neurons in the abdominal ganglia of *Aplysia*, or the segmental ganglia of the leech. But the architecture of these organisms' central nervous system is so different from mammals' that the probability of successfully using similar techniques for understanding humans is very low to zero.

In addition, the actual processing of information that goes on in those cells involves lots of different kinds of excitatory and inhibitory inputs from other areas in the brain stem, cerebellum, and cerebral cortex. The dorsal horn is supposed to integrate afferent nociceptive information from the periphery and pass it onto the motor system (among other things), but it does not do that segregated from the rest of the brain and what the brain is trying to do. It is integrating and passing as the organism is trying to pursue prey or flee from an enemy. Moreover, the brain regions that perform these tasks are often connected to the very area scientists are recording from. The motor system feeds back down into the dorsal horn, as do the thalamus and significant parts of cortex.

The impact on cognitive processing of such rampant feedback connections in the brain is only just now starting to be explored in neurobiological research, though exactly how to do this is a difficult question to answer. Of course, neurobiologists design their experiments keeping in mind the known anatomic connections between and among the relevant structures. At the same time, any actual experimental observations of all the remote influences on the dorsal horn, for example, are impossible, despite however many individual neurons scientists record from. They simply do not have any way of conducting such extensive, invasive tests on live animals. At best, the particular influences assumed in any particular recording series are a matter of previously accepted gospel, dogma, and faith.

Ideally, neurobiologists try to conjoin single-cell studies with some sort of lesion experiment. Once scientists construct a general flowchart of the relevant structures based on anatomy experiments, and they have estimated normal unit behavior from a series of single-cell studies, they then try to knock out the hypothesized functions by placing lesions in otherwise normal animals. They run their experiments on the basis of the assumption that these lesions, placed in regions known to be important, will change the unit behavior of cells they are studying in a consistent fashion. If they witness such a change, they use that information to explain the relative functional contributions of the lesioned region to the cells under scrutiny. In other words, they are using lesion studies to try to derive a functional boxology for the brain, just as cognitive psychologists use reaction time distributions and error measurements to find one for the mind.

But there is a larger theoretical concern. What neurobiologists know, but generally ignore, is that any functional change in the central nervous system will lead to compensatory changes elsewhere (e.g., Merzenich et al. 1983). Because the brain is highly plastic, lesioning it in one place will provoke it to react in some fashion in some other place. Usually these other places are not components in the system or region being studied. But even if they are, neurobiologists ignore plasticity of the brain in favor of assuming a consistent functional alteration as caused by the lesion and nothing more. How are investigators supposed to evaluate some observed functional change when the difference they see might have been evoked by the brain's attempt to compensate for its loss and not by any specific deficit induced by the lesion?

The short answer is that they cannot if they are restricted to single-cell recordings and lesion studies. To answer this question we need to be able to see the activity of the entire brain at once and over time. The excitement over functional magnetic resonance imaging (fMRI) and other imaging techniques concerns exactly this point: there is a way of looking at the activity of the whole brain at one time as tied to some cognitive activity or other. But magnetic resonance imaging, the best noninvasive recording device we currently have, only has a spatial resolution of about 0.1 millimeter and each scan samples a few seconds of activity. This imprecision forecloses the possibility of directly connecting single-cell activity – which operates three to four orders of magnitude smaller and faster – with larger brain activation patterns.

Methodological difficulties with current imaging techniques are now well known (Bechtel 2000, Cabeza and Nyberg 1997, 2000). Most center around the fact that MRI is a blood-oxygen-level-dependent (BOLD) measure, which can only be imperfectly correlated to brain activity. That is, MRI measures changes in the oxygenation level of blood; it does not directly measure anything about actual neuronal activity. Others are tied to the fact that the measure cannot differentiate between inhibitory and excitatory activation, and that can confound the way the images are interpreted. An area might be "read" as being part of the processing stream for some input, even though what is showing up in the MR analysis is that area actively damping down activity. A third set of limitations is tied to the sparse distribution of some processing systems. If a system – nociception in somatosensory cortex might be one example – is widely but sparsely distributed, then its activity level might never reach what is required for a BOLD measure to notice, given that cells surrounding the system are not activated by the particular stimulus in question.

The final set of concerns revolves around the subtraction method used in imaging studies to cull data. In brief, here is how that technique works. The experimenter picks two experimental conditions that she believes differ along only one dimension: they differ only with respect to the cognitive or perceptual process she wants to investigate. She then compares brain activity recorded under one condition with what happens in the second condition, looking for regions whose activity levels differ significantly across the two. These areas, she believes, constitute the neural substrates of the task under scrutiny. By subtracting one set of scans from the other, the hope is that one has removed activity not specifically relevant to the task at hand.

Let us set aside the fact that this method has no way of determining whether the differences found are actually tied to the cognitive process and not to something else occurring concurrently but coincidentally. Let us also set aside the fact that some activity might be both relevant to the task at hand and relevant to the baseline task. Notice that how well the subtraction method will work depends upon the sensitivity of the measuring devices such that the worse the instrument is, the better the method seems to be for localization studies. Low signal-to-noise ratio (SNR) means that scientists will find only a few statistically significant differences across conditions. And these are the sorts of results neurobiologists need in order to bolster any claims identifying particular cognitive processes with discrete brain regions.

But as the imaging technology improves and the SNR increases, scientists see more and more sites that differ across trials. The more sites they get, the more it appears that essentially the entire brain is involved in each cognitive computation. And the more it appears that the entire brain is involved in each thought, the less it is they can justify any assumption of functional specificity in the brain. If we extrapolate from what scientists might learn with more sensitive measures, we can easily see that there will be a time when this whole approach just will not work anymore. Put in the harshest terms, brain imaging seems to support reductionism because the science is not very good yet.

For example, Brodman area 6 appears significantly active after subtraction in studies of phonetic speech processing, voluntary hand and arm movements, sight-reading of music, spatial working memory, recognizing facial emotions, binocular disparity, sequence learning, idiopathic dystonia, pain, itch, delayed response alternation, and category-specific knowledge, to list only a subset of activities in which it is significantly and differentially active. It could be the case that if scientists keep on doing the sort of subtraction studies that they currently are doing, then eventually they will find a unifying and pithy way to describe what premotor cortex is doing in humans. In this instance, neurobiology would be on the right track to determining brain function, but they still have a long way to go. But it could also be true that how a region functions depends heavily on the "neural context." Its functional role in a cognitive economy depends on how it is connected to other areas and how those other areas are responding. (The function of these areas would also be dependent on their particular connectivity and the current patterns of activation. And so it would go.) If this is correct, then searching for "the" function of particular areas is misguided, for different brain regions play different roles depending upon the cognitive tasks at hand.

4.   NEUROETHICS

As progress is made into understanding how the brain works and how to influence brain functioning, serious ethical questions arise concerning how the medical, insurance, and governmental leaders should react to new information and possibilities (Marcus 2004). Neuroethics is a newly burgeoning area of research, with national attention only now being focused on the issues. Particular questions that philosophers of neurobiology will have to answer concern how and whether we should alter normal functioning brains, how and whether we should use brain technology to track individuals' social behavior, and how and whether what we learn about the brain changes the way we think of ourselves as human.

We know a lot about how memory works, and, more importantly, how it fails us. Seven basic ways in which memory can fail are decreasing accessibility to memories over time, lapses in attention, temporary inability to access stored information, false recognition of something, false memory of something, contamination of stored information by current beliefs, and remembering of items at inappropriate times. All of these processes are perfectly normal and occur in all of us at some time or another. Suppose we have some way of correcting some or all of these deficits. Should we? Or should we accept less-than-perfect memories as the way we are?

Neurobiologists are already tracking where and how moral decisions are made in the brain; they are also looking at brain differences between normal and sociopathic, psychopathic, and violently impulsive individuals. We know that such individuals respond to violent or otherwise disturbing situations with increased activity in the amygdala and decreased activity in the frontal lobes relative to normal individuals. We can now identify such trends in individuals before they actually commit any crime. Should we? And what should we do with such information once we have it?

If we come to believe that violence is biologically based, as are all other behavioral decisions, then what does this say about notions of self or free will? How might this alter our court systems, since they operate under the assumption that one is guilty if one could have done otherwise in a situation but chose not to? Similar questions arise with gender differences in the brains. We know that female brains differ from males'. What effect, if any, should this fact have on our educational systems, our social expectations of gendered behavior, or men's and women's professional lives?

We are only beginning to confront these sorts of questions, as our technology is only beginning to allow us to understand and change the brain to any significant degree. As our knowledge of the mind/brain continues to increase exponentially, these and other similar questions will only become more pressing.

15 Biological Explanations of Human Sexuality

The Genetic Basis of Sexual Orientation

"Do genes make people gay?" This is a question scientists who study human sexuality hear all the time from politicians, students, and people just curious about their own sexual orientation. Many people believe that the causation of sexual orientation has important implications for the moral status of homosexuality. Both common sense and moral theory tend to evaluate the acceptability of a person's behavior on the basis of the controllability of that behavior. One cannot be held morally blameworthy for things over which one has no control. Because individuals have no control over the genes they inherit, according to this theory they should not be held responsible for being gay if it is genetically determined. This view is apparent in surveys of Americans that have found an association between the belief that homosexuality is genetically determined and less negative attitudes towards gay people (Schmalz 1993, Tygart 2000).

The purposes of this essay are to review what the best science currently available says about the genetic causes of sexual orientation and to discuss some of the normative implications and, more importantly, nonimplications of that science. As with any issue in which there is substantial political and scientific controversy, there is ongoing debate about the very distinctions that lie at its foundation. Nevertheless, terms like "heterosexual" and "gay" are commonplace in both the popular and the scientific literature. My purpose here is not to debate the distinctions, but to discuss the best science we have available today as it is applied to the distinctions as commonly understood by both experts in the field and the public at large. I have structured this article around the sorts of questions students or people unfamiliar with recent work in the field might ask.

HOW DO SCIENTISTS DEFINE SEXUAL ORIENTATION?

The Kinsey Scale (Kinsey, Pomeroy, and Martin 1948) has become the standard way to operationalize sexual orientation in almost all scientific studies of the phenomenon. Study participants are asked a series of questions about their sexual desires, behaviors, fantasies, and self-identity. On the basis of the answers given to these questions, each participant's sexual orientation is classified somewhere along a unidimensional linear scale where 0 = exclusively heterosexual and 6 = exclusively homosexual.

Some researchers weight the answers to certain questions more heavily than others, depending on the nature of the study. Those who are interested in investigating sexual orientation as a set of erotic desires directing sexual attraction towards particular kinds of people usually weigh the answers about desires and fantasies higher than the answers about behavior. This would seem a perfectly reasonable bias. To say that virgins have no sexual orientation simply because they are virgins would be to fly in the face of hundreds of years of romantic literature as well as most people's personal experience. It is also obviously true that people who find themselves living or working within very homophobic environments might never act on their true sexual desires out of fear of reprisal or rejection.

Critics of biological explanations of sexual orientation argue that the conception of sexual attraction implicit in the Kinsey test is inadequate because it treats sexual orientation as a discrete trait with clearly defined boundaries (e.g., Stein 1999). In terms of the Kinsey test, everybody fits neatly into one of seven categories. Sexual orientation is clearly a continuous character. People are generally attracted to other people with varying degrees of intensity. One person might be highly attracted to other males and only slightly attracted to females. A second person might be highly attracted to both males and females. The Kinsey scale collapses attractions to both males and females into a single unidimensional scale and divides that spectrum into seven discrete categories. However, the points of division and the number of categories are arbitrary. In practice, since most researchers are only interested in people whose sexual orientation lies on one or the other of the extremes, they collapse the seven Kinsey categories into a more easily managed two or three.

It is possible to use the same Kinsey-style instrument focusing on attraction and behavior, but measure sexual orientation as a continuous variable (Ellis, Burke, and Ames 1987). This procedure results in a particular pattern of distribution consistent with the results of the more standard Kinsey test. Most people cluster around the "heterosexual" end of the distribution, while there is a smaller cluster at the other, "homosexual" end. Also as in the more standard Kinsey test, there are more males than females at the homosexual end of the spectrum. Females are also more widely distributed along the middle of the spectrum while males tend to cluster at the extremes.

If you want to turn a continuous character into a discrete one for some practical or heuristic reason, you can always create the necessary categories by dividing the spectrum into discrete hunks with boundaries. It is to be hoped that these boundaries would not be placed at completely arbitrary points in the spectrum. Instead, one would hope for some principled reasons behind the creation of the various categories. For example, with sexual orientation, most researchers take care to make these boundaries consistent with people's self-reported sexual identity.

IS SEXUAL ORIENTATION GENETICALLY DETERMINED?

This question most commonly takes the form of 'Is there a gay gene?' For several reasons it is important to realize there is no such thing as a 'gay gene'. First, the term 'gay gene' is a misnomer in the same way that 'straight gene' would be. The same gene variant that predisposes someone to homosexuality when present would predispose towards heterosexuality when absent. Thus, it is more appropriate to ask, 'Is sexual orientation genetic?' Second, because sexual orientation is a complex phenomenon, it is unlikely that any single gene determines sexual orientation with total certainty. Rather, it is more likely that several genes interact with each other and with environmental factors to influence the development of sexual orientation. If they exist, such genes will predispose someone towards a particular sexual orientation rather than determine his or her orientation with certainty.

Three research methods have been used to explore the existence of genetic influences on sexual orientation: family pedigree studies, twin studies, and molecular genetic studies. Each contributes distinct and useful information as well as having specific limitations.

Family studies are used to answer two main questions. First, does the trait of interest run in families? If the answer is no, it is pointless to continue to look for genes. Not all traits that run in families are genetically determined, but traits that do not run in families cannot be genetically determined. Second, by looking for patterns in the way a trait is passed from generation to generation it is possible to get an idea about where in the genome the predisposing genes might be located (i.e., on autosomes or sex chromosomes).

In order to demonstrate that genes influence a trait such as sexual orientation, one must show both that it runs in families and that it occurs in these families at levels higher than the base rate in the general population. Studies have found that, on average, the rate of homosexuality in men who have gay brothers is approximately 9 percent (see Bailey and Pillard 1995, Mustanski, Bailey, and Chivers 2002, for reviews). Female homosexuality also appears to run in families, although estimates of homosexuality among sisters of lesbians vary widely, ranging from 6 to 25 percent (Bailey and Pillard 1995). In order to compare these rates with the population base rate, data from a large sexuality survey are needed. The most recent large survey of sexual behavior, conducted in the same country where the family data were generated (United States) and using the same criteria (self-identification as gay), produced rates of homosexuality of 2.4 percent in men and 1.4 percent in women (Laumann, Michael, and Gagnon 1994). The fact that the rate of homosexuality in relatives with a gay family member is higher than the population base rate suggests evidence for the familiality of both male and female sexual orientation.

Family studies are also useful for determining whether the pattern of transmission of a trait is autosomal or sex linked. Humans have twenty-three pairs of chromosomes, twenty-two of them classified as autosomes and one pair classified as sex chromosomes. Sex-linked traits are those that have predisposing genes transmitted on the X or Y chromosomes, and they have distinct patterns of inheritance. Since the Y chromosome is small and is believed not to contain many genes, most sex-linked traits are influenced by genes on the X chromosome. Males inherit their Y from their father and their X from their mother; thus maternal transmission of a trait is the strongest evidence that a gene is located on the X chromosome.

The first family study to examine patterns of transmission found gay men had more gay uncles and cousins on the maternal side than the paternal side of their family (i.e., maternal transmission; Hamer et al. 1993). As previously mentioned, such a pattern of results suggests that a gene associated with male homosexuality may be located on the X chromosome. One subsequent study replicated the findings of maternal transmission (Rice et al. 1999), while two others have not (Bailey et al. 1999, McKnight and Malcolm 2000). More research will be needed to clarify this incongruity. If the maternal transmission hypothesis were confirmed, it would add to a growing body of evidence suggesting the genetics of sexual orientation are different in men and women.

Family studies can demonstrate both whether a trait is passed down in families and possibly the pattern in which it is inherited, but simply that a trait is passed down in a family does not necessarily indicate that it is genetic. Religious affiliation (e.g., Methodist versus Catholic), for example, tends to be passed down from parents to children, but is not genetic in nature. Culture and family tradition are confounded with genetic transmission in family studies. Twin studies overcome this issue by comparing the similarity of mono-zygotic (identical) twins who share 100 percent of their genes to dizygotic (fraternal) twins, who share 50 percent of their genes in common on average. Because both types of twins are raised in the same family and share the same environmental experiences (including sharing the same uterine environment) differences in their degree of similarity should be due primarily to differences in their degree of genetic similarity. Using statistical procedures to compare twin correlations, the heritability of traits can be calcu-lated. Heritability refers to the proportion of variability in a trait within the population that is due to genetic effects. It is typically represented numerically on a scale from $H = 0$ (no genetic influences) to $H = 1$ (totally genetically controlled).

Several recent twin studies have been conducted on large, systematically recruited samples (reviewed in Mustanski et al. 2002). One recent study of 756 pairs of twins in the United States estimated a heritability of .62 for self-reported sexual orientation in men and women (Kendler et al. 2000). Another recent study of 1,405 twin pairs identified through an Australian twin registry produced heritability estimates of .58 for men and .30 for women (Kirk et al. 2000). To put this heritability estimate in perspective, the heritability of body mass index has been estimated at approximately .80 (Pietilainen et al. 1999) and alcoholism at between .50 and .60 (McGue 1999).

Although twin studies are helpful in determining whether genes are important, they are not able to tell you which particular genes are at work. Molecular genetic research is needed for this purpose. One type of molecular genetic study, called linkage analysis, has been used to study male sexual orientation. Linkage studies look for specific chromosomal regions that are passed down in families along with a trait (e.g., sexual orientation) at probability levels greater than chance (50 percent for siblings). Because some family pedigree studies have suggested that sexual orientation is transmitted through the maternal line, the studies published to date focus only on the X chromosome. Hamer and associates (1993) reported the first linkage study in forty families of gay men and found an association between male sexual orientation and a region of the X chromosome named Xq28. This finding was replicated in two independent samples (Hu et al. 1995, Sanders 1998), and not replicated in another (Rice et al. 1999). Furthermore, the region was shown not to be an important determinant of female sexual orientation (Hu et al. 1995). Given the mixed evidence for an association between Xq28 and male sexual orientation, future research will be needed before any definitive claims can be made for its importance in helping to determine sexual orientation in males.

It is important to note that Xq28 is a chromosomal region, not a specific gene. Finding linkage to a chromosomal region tells you that a gene located within this region is related to a particular trait. If Xq28 proves to be associated with male sexual orientation, the next step will be to determine which gene within this region is important. Future research will likely explore this, as well as genes on other chromosomes. Researchers may look for genes on other chromosomes because the evidence for maternal transmission is mixed, and because even if genes on the X chromosome are important, that importance does not preclude genes on other chromosomes from also having influence.

Although no specific genes for male sexual orientation have currently been identified, and no chromosomal regions have been implicated in women, it is possible to draw several conclusions about the role that genes play in determining sexual orientation. Twin and family studies have demonstrated that genetic influences do explain a portion of the variability in sexual orientation within the populations that have been studied. Genes also seem to be more influential in male sexual orientation. Finally, there is some evidence that the chromosomal region Xq28 may play a role in male sexual orientation, but further research is needed before any definitive conclusions can be drawn.

HOW COULD HOMOSEXUALITY HAVE EVOLVED?

People who ask how homosexuality could have evolved are usually what philosophers of biology have come to call "empirical adaptationists" (Godfrey-Smith 2001). They conceive of natural selection as the only evolutionary process of any real consequence and wonder how the genes that contribute to the development of a trait such as homosexuality could be maintained in a population when they are likely to decrease rather than increase the reproductive success of the organisms that bear them.

First, the presence of genes that contribute to homosexuality in contemporary populations is explained by their relative fitness in prior generations. While it may seem intuitively obvious that homosexual humans have a lower fitness than heterosexual humans as measured by actual reproductive success, there are no good empirical studies that provide actual evidence that homosexuals have had substantially lower fitness throughout the relevant evolutionary past. Our intuitions about the reproductive success of homosexuals are based on our modern conception of "gay" and "lesbian" people. When we think of homosexuals, we think of the single urban gay man or woman, or we think of the nice gay couple living in the house down the block. The idea of organizing one's identity and one's life around homosexual desires is very contemporary (Halperin 1990) and largely confined to a very few cultures (Herdt 1997). For the vast majority of human history and even today in the majority of the world's cultures, people who have homosexual desires are far more likely to structure their lives in heterosexual-typical ways. If cultural mores require adults to marry and to produce offspring, then homosexual members of the culture are under as much pressure to accede to these expectations as heterosexual members. What little we know about the way homosexual members of the population have lived for the majority of human evolutionary history undermines the intuition that their actual reproductive success is substantially different from that of the heterosexual members of the population. Of course, it would be nice to have some actual empirical data on which to base an evolutionary hypothesis. Unfortunately, such data about the actual reproductive success of homosexuals throughout human history do not exist.

For the last thirty years, the kin selection hypothesis was the most widely held evolutionary hypothesis regarding the maintenance of homosexuality. In 1975, E. O. Wilson argued that instead of producing their own offspring, homosexual men reproduce "indirectly" by contributing their resources to the offspring of their relatives. Wilson hypothesized that homosexual male members of a family might have acted as helpers in hunting or in various other domestic chores. Others theorized that the homosexual family member's contribution might take the form of sharing income, bequeathing wealth, or assisting in child rearing (Ruse 1981, Weinrich 1987).

A very important part of the kin selection hypothesis is that homosexual males possess a set of characteristics that make them better suited to life as domestic helpers than as actual fathers. Ruse (1981) and Weinrich (1987) argued that homosexual males possess a set of characteristics that substantially reduced their chances of success in the competition for mates. There is a great deal of evidence to suggest that homosexual men tend to exhibit a range of gender-atypical behavioral characteristics that would have contributed to decreased mating success in the evolutionary past (Bailey et al. 1994). For example, studies of contemporary homosexual males show that they tend to be less physically aggressive than heterosexual males (Bailey and Zucker 1995).

Homosexual males also tend to be "later born sons" with a disproportionate number of older brothers (Blanchard and Bogart 1996). By the time the homosexual son came along, the family may have already invested the majority of its resources in the earlier born sons. Thus, the later born homosexual son may have had a greater chance of getting his genes in the next generation by contributing to the reproductive success of his older siblings who were better equipped to compete successfully for mates.

In addition to the presence of characteristics making them less suited for the competition for mates, the kin selection hypothesis includes the supposition that homosexual males had sex-atypical traits that made them more willing to contribute resources to their nieces and nephews than their heterosexual siblings. Females typically exhibit a greater degree of empathy than males. In 1995, Salais and Fischer reported evidence that homosexual men were on average more empathetic than heterosexual men. Proponents of the kin selection theory hypothesized that the increased level of empathy might translate into an increase in altruistic behavior towards relatives.

While the kin selection hypothesis has been the dominant evolutionary hypothesis with respect to homosexuality since its conception, there is no shortage of scientists who reject it as a viable alternative. Some reject it on theoretical or conceptual grounds (e.g., Dickemann 1995). Others reject it because the few empirical studies directed at testing the hypothesis have failed to yield any convincing supporting evidence. For example, Bobrow and Bailey (2001) investigated whether contemporary homosexual men were more or less likely than heterosexual men to contribute financial, emotional, and social resources to siblings, nieces, and nephews. They found that, contrary to the predictions of the hypothesis, heterosexual men tend to contribute more financial resources to kin than homosexual men. This would seem to be strong evidence against the kin selection hypothesis.

Despite its considerable conceptual and theoretical weaknesses and the empirical evidence against it, the kin selection hypothesis has not yet been rejected. I would argue that this is due in large part to the adaptationist bias prominent in sociobiology and evolutionary psychology. Another, much less well recognized reason for the persistence of the kin selection hypothesis is the fact that while most research on sexual orientation has been conducted within a highly homophobic context, this hypothesis treats homosexuality as an adaptation. On the kin selection hypothesis, homosexuality is not seen as a developmental flaw or disease, but as a "normal" variation in human sexuality that conveys a positive advantage to the families in which it is found. A preference for viewing homosexuality in this "positive" light is not so much a reason for believing the hypothesis as it is a motivation to resist the alternative nonadaptationist hypotheses.

One such nonadaptationist hypothesis developed out of the birth-order effect described earlier. The higher incidence of homosexuality in later born sons might be a result of a maternal immune response to one or more of the male-specific, Y-linked, minor histocompatability antigens (referred to collectively as H-Y antigens) carried by the male fetus (Blanchard and Klassen 1997). The mother's immune response would be triggered by the presence of the H-Y antigens of the first male fetus she carried and would become stronger with each subsequent pregnancy. There is good evidence, independent of this hypothesis, that H-Y antigens play some role in the sexual differentiation of vertebrates (Wachtel 1983). They are usually present in the heterogametic sex (male mammals) and absent in the homogametic sex (female mammals) and they have been highly conserved throughout vertebrate evolution (Wachtel 1983). For these reasons, Blanchard and Klassen (1997) hypothesize that the H-Y antigens are involved in the development of sex-typical psychological and behavioral characteristics. They go on to hypothesize that exposure to H-Y antibodies produced by the mother would affect the male fetus's development and possibly produce males with sex-atypical sexual behavior (Blanchard and Klassen 1997, 374). Thus, the interaction between the H-Y antigens of the male fetus and the immune system of the mother would produce male offspring with a range of gender-nonconforming behavior that might well include more female-typical sexual desires – that is, sexual desires directed at males.

IF SEXUAL ORIENTATION IS HERITABLE, DOES THAT MEAN IT IS COMPELLED, IMMUTABLE, AND INNATE? DOES THAT MAKE IT MORALLY ACCEPTABLE TO ACT ON HOMOSEXUAL DESIRES?

Despite common assertions to the contrary, evidence for biological causation does not have clear moral, legal, or policy consequences. The argument that biological causation does render homosexuality morally acceptable might proceed as follows: A person cannot be held morally blameworthy for a particular behavior if that behavior is a result of neurophysiology and if he or she possesses genes that predispose him or her to have such neurophysiology. Put this way, it is easy to see that such reasoning is irrational. In order for a person to be morally blameworthy for some action, that action would have to be not a result of genetically influenced neurophysiology. But there are no behaviors whose proximate cause is not neurophysiological and brains are all a product, at least in part, of genes. (For a more detailed version of this argument see Greenberg and Bailey 1993).

It is also frequently assumed that traits that have genetic under-pinnings are compelled, unchangeable, and innate. Questions along these lines usual take the form of 'Do people choose their sexual orientation? Can people change sexual orientation? Are people born gay?' While these are interesting questions in their own right, it is important to understand that knowing that a trait is heritable does not provide a clear answer to any of them.

A trait is said to be compelled if it is completely determined by factors over which one has no control. Traits that have both high and low heritability can both be compelled. Eye color is completely genetically determined and totally involuntary, while the language you are taught in childhood is totally socially determined and yet equally compelled. Environmental or genetic determination is uninformative about the extent to which a person chooses sexual orientation. For example, one hypothesis about the development of sexual orientation based on learning theory posits that women become lesbians because of negative experiences with men. Yet, if this were the case, homosexuality would be no more voluntary than if it were genetically determined because it was caused by events beyond the woman's control.

The fact that a trait is heritable does not mean that it is immu-table, or resistant to change. This is because heritability estimates can, and do, change over time and when the distributions of genes and/or environments change. For example, recent research has demonstrated that the percentage of young adults and the amount of migration in different communities strongly moderate the herita-bility of adolescent alcoholism (Dick et al. 2001). This suggests that communities with more young adults and greater social mobility allow for increased expression of genetic dispositions, whereas communities with fewer older peers and greater social structure facilitate greater social control over drinking. Such interactions, which scientists believe are important for most complex traits, demonstrate that heritability estimates should not be used to determine whether a trait is immutable. Instead, it is proper to review the evidence concerning attempts at changing it. The results of such reviews suggest no evidence that psychotherapeutic or religious conversion therapies are effective in changing sexual orientation (Haldeman 1994).

Perhaps the most commonly misunderstood association is between heritability and innateness. The general public thinks of a trait as innate if it is present at birth. The assumption seems to be that if a trait is present at birth it must be the product of genetic rather than environmental factors. Of course, biologists know that all traits are a product of the interaction between genetic and environmental factors and that a great deal of this interaction occurs before birth. Nevertheless, the distinction between genetically "determined" traits and "learned" traits continues to be strongly held by the general public. To illustrate the disconnect between the concepts of heritability and innateness consider the following example: The tendency to have two thumbs – one on each hand – is "innate" and most people have two thumbs. What variation does exist in thumb number is largely environmental (i.e., losing a thumb). Because heritability is defined as the proportion of variability within a population due to genetic factors, and variability in thumb number is largely environmental, the heritability of thumb number is low. But few people would argue that thumb number is not an innate characteristic. Most researchers in the field do believe that sexual orientation is influenced by prenatal factors, but not because it is heritable. Instead, they are convinced by evidence of associations between homosexuality and traits that either show up early in development (such as childhood gendered behaviors; Bailey and Zucker 1995) or are believed to be determined before birth (such as handedness and finger length; see Mustanski et al. 2002 for review).

People looking for answers to normative questions about homosexuality by examining its biological origins are focusing on the wrong issue. The value of a characteristic such as homosexuality depends on its effects rather than its causes. Homosexual relationships between consenting adults do not in themselves cause harm or at least do so no more often than heterosexual relationships do. Furthermore, attempts by homosexually oriented individuals to live heterosexually are fraught with harm, both to the self and to others, especially those who form heterosexual relationships with such individuals, such as their wives and husbands.

CONCLUSION

Scientists who study human sexuality are interested in a variety of questions about human behavior and the neurophysiology that underlies it. Nevertheless, the questions that are most interesting to the public at large are questions about the relationship between genes and sexual orientation. Although there are many questions that cannot be answered at this time, there are sufficient scientific data to allow some definitive statements to be made. Homosexuality does appear to run in families and studies have established that genes play a substantial role in explaining individual differences in sexual orientation. There is some evidence that a region on the X chromosome may influence male sexual orientation, but further research is needed and is under way. The sum of the evidence from these studies also strongly suggests that the genetics of homosexuality must be different for men and women.

The maternal immune response hypothesis is replacing the sociobiological kin selection hypothesis as the dominant theory regarding the ultimate cause of male homosexuality. The maternal immune response hypothesis provides a nonadaptationist explanation that is theoretically more coherent and more consistent with all of the empirical data we have at this point. However, it cannot explain the existence of female homosexuality. Given that the data collected to date strongly suggest that male and female homosexuality have different patterns of inheritance, one might expect there to be different causal explanations for the maintenance of the biological systems that produce them.

Interest in the genetic causation of human sexual orientation is driven in large part by the social and political controversy over the status of homosexuals and homosexual relationships within modern culture. Nevertheless, it is important to appreciate the fact that the heritability of a trait provides little information about the extent to which it is compelled, immutable, innate, or, most importantly, morally acceptable. All people should be treated with dignity and respect regardless of their sexual orientation.

16   Game Theory in Evolutionary Biology

1.   INTRODUCTION

Game theory is now a standard tool for explaining puzzling and counterintuitive behavior. But in spite of the fact that game theory was developed to investigate rational and economic behavior of modern humans, it has found equally valuable application in biology. For example, it has often been observed that when fights break out between members of the same species, the antagonists often display restraint by not inflicting serious injury on each other during the fight. To take another example, individual guppies will sometimes go out of their way to swim beside a larger fish that may turn out to be a predator (Dugatkin and Alfieri 1991a, b). And chimpanzees will often raise an alarm to the rest of their group when they spot a dangerous predator, in spite of the fact that an individual who does so will attract the attention of the predator.

These behaviors should be puzzling to anyone who thinks of evolution as 'nature red in tooth and claw'. However, they can all be explained in a very satisfying way by applying simple game-theoretic analyses. The purpose of this chapter is to illustrate enough game theory to show how it may be applied to the task of explaining such puzzling behavior. Although the range of such behaviors is extremely large, and the range of available game-theoretic techniques is equally large, I shall focus on game-theoretic explanations of one particular kind of behavior – namely, altruism.

In what follows, I shall refer to any behavior that lowers the fitness of its bearer, but raises the fitness of another individual, as altruistic. When biologists refer to a behavior as 'altruistic', they mean only that it has this effect on fitness – no claim is made about the psychological states of the individuals involved. Indeed, there is no assumption that the individuals even have psychological states at all. For example, it turns out that some viruses reproduce much more slowly than they could. Such low virulence thereby benefits other viruses in the host (by keeping the host alive longer) while lowering the fitness of the virus. We may therefore properly call the virus an 'altruist', even though it obviously has no capacity for having altruistic or self-sacrificial feelings and sentiments. Furthermore, the behavior of this virus is every bit as puzzling as the psychologically laden behavior of a chimpanzee or modern human. Explaining the evolution of altruistic behavior is commonly called the problem of altruism.

The problem of altruism is an excellent illustration of the power of game theory. This is because the very existence of altruism seems, at first glance, to be ruled out by evolution. After all, any genetic propensity to behave altruistically would, by definition, result in a lowering of that individual's fitness. Therefore, we should expect that evolution would relentlessly eliminate altruism from any population. It turns out, however, that natural selection's effect on altruism is much more subtle and interesting. Game theory throws these subtle features of natural selection into sharp relief.

2.   GAME THEORY AND THE PROBLEM OF ALTRUISM

Game theory is used to analyze situations in which two or more individuals have conflicting interests.[1] Typically, game theorists do not care about many of the features of the situation that the layperson would normally think are important. For example, suppose that Betty is deciding whether to pay back a loan that Al gave her last week. If a layperson were asked to predict Betty's behavior, the layperson would probably take into consideration such facts as whether Betty has promised to pay it back, and whether she appears to be an honest and trustworthy person.

In contrast, a game theorist does not normally care about such things. When we subject a situation to a game-theoretic analysis, we strip away most of the contextual features of the situation, focusing

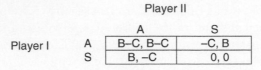

Figure 16.1 The problem of altruism as a game

only on a few key facts. First, we want to know what behaviors are available to the individuals. In analogy with ordinary games such as checkers or chess, we refer to the available behaviors as strategies and we refer to the individuals involved as players. So Betty has two options – she can pay back the loan or not. Accordingly, we say that her strategy set consists of the two strategies 'pay back' and 'do not pay back'.

The strategies selected by the players will lead to an outcome; some of the outcomes may be more beneficial to some players than to others. We measure the desirability of an outcome by its payoff to each player. For instance, if Betty repays the loan, then one result of her behavior is that she gives Al some money. Thus, if we are interested only in the amount of money that Betty and Al have at the end of the day, then Betty's payoff will be negative if she plays the strategy of repaying the loan.

Earlier, we defined altruistic behavior as behavior that raises the fitness of another individual at some cost to oneself. In game-theoretic terms, we say that the altruist raises the payoff of another individual while lowering her own payoff. Let us say that an individual has a choice as to whether to bestow a benefit B onto another individual at some cost C to herself, and that the potential benefit is greater than the cost – in other words that $B > C$. If two individuals – call them Player I and Player II – are to interact with each other, then we may represent their situation in the simple matrix shown in Figure 16.1. There, the players have a choice between behaving Altruistically (A) or Selfishly (S).

If a player chooses strategy A, then she incurs a cost to herself of $-C$; the strategy S incurs no cost. Playing strategy A bestows a benefit of B to one's partner; playing S does not.

This game is probably the most famous one in game theory. It is usually called the 'Prisoner's Dilemma', because the following story usually goes along with it. Two bank robbers have been arrested by

Player II

		D	R
Player I	D	1, 1	10, 0
	R	0, 10	4, 4

Figure 16.2   The Prisoner's Dilemma. The numbers represent the jail sentences for each player, so larger numbers are worse

the police, but the police do not have enough evidence to convict them of the robbery. So the police put the two suspects in different cells where they cannot talk to each other, and the police make the following speech to each suspect:

We know you robbed the bank, but we cannot convict you with the information that we have. So if you agree to rat out your partner, and your partner does not rat you out, then we will let you go free while we give your partner ten years in prison. On the other hand, if you do not rat out your partner, but your partner rats you out, then you will go to prison for ten years while your partner goes free. If you both rat each other out, then you will both get four years in prison. If neither one of you rats the other out, then we will make sure that you both go to prison for one year on trumped-up charges.

We represent the dilemma faced by the robbers as in Figure 16.2. If the prisoners are fairly sophisticated, then they could reason as follows:

Suppose that my accomplice rats me out. If so, then I am better off ratting him out, since I would rather get four years in prison instead of ten. On the other hand, suppose that my accomplice does not rat me out. In that case, I am still better off ratting him out, because I will get no time in prison rather than one year. So no matter what my accomplice does, I am better off ratting him out.

The trouble is that both prisoners go through exactly the same line of reasoning. So we expect each to wind up with four years in prison as a result. But this seems paradoxical, for if they can anticipate getting four years in prison, why not both remain silent and get one year instead?

A quick comparison of the Prisoner's Dilemma and the problem of altruism reveals that they are really the same situation. The only difference between the two matrices is that we have represented positive numbers as payoffs in the first figure, and as penalties in the second. But clearly, this makes no substantive difference.

As we have seen, the prisoners have a convincing line of reasoning that leads them toward behaving selfishly, ultimately to their own detriment.[2] But if the Prisoner's Dilemma is really just a restatement of the problem of altruism, then we have a puzzle: how is it possible for natural selection to lead to the evolution of any altruistic tendencies at all? For if we have a sound argument in favor of ratting out one's accomplice in the Prisoner's Dilemma, then the same argument should favor selfishness in the problem of altruism.

3.   TWO GAME-THEORETIC APPROACHES

As is well appreciated by philosophers of science, the best explanations frequently display a great deal of generality, applying to a wide variety of different cases across many different circumstances.[3] Indeed, it is one of the strengths of game theory that it shows us what relevant similarities exist across different phenomena. Thus, game theory holds out hope of constructing general explanations that apply to many diverse phenomena.

Game-theoretic explanations may be divided into two categories. On the one hand, we might hope that there is some characteristic of the strategies that makes them much more likely. If so, then we may largely ignore the question of how the plays settle on those strategies. I shall call this the static approach. On the other hand, we might focus on modeling the process by which agents determine their strategies and largely ignore the characteristics of the strategies upon which they settle. I shall call this the dynamic approach.

The advantage to the first method is that various situations will follow different processes and dynamics. Thus, if we can construct a satisfactory explanation for an observed set of behaviors that ignores those processes, then the explanation may apply to a wide range of other cases. Of course, this simultaneously suggests the biggest drawback to the equilibrium approach – it may turn out that there is no general explanation, and that we have no choice but to be drawn into an analysis of those evolutionary processes.

3.1   The Static Approach

Historically, the static approach is the one that has drawn the most attention from biologists, philosophers, and economists. Indeed, it has found application not only in biology and economics, but also in philosophical discussions concerning the origins of communication and coordinated conventions (for which the classic source is David Lewis 1969).

One way intuitively to motivate the static approach is to consider a strategic situation faced by two or more players as a system in which the game is repeated an indefinite number of times. Sometimes, the players will settle upon an outcome in which they have an incentive to change strategies; sometimes the outcome will be stable, leaving the players with no incentive to change their strategies. We say that such a stable outcome is an equilibrium.

This simple and elegant idea was pioneered in economics by John Nash (1950, 1951). Nash's insight gave an intuitively compelling and mathematically precise characterization of the sets of strategies that make up a so-called Nash equilibrium. According to Nash, we should consider a set of strategies to compose an equilibrium just in case no single player could do better than her current payoff, given the current strategies played by the other individuals. In other words, a Nash equilibrium exists when no player has an incentive unilaterally to switch strategies. In this way, the Nash equilibrium concept is a predictive tool that tells us that, all other factors being equal, if a game were to be repeated an indefinite number of times, and we randomly select a time at which we observe the players' choices, there is a high probability that we will observe a set of strategies in Nash equilibrium.

Nash's original treatment was concerned with bargaining situations between rational agents; in fact, in Nash's original conception, the preequilibrium 'experimentation' phase was to be conducted only hypothetically, in the minds of rational agents who were to bargain only once. Thus, the very term 'equilibrium' was only a metaphor in Nash's intended context, because there is no dynamic process that could be in equilibrium in the first place. However, even in a more literal context, in which the 'experimentation' is guided by trial and error or by evolutionary processes, we should still expect to observe sets of strategies in Nash equilibrium.

It was John Maynard Smith who gave a biological motivation for a closely related approach (1973, 1974, 1982). In his formulation, we may explain the evolution of a behavior by showing that it is stable in a slightly different sense. His sense of 'stable' is that the behavior, once it has been adopted by a population, cannot be invaded by a small number of 'mutants' who behave differently. More precisely, we consider a population whose members exhibit a particular kind of behavior. That behavior will in the population entail that the individuals have a particular fitness. The behavior is 'uninvadable', according to Maynard Smith, if a small group of mutants who behave differently cannot have a higher fitness if they were to interact with that population. Any strategy having this stability property is called an 'evolutionarily stable strategy', or an 'ESS' for short.[4]

We may illustrate the ESS concept using the Prisoner's Dilemma. Let us suppose that a population of players faces the Prisoner's Dilemma, and that everyone in the population plays Don't Rat. This is not an ESS, because a mutant who plays Rat will enjoy a much higher payoff. Thus, the strategy of playing Don't Rat is invadable by mutants who Rat. Thus, Maynard Smith's ESS concept predicts that populations will play Rat in the Prisoner's Dilemma, and that therefore (because it is really the same game), they will behave selfishly rather than altruistically.

The observation that the strategy Don't Rat is not an ESS is puzzling because we still have the observation that altruism does exist in nature. So our model must be flawed or incomplete if it predicts only selfishness; in other words, there must be important biological facts that we are overlooking. Fortunately, game theory provides a flexible enough framework for including other biological facts. But this requires us to adopt (what I have called) the dynamic approach.

3.2   The Dynamic Approach

In order to introduce the dynamic approach, consider the following simple game, which is sometimes called 'Hi-Lo'. In Hi-Lo, the two players have a choice between playing strategy A and strategy B. They receive a payoff only if they play the same strategy, but they both receive a much higher payoff if they both play A. For instance, we could suppose that the payoff if they both play A is 10, while their payoff if they both play B is merely 1.

In this game, it is obvious that we should expect the players to converge on the A outcome. However, closer inspection reveals that neither the Nash equilibrium concept nor the ESS concept makes this prediction. This is because the outcomes in which they both play A and the outcome in which they both play B are Nash equilibria and evolutionarily stable strategies. After all, a population of B players cannot be invaded by a mutant A player, because the mutant will receive a payoff of 0, which is lower than the population's payoff of 1.

So we would like to have a principled test that would tell us (in this case) that the players are more likely to converge on the B strategy. Criteria that eliminate some equilibria from consideration are commonly called 'equilibrium refinements', and there is a large literature on these that may be found in economics. However, as most of these refinements make recourse to the characteristics of rational decision makers, they are largely outside the scope of our consideration here.[5]

If we are working within a biological context, we need to refocus our attention on the evolutionary processes that yield determinate behaviors among populations of nonrational agents. To return to our earlier coordination example, it is implausible to suppose that most evolutionary processes will equally favor coordination on A and coordination on B, given the radically unequal payoffs in those two coordinated states. Rather, it seems clear that most evolutionary processes will tend to favor coordination on the more profitable behavior.

Thus we are led to consider the evolutionary process itself. So let us suppose that we have a very large population of nonrational agents who pair up periodically and at random to play this coordination game. To simplify matters, imagine that initially, the population is divided roughly evenly between A-players and B-players. Since we are working within a biological context, we interpret payoff as fitness, or expected number of offspring. As is the usual custom with this sort of example, we assume that each agent's behavior (that is, the choice of strategy) is determined genetically, so that the offspring of A-players will tend to be A-players, and similarly for the offspring of B-players.

It is clear that this simple evolutionary process will dramatically favor the evolution of B-players, in agreement with our expectations for this example, for suppose that you are an A-player. About half the time, you will meet another A-player and receive a small payoff; the other half of the time, you will meet a B-player and receive nothing. But if you are a B-player, you will meet another B-player about half the time and receive a much larger payoff (since a pair of B-players will have a higher payoff than a pair of A-players). Thus, on average, B-players will enjoy a much higher fitness than A-players under this dynamic, as the following simple equations show:

expected payoff for A = .5(1) + .5(0) = .5
expected payoff for B = .5(0) + .5(10) = 5

Since payoff is to be interpreted as the expected number of offspring, there will be more B-players in the population in the next generation. This is where the dynamic begins to magnify the positive effect of being a B-player and leads to the spread of B throughout the entire population, for if there are more B-players than A-players in the population in the next generation, then A-players will receive even lower payoffs than they did during the first generation. If they are in the minority, then they are less likely to meet each other and therefore are less likely to receive any payoff at all. The opposite is clearly true of those who are fortunate enough to be born with a tendency to play B. They will be more likely to meet each other in the next generation and will therefore be more likely to enjoy high payoffs. In this way, the spread of the B strategy throughout the population will accelerate from one generation to the next. If there is no other process to stop it, we should expect that within relatively few generations, the population would be composed almost entirely of individuals who play strategy B.

However, it is important to note that these considerations do not imply that the B strategy will always be observed, for suppose that chance events cause the population to start in a state in which 95 percent of the individuals play A. Then an individual will tend to encounter an A-player 95 percent of the time, and we can therefore calculate the average payoff for each type of player as follows:

expected payoff for A = .95(1) + .05(0) = .95
expected payoff for B = .95(0) + .05(10) = .5

Thus, if the population happens to begin in a state that is heavily skewed toward the A strategy, then the population will tend to move to a state in which everyone plays that strategy. So what these considerations really say is that most of the time we should observe the population playing strategy B, but that a small percentage of the time, populations will play A.

This simple example is an informal illustration of a formal model called the replicator dynamics, which is due originally to Taylor and Jonker (1978), and which has played a dominant role in so-called evolutionary game theory.[6] In contrast to the equilibrium approach invented by Nash, the dynamic approach of evolutionary game theory is concerned with the process by which a population of agents settles upon a behavior. In the replicator dynamics, which is the simplest and most common of such models, that process is taken to be evolution and natural selection, where strategies compete with each other as replicators. In contrast to traditional game-theoretic models in which payoffs are interpreted as money or welfare, the replicator dynamics interprets payoffs as fitness. Thus, individuals whose strategies tend to yield higher payoffs in the population will enjoy higher fitness and will thereby tend to have a greater number of offspring in subsequent generations. So if we were to look at a population of agents over many generations, we should expect the most successful strategies eventually to predominate in the population. But what of the problem of altruism?

Unfortunately, it turns out that the selfish strategy is favored by the replicator dynamics. To see this, we simply calculate the expected payoffs of altruism and selfishness as follows. Let us suppose that the population is randomly divided between altruistic and selfish types. Call the proportion of the population that is altruistic PA, and the proportion that is selfish PS. Because PA + PS = 1, we can replace PS with (1 − PA). Using the payoff matrix from Figure 16.1, we may calculate the expected payoff of altruists and selfish types:

$$\text{payoff to altruists} = P_A (B - C) + (1 - P_A) (-C)$$
$$\text{payoff to selfish type} = P_A (B) + (1 - P_A) (0)$$

which simplifies to

$$\text{payoff to altruists} = P_A (B) - C (1 + 2P_A)$$
$$\text{payoff to selfish types} = P_A(B)$$

Clearly, it is better to be selfish – after all, selfish types get all the benefits of interacting with altruists, but never incur any costs.

Thus, the replicator dynamics predicts that populations will move inexorably toward a state in which everyone is selfish.

4.   ADDITIONS TO GAME-THEORETIC EXPLANATIONS

So it begins to appear that the problem of altruism is a very thorny one, and that the game-theoretic tools suggest only what we knew already – namely, that we should expect individuals to be selfish. Specifically, we find that selfishness is the predicted behavior on both the static and dynamic approaches, for a population in which everyone is selfish is more stable than one which has altruists, and the replicator dynamics shows us that the simplest population dynamic favors the spread of selfishness in any mixed population.

However, we must return to the empirical fact that many species do exhibit altruistic behavior, so there must be more to the story. In the remainder of this essay, I shall discuss two additional features that have been prominent in explaining the evolution of altruism. Then I shall offer some speculations about promising areas for future research.

4.1   Iteration

One of the most influential proposals for explaining the evolution of altruism relies upon the commonsense observation that many individuals interact with each other repeatedly. As we all know, it is easy to acquire a reputation for interacting in a particular way, and one can carry that reputation to future encounters. Thus, the suggestion is that we should explain the evolution of altruism by considering iterated interactions instead of the simpler 'one-shot' interactions.

So we need to differentiate between two different versions of the Prisoner's Dilemma. On the one hand, we have the 'one-shot' Prisoner's Dilemma in which individuals pair up, play the Prisoner's Dilemma once, and never interact again. On the other hand, we have the so-called Iterated Prisoner's Dilemma, in which individuals pair up and play the Prisoner's Dilemma repeatedly with the same partners.

It is important to note that the one-shot Prisoner's Dilemma and the Iterated Prisoner's Dilemma are very different games. In

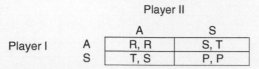

Figure 16.3 The Prisoner's Dilemma in general form

particular, the Iterated Prisoner's Dilemma has a much richer set of available strategies. For example, one can play the relatively simple strategies 'Always Be Altruistic' (henceforth, AA) or 'Always Be Selfish' (AS). But one can also play a strategy that takes one's partner's past behavior into account. For instance, one can play the appropriately named 'Tit-for-Tat' strategy (TFT). When playing TFT, one begins by behaving altruistically toward one's opponent. However, the TFT strategy retaliates against a selfish opponent by behaving selfishly on the round after the opponent's selfish behavior. Conversely, the TFT strategy rewards an altruistic opponent by behaving altruistically in the round after that behavior. Let us analyze the problem of altruism using the three previous strategies.

For simplicity, I will follow the standard practice of rewriting the matrix in Figure 16.1 using these shortcuts:

$$B = T \text{ (Temptation to defact)}$$
$$B - C = R \text{ (Reward for mutual cooperation)}$$
$$0 = P \text{ (Punishment for mutual defection)}$$
$$C = S \text{ (Sucker's payoff)}$$

As Figures 16.1 and 16.2 make clear, we should require that $T > R > P > S$ so that the best possible result for a player is to be selfish against an altruist; the next best is to be altruistic with another altruist; and so on.[7] For easy reference, I place an equivalent version of Figure 16.1 using the new abbreviations as Figure 16.3.

Consider two players who pair up to play the Iterated Prisoner's Dilemma for ten rounds using the payoff matrix from Figure 16.3. Suppose that the players both play the selfish strategy AS. Then in every round, both players defect, each receiving a payoff of P. Thus, their overall payoff for ten rounds will be 10P. On the other hand, suppose that a TFT player interacts ten times with a selfish AS player. In the first round, the TFT player will be altruistic and the selfish player will be selfish, so the payoffs to the two players will be

	TFT	AA	AS
TFT	10R, 10R	10R, 10R	S+9P, T+9P
AA	10R, 10R	10R, 10R	10S, 10T
AS	T+9P, S+9P	10T, 10S	10P, 10P

Figure 16.4   Payoffs for the Iterated Prisoner's Dilemma

S and T, respectively. But after that point, the TFT player will retaliate against the selfish AS player by playing the selfish strategy.

Thus, for the remaining nine rounds, both players will receive the payoff of P. So over the course of ten rounds, the TFT player's payoff will be S + 9P, and the payoff to the AS player will be T + 9P. So again, the more selfish player will do better (since T > S).

Obviously, when a highly altruistic AA player goes up against a selfish AS player, the altruist will be at a severe disadvantage. The AA strategy gives a payoff of 10S, while the AS strategy receives the vastly higher payoff of 10T. On the other end of the spectrum, when two AA players, two TFT players, or an AA/TFT pair interact with each other, no player is ever selfish, for the TFT strategy only retaliates against selfish behavior.[8] So in each of these cases, the payoff to both players is 10R.

Keeping in mind that T > R > P > S, it is clear that in any iterated interaction, the more selfish strategy does better than its opponent. We may summarize the payoffs from the last two paragraphs as in Figure 16.4. But surprisingly, the fact that AS does better in every interaction does not imply that it does best over the long run. To see this, we have to adopt a dynamic perspective on the evolutionary process. Imagine that we begin with a population that is divided roughly evenly among those three strategies, and that the individuals pair up randomly to play the iterated game for ten rounds. As in the replicator dynamics, we assume that the players' fitness depends on their total payoff for those ten rounds.

Intuitively, the problem faced by an AS strategy over the long run is that its success depends upon victimizing altruistic AA players in the population. However, the AS players dramatically lower the fitness of those AA players – for recall that in an AS/AA interaction, the AA player gets the lowest possible payoff of 10S. So over the long run, the AS players will tend to drive the AA players to extinction. Once those altruistic AA players have been driven out of the population, the selfish AS types no longer have those highly profitable interactions. So in this way, the most selfish strategy is self-defeating over the long run.

On the other hand, this is not the case for TFT or AA players. They benefit from interactions with other 'nice' strategies. But most importantly, they do not harm those players upon whom they depend for high payoffs. So in contrast to the selfish AS players, the nicer TFT and AA strategies have the potential to do well over the long run, by helping their partners to remain in the population.

In fact, these informal considerations have been verified in a famous computer tournament organized by Robert Axelrod at the University of Michigan (1984, 1997). He invited people to submit strategies for the Iterated Prisoner's Dilemma and programmed a computer to play each of the strategies against every other strategy in a round-robin competition. Although a variety of highly sophisticated strategies were submitted, the strategy that yielded the highest payoff in the tournament was none other than TFT. Axelrod's analysis of the results parallels the preceding considerations. In order for a strategy to do well over the long run, it must thrive once it has spread through the population. And in order for that to happen, it must do well when it plays against itself. Therefore, strategies such as AA and TFT have a distinct advantage – they yield the relatively high payoff of R in each round when they play against themselves. But selfish strategies are more likely to yield the lower payoff of P in any given round, because they tend to fall into a pattern of mutual defection.

However, strategies such as TFT are superior to more 'naive' strategies like AA in the following respect. The AA strategy can be victimized by selfish strategies to a much greater extent than TFT, for when a TFT player interacts with a selfish type, it begins to behave selfishly itself. Thus, it guarantees that it will receive the lowest payoff S less often; on the other hand, a 'naive' strategy like AA may receive a payoff of S repeatedly in any given interaction.

So we learn a few important lessons when we consider repeated interactions. In order to be successful, a strategy must exhibit two key features. First, the strategy must do well when it plays against itself – for if it does not, then it will be unlikely to predominate in the population. Second, the strategy must protect itself against highly selfish or opportunistic strategies by retaliating. Furthermore, the advantage of a strategy like TFT is increased if the number of repeated interactions is higher. For example, if there were 1,000 repeated interactions in each round, then there are more opportunities to receive the 'reward' payoff of R. Similarly, the highly altruistic AA players will be driven to extinction faster, followed by the demise of the selfish AS types who depend on the presence of naive altruists for their survival.

Finally, it appears as though we have made some progress on the problem of altruism. By considering the effect of repeated interactions on the evolution of altruistic behavior, we can make some definite predictions that can be confirmed or disconfirmed in a biological context. Let us go through some of these lessons individually.

First, the preceding considerations suggest that we are more likely to observe altruistic behavior when individuals have the opportunity to interact with each other repeatedly. In a biological context, this implies that altruism will be observed more often in species in which individuals live together in small groups over a long period. And indeed, this is confirmed by observation. Consider, for example, our evolutionary cousins among the great apes. Chimpanzees, bonobos, and gorillas each live in relatively small groups over much of their lives. Moreover, in each species, one gender leaves the natal group upon reaching sexual maturity in a migration pattern known as philopatry. Hence, the gender that remains in the natal group will have increased opportunities for repeated interaction. Thus, we expect to find a greater degree of altruism among the gender that remains in the natal group.

Indeed, this is the case. If the females remain in the natal group after reaching sexual maturity, then anthropologists refer to that species as 'female bonded'. Similarly, species are 'male bonded' if it is the males who remain in the natal group. It is a commonplace observation that in female bonded species, cooperation and altruism are much more likely to be observed among the females, and conversely if the species is male bonded.

Second, the game-theoretic analyses suggest that we are unlikely to observe 'pure' altruism – instead, we should observe a more 'restricted' form of altruism such as Tit-for-Tat. Successful individuals will retaliate against selfish individuals, so we should expect an 'ethic' among various species according to which selfish individuals are punished or isolated from the rest of the group. Among nonhuman primates, this behavior is well documented. In the most famous of such behaviors, individual hunters who successfully obtain meat will share it with the rest of their group in an overtly altruistic gesture. But successful hunters who fail to share are ostracized from the group and generally do not benefit from the altruism of others.

Third, we expect that there should be strong effects of reputation within a social group, for if successful strategies will be altruistic toward altruists, then it becomes important to establish a reputation for being altruistic. In cognitively sophisticated species, we would expect to see strong effects of reputation within a social group. Among humans, this effect has been documented by experimental economists, who have performed experiments in which individuals are observed after being given the opportunity to establish a reputation (Camerer and Thaler 1995, Frey and Bohnet 1980, Thaler 1988). Perhaps more surprisingly, it appears that this same effect occurs even among guppies. Experiments have shown that guppies prefer to associate with other guppies that exhibit the altruistic behavior of inspecting potential predators (Dugatkin and Alfieri 1991a, b, surveyed in Sober and Wilson 1998). Thus, we may have a form of 'reputation' even within species that have severely limited cognitive powers.

4.2   Correlation

Another highly influential consideration for game-theoretic analyses is the effect of correlation upon a population. By 'correlation', we mean a tendency for individuals with similar strategies to interact with each other. Applied to the problem of altruism, any assortative mechanism that brings about correlation will increase the probability that an altruist will interact with another altruist, and that selfish types will tend to meet each other.

Correlation is closely related to some of the considerations that we have made earlier. For as we have seen, altruists will benefit from interacting with each other, while selfish individuals are harmed when they interact with each other. The easiest way to see the effect of correlation upon the evolutionary altruism is to consider an extreme case of perfect correlation. Suppose that individuals in a population are sorted perfectly according to whether they are altruistic or selfish. In such a case, the altruists will always receive a payoff of R, while the selfish types will always receive the lower payoff of P. Thus, if there were perfect correlation, then an individual would be better off as an altruist.

Of course, correlation will rarely be perfect. There will always be some positive probability that a selfish type will interact with an altruist. But correlation need not be perfect in order for altruism to have the advantage. Given any values of T, R, P, and S, one can easily compute the degree of correlation that is required if altruism is to win out over selfishness. We will omit that computation here.[9] But it turns out that there are a variety of well-known mechanisms that increase the probability that like interacts with like. Perhaps the most important of these is simply the tendency of individuals to interact with genetic relatives, for if genetic relatives tend to be born in the same geographic area, and genes play a role in determining behavior, then individuals will tend to meet similarly behaving individuals to the degree to which their interactions are 'local'.

Models along these lines have been explored in a biological context by Sewall Wright (1943, 1945, 1969). In Wright's model, a population's tendency to remain within a particular geographic area is its 'viscosity'. High population viscosity entails that the members of a population are very likely to remain in a small area, while populations with lower viscosity will tend to wander across a larger region. In philosophical work, the effects of local interaction have been explored by Brian Skyrms (1999, 2004) and J. McKenzie Alexander (2000), who have shown that the effects of local interaction may be surprisingly powerful.

Thus, we are led to another specific prediction that emerges from the game-theoretic analysis. Populations with high viscosity should be expected to exhibit higher degrees of correlation, and hence, more altruistic behavior. Indeed, precisely this phenomenon is observed among nonhuman primates. Among the great apes, the species with the lowest population viscosity is the orangutan, and it is the orangutan that exhibits the least altruistic behavior.

On the other end of the spectrum, chimpanzees and bonobos have very high population viscosity, and they exhibit a great deal of altruistic behavior.

5.   FUTURE DIRECTIONS

So far, I hope to have motivated the conclusion that game theory is a valuable tool for gaining insight into the evolutionary origins of altruistic behavior. In spite of its abstractness and great generality, it yields a series of specific predictions that can be confirmed or disconfirmed by empirical observation. Thus, game-theoretic models are not 'just-so stories' – they are testable models capable of yielding insight into the evolutionary origins of puzzling behaviors. But as provocative as the preceding results are, however, I believe that the most fascinating work in this area is yet to be done.

As is illustrated by the preceding discussion of iteration and correlation, one of the most valuable features of game theory is that additional relevant features of the interaction may be integrated into the game-theoretic models. For example, although game theory does not necessarily take iteration and correlation into account, it is a very simple matter to build those features into the models. We should be looking for other features that are likely to have played an important role in the evolution of altruism, as well as other important social behaviors. To conclude this survey, I will briefly indicate two areas where important work remains to be done.

First, although current work on local interaction is highly provocative, there is a wealth of relevant population structures that should be studied further. For example, anthropologists are very confident that our evolutionary ancestors lived in a specific type of metapopulation structure in which large populations were divided into small bands of genetic relatives (Pusey and Packer 1986). Members of those bands would tend to migrate to other nearby bands, and their migration was heavily biased according to their age and gender (Cheney 1983, 1986). On the face of it, this population structure contains all of the features that would tend to favor the evolution of altruism (Ernst 2001). But much more work needs to be done on the evolution of large metapopulations such as these.

Second, interactions among the nonhuman primates are highly structured in a manner that is much richer than the simple correlation models that have been discussed. In most nonhuman primate species, there is a powerful dominance hierarchy that dictates the type of interactions that are observed (Colvin 1983, Essock-Vitale and Seyfarth 1986). Individuals that are highly ranked in the hierarchy tend to interact primarily with other high-ranking individuals, and the roles that each rank plays in the group are importantly different. Anthropologists are well aware of the fact that rank plays a crucial role in determining the structure of primate societies. However, there has been little work on this from a game-theoretic perspective.[10]

Although it would be a significant complication to the game-theoretic models, we could assume that each player in the population has an associated rank that affects the payoff structure of an interaction. It is generally believed that the existence of dominance hierarchies explains why primates form coalitions and cooperate within those coalitions to defend themselves against high-ranking individuals. It has been suggested that this coalition formation behavior is qualitatively similar to the sophisticated politics that are observed in modern human societies. As a working hypothesis, it may be reasonable to suppose that coalition formation (as well as its associated cooperative and agonistic behaviors) emerged as a direct result of differences in rank between individuals. A game-theoretic model that showed this analytically – or refuted this conjecture – would constitute an important advance.

NOTES

1. This is so-called noncooperative game theory. There is also 'cooperative' game theory, in which the players are able to make binding agreements before they play their strategies. However, I shall ignore cooperative game theory in this essay.
2. It is not unusual to come across arguments that their line of reasoning is unsound. But see Binmore (1998) for a compelling survey and refutation of such arguments.
3. Kitcher (1988) is the canonical expression of this 'generalist' view of scientific explanation.
4. For a thorough review of the ESS concept, see W. G. S. Hines (1987).
5. See, for instance, Bergstrom and Stark (1993); Bernheim (1984); Camerer and Thaler (1995); Güth, Schmittberger, and Schwarze (1982); Krepps (1990); McKelvey and Palfrey (1992); Ortona (1991); Rabin (1993); Selten (1975); and Sopher (1993).
6. An excellent formal introduction to evolutionary game theory is Weibull (1995), and a brief informal introduction may be found in Sober and D. S. Wilson (2000). For those who are interested in a survey from an economic perspective that combines technical sophistication with thorough informal motivation, I suggest Larry Samuelson (1997).

7. Technically, we should also require that $(T-S) < 2R$, for if that condition is violated, then individuals who play an Iterated Prisoner's Dilemma do best by taking turns 'exploiting' each other. However, I will not be concerned with this condition in what follows.

8. Axelrod (1984) refers to a strategy with this property as 'nice'.

9. See Skyrms (1994) for an excellent discussion of correlation in simple games from the perspective of decision theory, and Sober (2000) for a straightforward demonstration of how the cost of altruism is related to the level of coordination required for altruism to evolve.

10. Kitcher (1999) also makes a suggestion along these lines, although his proposed model bears little resemblance to the observed behavior of primates. Yasha Rohwer (forthcoming) has also argued convincingly that rank may play an important role in the evolution of so-called altruistic punishment, in which one individual punishes another at a cost to the punisher, and to the benefit of the rest of the group.

17   What Is an 'Embryo' and How Do We Know?

Because of recent public excitement about cloning and embryonic stem cell research, more people than just developmental biologists are busily talking about embryos. Human embryos are central players in proposed legislation at state, federal, and international levels. But what is meant by an embryo? Rarely is the term defined or defined clearly. Yet the term is used in quite different ways and has evolved over time.

How have meanings changed, and for what reasons? What is the relationship between public and scientific understandings of embryos? Here, the focus is most directly on evolving understandings of the biological embryo, including recent shifting public meanings. In each case, both metaphysical and epistemological considerations are important. Yet only after the emergence of in vitro fertilization did the embryo become an object of significant ethical concern, and only with cloning and human embryonic stem stem cell research was it widely seen as an object of social concern. This essay considers the changing understandings of embryos.

Since at least 1771, with the appearance of the first edition of the *Encyclopedia Britannica*, the embryo has been seen as the earliest – and undifferentiated – stage of an individual organism's development. The embryonic stage was clearly separated from the fetal stage, with the first giving way to the second as form gradually emerged from unformed matter. Specifically, the 'embrio' was "in physiology, the first rudiments of an animal in the womb, before the several members are distinctly formed; after which period it is denominated a foetus. The 'foetus' denotes the child while it is contained in the mother's womb, but particularly after it is formed, til which time it is more properly called embrio" (*Encyclopedia*

Britannica 1771). In the eighteenth century, the change was taken as occurring at "quickening" or after, while today the shift from embryo to fetus is defined for humans as occurring at roughly eight weeks.

Clearly, the embryo has long been seen as unformed, as undifferentiated, and (following Aristotle) as having the potential to become an individual of the appropriate type but as not yet having been actualized as such. The *Oxford English Dictionary* offers a similar picture. Early usages of the term in the seventeenth century emphasized the 'Embryo' as "A thing in its rudimentary stage or first beginning; a germ; that which is still in idea as opposed to what has become actual in fact." It remains "in an undeveloped stage, 'that is to be'," presumably from the Latin 'Embryon' with a suggestion of "swelling within."

The past century of embryology textbooks has continued to provide similar interpretations, with emphasis on the coming into being and gradual emergence of form through the process called epigenesis. Even when embryology had begun to be called developmental biology, and even as the presumed efficacy of genetic inheritance began to overwhelm the previous presumed causal force of epigenetic emergence, the 1961 *Britannica* captured a typical understanding of the embryo. The 'Embryo' entry was written by Aute Richards, an emeritus University of Oklahoma zoologist who had been director of the School of Applied Biology and had written the widely used 1931 *Outline of Comparative Embryology*. He portrayed the embryo as beginning with the biological action of fertilization and existing through the process of cell division, through the cleavage of one fertilized egg cell into many. The early developmental processes occur with largely undifferentiated cells, and only gradually do they become separated histological types. Richards summarized that "it is not until these histological changes are accomplished that the young embryo is ready to function fully and to take an independent place in the world" (*Britannica* 1961, "Embryo") Of course, some species move from the embryonic form through a fetal or larval stage, but the earliest embryonic form of all species consists of undifferentiated cells and is unformed.

This was the epigenetic view that the embryo is the stage when form emerges gradually from the unformed matter, where 'unformed' means lacking in organic differentiation and without the body parts and systems that will arise later. Epigenesis was the dominant interpretation of the embryo into the twentieth century. The alternative was preformation, with form presumed to be already present from the beginning (see Maienschein 2003). To document the shifts in epigenetic and preformationist thinking, and to understand the changing patterns of debate, it is useful to focus on a sequence of six selected historical episodes. Underlying metaphysical questions about the nature of life, organisms, and parts provided one focus, while epistemological questions about how best to understand the emerging organism provided another. What was thought to be at issue shifted over time and in different contexts in ways that are instructive for our understanding of current debates about embryos.

HISTORICAL INTERPRETATIONS

Of course, there are many ways to divide up the millennia starting with Aristotle, but the following selected slices capture the range of shifting central issues. Each episode raises new questions and introduces new relevant factors, but each case also reveals instructive decisions about what is being studied, how to do the studying, and what relevant factors should be brought to bear in interpreting the results.

KEY PERIODS IN UNDERSTANDING EMBRYOS

The *Hypothetical Embryo* remained largely invisible and a matter for theoretical interpretation from Aristotle to the Enlightenment. Eighteenth-century debates laid out the traditions of preformationism and epigenesis that have continued.

The *Physical Embryo* of the midnineteenth to early twentieth century introduced comparative study of embryos in many species to describe the details of organismal change. When the work was done in the context of evolutionary theory it influenced the interpretations of developmental stages in important ways.

The *Biological Embryo* gained attention in the 1920s–30s, the embryological "golden age," with emphasis on the "organizer" and processes and causes of differentiation.

The *Inherited Embryo* of the 1950s–60s, with frog cloning and nuclear transfer, appeared with an enthusiasm for geneticism and eventually genomicism, and for reductionist methodologies.

The *Visible Human Embryo* started in the 1970s with in vitro fertilization (IVF) that took embryos out of the womb, with Nilsson's widely published photographs of fetuses that came (inaccurately) to represent embryos, and with use of other imaging techniques, all in the context of abortion politics.

The *Constructed Embryo* arrived with genetic recombination, cloning, and stem cell research that have allowed researchers to construct, deconstruct, and reconstruct embryos. Because of the fears and promises, the embryo becomes a public as well as a biological object.

We might also point to a seventh period, of the *Computed Embryo*, with an emphasis on collected data management and informatics, but that raises different issues and therefore will not be considered in further detail here.

The Hypothetical Embryo

The Hypothetical period, drawing more on theory and inference than on observation, provides important background about the interpretations that dominated thinking for more than two millennia. Aristotle outlined an epigenetic hypothesis for embryonic development that remained the only serious interpretation until the eighteenth century. According to Aristotle, embryo development was part of the natural processes of generation and corruption. Generation of animals, through sexual reproduction, involves combining fluids (or "semen") from both parents. This mingling of male and female fluids provides the material and the motive force for development.

More specifically, the female contributes the material cause that resides in the menstrual blood, and after "the discharge is over and most of it has passed off, then what remains begins to take shape as a fetus." This menstrual blood is not pure, however, and is simply "that out of which it generates." The material must be acted upon by the male fluid, which provides the formal cause and initiates the efficient cause for the development that follows. The formal and efficient causes therefore both act through the joining of the male and female fluids. Only then can the final cause serve as the telos for the living organism. Aristotle's four causes together bring about generation of each individual organism. "Thus things are alive in virtue of having in them a share of the male and of the female." The male and female serve as the "principles of generation" (Aristotle 1979, 99, 111, 133, 129).

Aristotle urged that the form must be guided by internal and not outside causes. "From the outset," an individual life requires a "soul" that guides the gradual unfolding of form (or epigenetic process) from the unformed matter. This soul – consisting of vegetative soul for all living beings, plus locomotory soul for all ambulatory beings capable of picking themselves up and moving around, plus rational soul for humans alone since only humans have the power to reason – gives the potential to become actualized as an individual organism.

Aristotle did not picture an embryo in our sense of a material cell that is fertilized by another cell to form a new union. Rather, his embryo was more process than object, and it was theoretical rather than observed for most species. He would have been able to see eggs only in such nonplacental animals as chickens, frogs, or insects. Yet what Aristotle did see, especially in chicks, looked initially unformed and only gradually actualizing the potential through the formation effected by epigenetic emergence.

As usual, Aristotle's interpretation was reasoned, accorded with available observable evidence, and provided an explanation for the manifest developmental processes of growth and differentiation. His epigenetic interpretation dominated into the eighteenth century and found resonance with leading Catholic thinkers. Saint Augustine held that the process of giving rise to a human life was gradual and that the human only becomes human after the fetus is formed and growing, after quickening. Abortion was considered a sin, but not homicide until after full "hominization" had occurred. An embryo was material and was alive, but it was not yet a human. For Saint Thomas Aquinas, the fetus first acquires a vegetative soul and begins to grow larger. Then later it acquires an animal soul and begins to become differentiated with all its animal parts. Only then does it acquire its intellectual or rational soul, and only then is it fully ensouled and human. This interpretation of "delayed hominization" dominated early Catholic history.

Epigenesis prevailed despite influences such as Pope Sixtus V, who issued the first papal declaration on the subject in 1588 apparently because of worry about rising prostitution in Rome. Sixtus decreed that contraception and abortion at any stage constituted homicide. Yet three years later when he died, Pope Gregory XIV pointed to the standard theological understanding that hominization occurs only gradually and returned to the long-standing interpretation that development is epigenetic and that the human emerges only later with full ensoulment. With time, additional observations, and additional philosophical reflection, however, other interpretations emerged from within natural philosophy to challenge Aristotle's epigenesis.

Some researchers were pushed toward an alternative because of their metaphysical materialism. If all that exists are matter along with the motion it experiences, then they asked how epigenetic development could yield form from nonform. Surely, such emergence requires some unacceptable vital force or directive, like Aristotle's hypothetical causes. The demand for materialistic metaphysic therefore led to preformationism. The form simply must be already in the very earliest moment of an individual's life. Otherwise, how could the necessary sorts of change occur (Roe 1981)?

This led to heated eighteenth-century arguments about whether an organism begins more or less literally preformed and just unfolds (or "evolves"), or whether it arises gradually and epigenetically through a process of embryonic development. While materialists emphasized the metaphysical unacceptability of hypothetical and apparently nonmaterial or vitalistic causes of emergence of form from nonform, however, epigenesists insisted on the primacy of an epistemology based on observation. And observation did not reveal tiny little already formed beings from the beginning. In important ways, this debate about the relative primacy of metaphysics or epistemology, about unfolding or emergence, about preformation or epigenesis, has informed all discussions since and even lies at the root of today's heated debates (see Pinto-Correia 1997).

The Physical Embryo

Debate began seriously in the late eighteenth and early nineteenth centuries, when embryos became Physical and visible in nonhuman species. Microscopes and an ethos of natural philosophy encouraged observation. The questions were, What could observations reveal, and what did the observations mean?

This debate played out, for example, in the work of Caspar Friedrich Wolff and Charles Bonnet, both looking at chick development. They looked at the same thing and even agreed about what it was that they saw, but their conclusions differed. Wolff was an epigenesist, for whom form emerges only gradually; Bonnet was a preformationist, who insisted that form must exist from the beginning of each individual organism. This is a story about competing metaphysical and epistemological convictions (Roe 1981, Maienschein 2003)

In 1759, Wolff studied the chick egg from fertilization through the twenty-eight-hour stage, which is shortly before the heart becomes clearly visible and begins to beat. As Wolff looked at chick after chick, hour after hour, he saw change, yes, but no chick. He did not see the chick form, a beating heart, or any small preformed chick. Instead, he witnessed movement and gradual change. He did consider that perhaps the form was just not visible yet because it was so tiny, but then stronger microscopes should reveal more detail and they did not. Wolff concluded that we should trust our observations. If we cannot see something, then we can legitimately assume that it is not there. This is a strong epistemological assumption about the nature of knowledge and justification, and it helped that Wolff's interpretation also conformed to the standard Arisotelian epigenetic interpretations.

Ten years later, Bonnet also understood the power of empirical observation. He also looked closely at many chicks, and he agreed that he did not see the formed chick before the twenty-eight-hour stage. He agreed with Wolff that they are not visible yet. But Bonnet concluded that the preformed form of the chick must be there, just somehow hidden in the egg. Since we know that form exists later, and it must arise through the actions of matter and motion, therefore it must be present at the beginning. Since vital forces were unacceptable to this materialist and since there was no other explanation for the gradual emergence of form, Bonnet concluded that the form had to exist already. If natural knowledge is to rely on observation, on logic, and on a proper materialistic metaphysics, then there could be no further question. An organism must be preformed in some way.

In retrospect, we see other alternatives. Bonnet might have said, as Newton did about gravity, "I do not know how form arises. Hypotheses non fingo." But Bonnet wanted an explanation of the origin of individual form and concluded that it must reside in preformation. The results were debates on several levels and the coexistence of competing interpretations of individual development with epigenesis and preformationism.

Further observations of embryos introduced new grounds for debate. So far, observers had seen chick, frog, and a variety of insect eggs, but it was not yet clear whether mammals also have eggs. Some assumed that all animals share the beginning as an egg. In 1827, Karl Ernst von Baer announced his discovery (in a friend's dog, killed for the purpose of experimental study) that even mammals have eggs, though it is difficult to observe their normal development since the eggs remain inside mothers. This led to enthusiastic study of the developmental stages of embryos in as many species as it was possible to study. Improved microscopes and microscopic techniques play a central role here as they made embryos in a growing number of species, and at increasingly earlier stages, more visible. Representations in illustrated plates and in wax models were important in presenting the embryo to other researchers and to the public (Hopwood 2002).

The last half of the nineteenth century was also dominated by the importance of the embryo for evolutionary theory. Darwin pointed to embryology as fundamental for interpreting relationships. In chapter 13 of the *Origin*, he asked:

How, then, can we explain these several facts in embryology, namely the very general, but not universal difference in structure between the embryo and the adult; of parts in the same individual embryo, which ultimately become very unlike and serve for diverse purposes, being at this early period of growth alike; of embryos of different species within the same class, generally, but not universally, resembling each other; of the structure of the embryo not being closely related to its conditions of existence, except when the embryo becomes at any period of life active and has to provide for itself; of the embryo apparently having sometimes a higher organization than the mature animal, into which it is developed.

We know this was a rhetorical question, and sure enough he concluded, "I believe that all these facts can be explained, as follows, on the view of descent with modification." And that furthermore,

the leading facts in embryology, which are second in importance to none in natural history, are explained on the principle of slight modifications not appearing, in the many descendants from some one ancient progenitor, at a very period of the life of each, though perhaps caused at the earliest, and being inherited at a corresponding not early period. Embryology rises greatly in interest, when we thus look at the embryo as a picture, more or less obscured, of the common parent-form of each great class of animals.

In his *Generalle Morphologie der Organismen*, Ernst Haeckel went further. He saw "ontogeny as the brief and rapid recapitulation of phylogeny" and saw each individual's development as following the sequence of, and indeed being caused by, the evolutionary history of that individual organism's species. In his highly popular and widely translated books, Haeckel offered pictures of comparative embryology. "See," he seemed to suggest, "the human form emerges following the evolutionary development and adaptations of its ancestors." Form arises from form of the ancestors and unfolds following prescripted stages.

Darwin was not an embryologist; nor did he contribute directly to our understanding of the embryo. Nor did Haeckel. But while Darwin's use of the embryo in supporting evolutionary theory and in helping to interpret evolutionary relationships was consistent with various versions of either epigenetic or preformationist development, his view was decidedly preformationist. His was another preformationist interpretation based not on observations but on the metaphysical demands of his form of monistic materialism and motivated by his desire to provide evidence for evolution. This provided the context in which those studying cells and embryos worked at the end of the nineteenth century.

Only in 1869, that is, shortly after Darwin, did the Catholic Church alter its long-standing epigenetic interpretation, when Pope Pius IX decreed that hominization is immediate and begins implicitly at "conception." Unfortunately, his *Apostolicae Sedis* gives few clues about what led to his interpretation, which overthrew centuries of Aristotelian thinking, nor whether he was drawing on the recent biological discoveries of fertilization and of the mammalian egg.

Meanwhile, in Germany the anatomist Wilhelm His turned to human embryos. He collected every human embryo he could find and set up networks of physicians to contribute, seeking to establish the patterns of human development. The American anatomist

Franklin Paine Mall studied with His and carried human embryology to the United States. When His died, his collection went to Mall, then at the Johns Hopkins University. In 1914, Mall persuaded the Carnegie Institution of Washington to support his growing embryo collection, that led to the national Human Embryo Collection, which is still the most important source of human embryo material and history (Maienschein, Glitz, and Allen 2004).

His and Mall's embryos were all necessarily dead, since there was no way to study living human embryos inside the mothers. The embryos were seen as material objects, without questions about the appropriateness of collecting and studying them. Evidence that the public largely agreed includes the fact that the state of Maryland's Department of Public Health urged physicians to contribute. There were no known complaints about the project, perhaps in the belief that improved understanding of human embryonic development would have medical therapeutic value, over time.

The Biological Embryo

The 1920s–30s brought, as the Yale embryologist Ross Harrison put it, a "gold rush" of studies of causes and processes of differentiation. What causes the unformed to become formed: material mechanical changes within the embryo itself, as His had argued, or some set of special directive forces within or outside the embryo? Was there something unique about the living organism? Did we need what biologists by the twentieth century regarded as a metaphysically questionable vital force to explain emergence of form (and Hans Driesch did give up embryology and take up metaphysics precisely on this assumption), or is it something about the nature of the organic matter and its organization that allows development and differentiation of complex forms? How should embryos be studied?

Hans Spemann theorized that the tissue from the dorsal lip of the blastopore in amphibians has special powers to "induce" the rest of the embryo and to serve as a material "organizer" to produce differentiation and morphogenesis (Hamburger 1988, Maienschein 1991). Dozens of researchers took up the challenge to find the precise nature of the organizer. There is little evidence that this theory reached the general public or even biologists in other specialities often or in much detail, but an educated lay audience did ask whether such organization resulted from mechanistic or vitalistic forces. The physicist Erwin Schrödinger reached this wide audience when he asked, "What is Life?" and that discussion was clearly informed by embryo research.

The episode has an internal logic but also raises questions about our selective historical memory. This episode started with transplantation experiments of the 1890s and into the early twentieth century. These included the first stem cell experiments by Harrison with neuroblasts in tissue culture, and other experiments with nuclear transplantation. Today's stem cell researchers know little of this history of their own research, which leaves them – and the wider public – with the impression that something new and amazing (or horrifying) has been invented just in the last few years. Instead, stem cell and cloning research today is rooted firmly in traditions carrying back to Aristotle, through the work of the transplantation researchers of the late nineteenth and early twentieth centuries.

Harrison, Spemann, and their leading contemporaries assumed a metaphysical materialism. What exists are matter, its patterns of organization, and change over time. There is no room for vital forces or fluids, and they assumed that development is epigenetic. There is no form from the beginning, but it emerges gradually, over time, and guided by internal forces and factors. But how? That was the research program, focused on discovering the material processes that shape the embryo into an individual organized organism of the right sort. Epistemologically, they assumed that experimentation was the appropriate approach. Since it was not possible simply to observe natural processes and to see inside the egg and embryo, it was necessary to contrive experimental conditions. Manipulation of conditions, carefully controlling the environment as much as possible, could produce new knowledge. It was that new knowledge, taken collectively, that would reveal the patterns and processes of what came to be called morphogenesis, or the appearance of form.

The most important of the experiments were Harrison's on nerve fibers and Spemann's on the organizer. Both began with the idea that transplanting pieces of a developing embryo from one organism to another could reveal the relative contributions from the donor and the host. Working with frogs, which are abundant, have large eggs, and are easy to manipulate experimentally, they both saw the power of "heteroplastic grafting," or taking and recombining pieces from animals that look different – with different colors or sizes, for example. This made it easy to tell which tissue was from which organisms.

In 1907–10, Harrison refined the first tissue culture technique with neuroblast cells (essentially today's neural stem cells, known to give rise to nerve fibers). He "explanted" these cells, transplanting them out of a developing frog into a culture medium in a dish. They grew out as nerve fibers, apparently just as they would have done under normal conditions. This suggested that the cells contained an intrinsic capacity for differentiation, yet under normal conditions that differentiation would also be constrained and directed by the environmental conditions for each particular cell. The conclusion was clear: an embryo has internal capacities for development, and it also depends on cues and input from factors external to the egg and embryo itself.

Spemann focused on the earlier stages of development, on the stage when the blastocyst undergoes gastrulation. That is the first stage when the embryo begins to become visibly differentiated. The undifferentiated clump of cells undergoes rapid cell movement, with a flowing of cells into what is called the blastopore and the formation of germ layers. The clump becomes an organized ball of three layers that will become different parts of the organism. This is the first time when there is clear organization. Spemann asked, How does this happen? What causes the apparently undifferentiated mass of cells to become organized?

Perhaps, Spemann hypothesized, there is an "organizer." This must be material and it should be accessible through experimentation. Indeed, he found that the dorsal lip of the blastopore (that is, a particular set of a few cells at the still unorganized and undifferentiated blastocyst stage of development) induces organization of the cells. This particular material seemed to set up the layering of cells into the three layers. Other research suggested that the process is much more complicated, but during the 1920s and 1930s there was tremendous excitement about what seemed to be discoverable material causes of the production of form from the unformed cells.

Continuing research has reinforced this early conviction that it is the blastocyst stage of development, around days five to fourteen in humans, that is the beginning of organization and differentiation. This is the stage at which, in humans and other mammals, the preimplantation embryo must be implanted in the mother and must begin to exchange nutrients with the mother, or it will not survive. It is also the stage at which the embryo begins to grow, as it absorbs nutrients from the mother. And it is the stage when the clump of largely identical cells, now called embryonic stem cells, begin to undergo differentiation and development.

In addition to the biological scientific research, the social and cultural context began to have some influence. Harrison chaired the U.S. National Research Council (NRC), which promoted blood transfusions during World War I, and his tissue culture discovery inspired Rockefeller University's Alexis Carrel to expand tissue culture study for therapeutic applications. It is this research and the assumptions that underlie it that have provided the tradition of cell line development and application that have led to today's hopes for effective stem cell therapies. Harrison and Spemann were studying frogs, while Carrel and his medical colleagues worked with humans (Landecker 2004).

It seems likely that even if biologists had been able to study human embryos, the public response would have been positive. There is no evidence of early twentieth-century social concern about embryos. Human embryos were invisible, inside women. Because mammalian embryos remained hidden, early development remained a mystery to most people. The assumption was that human embryo development was similar to that of other animals, but even the experimental study of primate reproduction that began in the early twentieth century in places like the Carnegie Institution of Washington was slow to reveal insights about the earliest developmental stages (Maienschein, Glitz, and Allen 2004). Embryos must develop gradually, epigenetically, with form emerging through a process of stages. But how, and what directs the development? Is there really an "organizer," and if so what is it and how does it work? These questions remained.

The Inherited Embryo

The 1950s yielded one sort of answer with the Inherited embryo. The discovery of the structure of DNA, the stuff of heredity, has been well documented (Olby 1974, Judson 1979). Genetics had emerged as a field of study, and researchers had been exploring both the effects

of the theoretical units called "genes" and the chromosomal structure presumed to contain the genes. Yet the study was largely
abstract, and biologists such as Harrison did not see genetics as
contributing anything to the study of development. As department
chair at Yale, Harrison saw no point in hiring a geneticist, for
example, since it was the study of organisms and their processes that
he saw as the important work for biology.

Only in the 1950s did genetics begin to link effectively through
molecular biology such that it was possible to imagine the DNA and
its presumed genes as the concrete material basis for heredity. This,
in turn, suggested that heredity provides the underlying causal
shaping of developmental processes. Since every cell contains the
same DNA and genes, and yet the cells begin to differentiate and to
produce morphogenesis over time, it might be that the genes and
DNA carry the necessary information to guide development. Genes
could be the organizer that Spemann had sought.

In 1938 Spemann had suggested a "fantastical experiment" that
would get at the relative contributions of the nucleus – with its
chromosomes and genes – and cytoplasm. It was a conceptually
simple transplantation experiment, removing the nucleus of one egg
and replacing with another nucleus. He did not carry out this
experiment, or not successfully, but in 1951 Robert Briggs and
Thomas King did. They transferred frog nuclei to produce a new kind
of hybrid. Cloned frogs graced the cover of popular magazines such
as *Time* and *Newsweek*. John Gurdon went further in the 1960s,
transplanting donor nuclei from frogs in later developmental stages
into the eggs. He transplanted donor nuclei from an albino frog into
the egg of a normally pigmented mother, and the offspring turning
out to be like the donor nucleus, all albino. This suggested the very
strong predominant influence of nuclear inheritance over development. Yes, epigenetic development might occur gradually, through
time, as form emerges from the unformed material. But the guiding
direction seemed increasingly to originate in the nucleus and the
gradual expression of genetic information coded there. Accumulating evidence of this sort reinforced the idea that development is
not only loosely directed but actually caused by the genes. And if
so, then the information and determinants for development and
differentiation are already present at fertilization. It seemed that
development occurs by preformation after all, or at least by some

version of genetic predetermination. To what extent is development actually determined or fixed, then, and to what extent can it respond to changing environmental conditions? Opinion shifted toward increasing determinism, reinforced by the enthusiasm for the Human Genome Project of the late 1980s and 1990s.

Public response to discussions of development during this time was limited, but what reaction there was remained largely curious rather than critical or concerned. The research might have raised new philosophical and biological questions about individuality and identity? After all, if an egg did not need its own nucleus, but another would do, what is the biological basis for concepts of self and other, of identity and autonomy? But this questioning seems not to have occurred to any significant extent. This research was largely taken as just that, biological research, to be carried out without much thought about its interpretations or implications.

The Visible Human Embryo

It has been only since 1978, when the human embryo became literally visible, that embryos have become an object of wide public interest. The birth of the first "test tube baby," the lively and normal little Louise Brown, first took human embryos out of the mother and into the public eye. Socially and culturally, we are still sorting out the implications of this discovery that the earliest human developmental stages can take place separately from the mother, and the questions raised about "what's in the dish." An embryo is still identified as beginning with the process of fertilization until, in humans and other mammals, the developing and differentiating organism gains all its organ systems and becomes a fetus. What happened after 1978 was initially not about changing scientific definitions of the embryo, though the emphasis shifted from animal studies to human embryo research for the first time. The primary changes right away were that the embryo gained individual meaning for prospective parents as well as public meaning, especially in the context of abortion politics.

Clearly, technology and images have been very important in introducing this scientific research to the public. Lennart Nilsson's photographs had provided a background of assumptions and interest. His stunning pictures were taken with a scanning electron microscope and endoscopes and showed the fetus in the womb.

The *Life* magazine presentation in 1969 gave most people their first images of the developing human. The fact that these were fetuses, and often later-stage fetuses, and the fact that these were highly colored and contrived pictures were not part of the public impression. Instead these images of little persons, sucking thumbs and looking innocent, became the public image of embryos. Many people, and perhaps a significant majority, still imagine embryos as these tiny clearly formed beings floating in the womb. This is not the biological embryo, but since it has become the public embryo for many people, any attempt to understand shifting meanings must take this misconception into account.

Another important social shift in meaning of the embryo comes with the recognized clinical importance of that clump of undifferentiated cells in a dish for patients. At significant personal cost in the United States and with significant personal investment in any case, those individuals engaging in IVF have come to see the embryo as the beginning of their baby. The biological fact that before implantation, this clump of cells is really just that, a clump of undifferentiated cells, does not take away the social and medical meaning invested in those cells.

Making the human embryo visible, with its potential for medical advantage, has produced increased funding, as well as increased public nervousness. Was IVF safe? For whom and under what conditions was it desirable, and who should pay for this in vitro process if it was considered a medical treatment? Countries have had different responses to embryo research. In the United States, IVF and embryo research generally has remained unregulated and largely privatized. What regulation exists is at the state level through laws and court case rulings, with overlapping and often contradictory results. The only federal restrictions are on funding through the appropriations bills for the National Institutes of Health. Canada, Australia, and the United Kingdom had different responses in 1978 and since, accepting IVF as a public good with public funding and public regulation. As a result, the human embryo is differently visible in each country.

The Constructed Embryo

In 1997, Ian Wilmut announced the first cloned mammal, Dolly the sheep (Wilmut 2000). As with Briggs and King's frogs, Dolly was produced by transplanting a donor nucleus into a host egg from which the nucleus had been removed. In this case, however, the donor nucleus came from an adult, and the result was seen as an "unnatural" hybrid that challenged assumptions about what is possible. Wilmut and his team brought experimental embryology to the public very dramatically and thereby first made the Constructed Embryo public though such research had already been long under way inside research laboratories. That Dolly existed, suddenly and surprisingly, as Louise Brown had existed suddenly and surprisingly out of IVF, challenged treasured assumptions within society as well as within science. Perhaps clones would develop differently than normal organisms, or perhaps they would have a diminished (or enhanced) life. What is normal and what is acceptable for embryos? the public began to ask.

Embryonic stem cell research, which reached public attention just a year later, raised new questions – both scientific and public – and generated prospects for regenerative medical therapies. But this research can (at least for now) best be carried out by harvesting undifferentiated embryonic stem cells from embryos. As we know from the heated debates, some members of the public find this unacceptable because they make the assumption that an embryo is already a person (or at least a potential person) and therefore we do not have the right to harm it.

This discussion gets right at the heart of what we mean by an embryo, and how we know. Scientists urge that we draw on scientific understanding of the embryo. A human embryo, especially before it is implanted in the mother, is really just a bunch of undifferentiated cells. To the best of available knowledge, no significant gene expression has begun; there is no differentiation; there is no significant growth. This is a bunch of cells dividing and dividing, at least up through the blastocyst stage. Only at that point, as Harrison and Spemann recognized, do differentiation and morphogenesis begin to occur. Only with implantation and gastrulation does the embryo begin acting and begin the epigenetic processes of development, informed by the heredity carried in the genes.

CONCLUSIONS

What, then, is an embryo? Some biologists prefer to call the preimplantation stages a "preembryo." Others urge that it would be easier and politically safer to drop the term 'embryo' altogether, though that is surely politically naïve. Who gets to decide what an embryo is? On the face of it, this is a biological question since embryos are biological objects. Therefore, biologists should at least have a say. They are quite clear on the matter. An embryo is not yet formed in the sense of structured with functioning differentiated parts, and a preimplantation embryo is really little more than a bunch of undifferentiated cells. A two-, four-, or even eight-celled preimplantation embryo can become twins, quadruplets, or even octuplets. Up through the eight-cell stage, the cells can even be pulled apart in the lab and the separated cells can develop individually. Or one or two or more cells can be removed (perhaps, as is commonly done now in fertility clinics, to test one of the cells genetically), and the rest can develop normally. This is clear. Biologically, an embryo is not yet formed, not yet differentiated, not yet recognizably human, and indeed not even unalterably an indivisible single individual.

Yet biologists do not alone "own" embryos. As public objects, embryos are much more complicated. A large political group would like to define embryos as beginning with fertilization, and as having status as embryonic human persons at that time. What we must realize is that in doing so, they are invoking metaphysical assumptions that lie outside science and that often depend on religious assumptions that are not shared by the larger community and cannot be justified on scientific or any other clear-cut grounds. Their epistemic warrant comes from such claims of "intuition" or divine knowledge or pure conjecture. Such meanings are, of course, highly problematic. We have yet to work through ways to deal with cases where the biological and social become so intertwined as in this case.

We need to ask what authority and what processes we have for carrying political and social decisions to restrict research back into the laboratory. Does the loboratory have any sort of protection, as some argue; is there a right to carry out scientific research? If so, when can there be limits and how are they to be imposed? The questions 'What is an embryo?' and 'How do we know?' remain works in progress – biologically, philosophically, and publicly.

18 Evolutionary Developmental Biology

Evolutionary developmental biology is an emerging new research area that explores the links between two fundamental processes of life: development of individual organisms (ontogeny) and evolutionary transformation in the course of the history of life (phylogeny). For some of its more ardent proponents evolutionary developmental biology, or evo-devo for short, represents a new paradigm that completes the "Modern Synthesis" of the 1930s and 1940s, while others, often those with a more astute sense of the history of biology, have emphasized the long-standing connections between these two areas of study. But all agree that evo-devo offers some of the most promising theoretical perspectives in evolutionary biology at the beginning of the twenty-first century (see for example Amundson 2005, Carroll 2005, Carroll et al. 2005, Hall 1998, Kirschner and Gerhart 2005, Laubichler 2005, Müller 2005, Wagner et al. 2000). In this essay I will first sketch the emergence of present-day evolutionary developmental biology during the last decades of the twentieth century followed by a brief overview of the central questions and research programs of evo-devo. I will conclude with a discussion of the one problem – the issue of how to explain evolutionary innovations and novelties – that has the most profound implications for the philosophy of biology. As this example illustrates, the theoretical promise of evo-devo lies in the integration of different explanatory paradigms, those of evolutionary biology and population genetics, which are based on the analysis of ultimate causes in the sense of Ernst Mayr, and of developmental genetics and physiology, which attempt to give a mechanistic account of the origins of organismal structures in form of proximate causes, that is, of molecular and cellular mechanisms (Mayr 1961). As we will see,

342

such integration is not possible without first overcoming serious conceptual and theoretical difficulties. Therefore it should become clear in the course of this essay that any future synthesis of evo-devo will be conceptual rather than simply data driven.

The concept of regulation has, somewhat surprisingly perhaps, emerged as a central topic in evo-devo (see, e.g., Carroll 2005 and Carroll et al. 2005). The idea of developmental regulation has, of course, a long history dating back to early conceptions of the organism as a self-regulating individual, of regeneration and the maintenance of a *milieu interieur*, and of course to Hans Driesch's famous experiment and infamous idea that organisms are self-regulating equipotential systems governed by entelechy, which became the foundation of neovitalism at the beginning of the twentieth century. While the mystery behind the phenomenon of individual development has always been how it is possible that such a complex and fragile process generally leads to a predictable outcome – human eggs develop into recognizable humans and seas urchin eggs into sea urchins – evolution has been perceived as an open-ended process of constant transformation, thus in many ways as the exact opposite of regulation (see Canguilhem 1979). However, development and evolution are more closely linked than one would expect. As the German evolutionary biologist Günter Osche once dryly remarked, evolution cannot put out a sign "Closed Because of Reconstruction." Indeed, all evolutionary transformations, even though they manifest themselves on a population level, need to materialize within the constraints and possibilities of a functioning developmental system of individual organisms. And the more we learn about the details of these developmental systems – about the intricate regulatory networks and linked pathways that control gene expression and thus differentiation during ontogeny – the more we realize that evolutionary transformations are a consequence of changes in these regulatory systems and other developmental processes. Regulation has thus become a concept that allows us simultaneously to understand the stability of individual development and the possibilities of evolutionary transformations.

But before we continue these discussions of the conceptual structure of evo-devo let us first briefly explore its history (see Laubichler and Maienschein 2007, Amundson 2005, Laubichler 2005, Love and Raff 2003).

1.   EVOLUTIONARY DEVELOPMENTAL BIOLOGY:
A NEW SYNTHESIS OR ''OLD WINE IN
NEW BOTTLES''?

The study of individual development of organisms has a long history that dates all the way back to Aristotle. Theories of generation, preformism, and epigenesis had a profound impact on the way life was understood as either creative and vital or mechanical as well as on the interpretation of organic forms and their history. And indeed it was in the context of discussions about the development of organic forms that ideas related to the evolutionary history of life first emerged. Darwin, of course, put these questions in focus when he laid out the following problems and observations: (1) Both organisms and species vary. (2) The variation of both organisms and species is not arbitrary, but clustered. In other words, parents and offspring closely resemble each other, but not completely, and different species can be grouped together into higher systematic groups that are united by common features. (3) The specific causes for the similarity and differences between parents and offspring were unknown, but it was clear that these causes had to act in the course of individual development (ontogeny). (4) The principles for grouping individual species together into higher systematic groups were based on comparison and the establishment of homologies. Homologies – the same organs in different individuals irrespective of form and function – were considered the basis for the hierarchical system of classification; it was, however, not always clear how homologies could be established; (5) Embryological observations (*Entwicklungsgeschichte*) had revealed that earlier (less complex) developmental stages of different species more closely resembled each other than did adult stages; (6) The genealogical perspective and the geological record suggested that less complex forms of life emerged earlier in the phylogenetic history (*Stammesgeschichte*) than more complex forms (an argument that had already been made in the context of cultural stages in the history of humankind); (7) It was clear that all explanations as well as the specific details of this history needed to be inferred, often, as with the geological record, from rather incomplete data sets, as direct observation was not an option (Darwin 1859).

Darwin himself, honed by detailed observations of the developmental stages of barnacles, carefully suggested a connection between ontogeny and phylogeny: "Thus community in the embryonic structure reveals community of descent" (Darwin 1859, 449). Others soon followed suit, exploring how the observable patterns of ontogeny could help to reveal the hidden patterns of phylogeny, most famously Ernst Haeckel, who formulated the biogenetic law, first called that way in 1872. The idea that ontogeny recapitulates phylogeny and the developmental perspective implicit in the biogenetic law also seemed to offer insights into one of the major problems of comparative biology, the establishment of homology relations. A shared developmental history, so Haeckel, should be a solid basis for the assessment of homologies. This later assumption became one of the foundations of evolutionary morphology, a program initiated by Haeckel and his close friend Carl Gegenbaur, which focused on the establishment of a phylogenetic system as the goal of comparative anatomy and embryology (see Nyhart 1995, 2002, Laubichler 2003, Laubichler and Maienschein, 2003).

This new orientation brought about by the emphasis on the relations between ontogeny and phylogeny led to some spectacular insights. In the late 1860s the Russian embryologist Alexander Kowalevsky discovered similarities between the ontogenetic sequences of *Amphioxus* and vertebrates as well as the existence of a *chorda dorsalis* in the larvae of ascidians (Kowalvesky 1867). These discoveries suggested that vertebrates were derived from the larvae of ascidians, a theory that was soon challenged by, among others, Anton Dohrn, the founder of the zoological station in Naples, who proposed an annelid ancestry for vertebrates (Dohrn 1875). This and many similar examples illustrate how, in the decades after the publication of the *Origin of Species*, the scientific problem of ontogeny and phylogeny was primarily determined by the context of evolutionary morphology (phylogeny).

Matters changed around the turn of the twentieth century, when evolutionary morphology gradually disappeared. More experimentally oriented research programs, such as genetics and developmental mechanics and physiology (*Entwicklungsmechanik*), carried the day. *Entwicklungsmechanik* initially emphasized the role of mechanical and physico-chemical causes in explanations of development; only later, in the context of physiological and genetic approaches to development, did evolutionary questions return. These concerns had, in the meantime, been transformed by the results of experimental genetics. Both Richard Goldschmidt and Alfred Kühn attempted to integrate genetics, development, and evolution; Kühn with his concept of the "Wirkgetriebe der Erbanlagen," a mechanism for realizing hereditary dispositions, and Goldschmidt with his conception of physiological developmental genetics and his discussion of chromosomal rearrangements, macromutations, and "hopeful monsters" (Goldschmidt 1940, Kühn 1955).

The most inclusive research program devoted to questions of ontogeny and phylogeny was directed by Hans Przibram at the Vienna Vivarium, a privately owned and funded research station devoted to experimental biology. Przibram and a group of like-minded and well-to-do scientists established what was at the time the most sophisticated institution for experimental research that focused on the study of development, regeneration, endocrinology, experimental evolution, and the life history of organisms (both plants and animals). In addition, this group was interested in the integration of experimental and theoretical approaches. Their work was, in many ways, the most direct forerunner of current evo-devo. Unfortunately the political turmoil of the 1930s and 1940s put an end to this unique research environment (see, e.g., Przibram 1907).

The two unifying proposals of mid- to late twentieth-century biology were the Modern Synthesis and the emerging molecular biology (e.g., Mayr and Provine 1980, Morange 1998). The former was based on the conceptual integration allowed by the mathematical theory of population genetics, while the latter was based more on a unifying level of analysis and a shared repertoire of experimental approaches than an integrative conceptual framework. While these two approaches and their sometimes acrimonious relations dominated midtwentieth-century biology, experimental embryology, rechristened as developmental biology, continued to flourish in its own scientific niche. But there were always some, such as Conrad Waddington, who emphasized the unity of the evolutionary and developmental research programs. His idea of an epigenetic landscape and the concept of genetic assimilation both represented attempts to incorporate developmental mechanisms into evolutionary explanations (Waddington 1940). By the early 1970s the main focus of developmental biology had turned toward questions of developmental genetics. This was also the time when a renewed interest in the problem of "ontogeny and phylogeny" began to emerge.

Early contributors to what would later become evo-devo focused on the problem of development and evolution because they were dissatisfied with the explanatory framework of the Modern Synthesis. Several related issues, in particular, were at the heart of their critique. For one, the assumption of the Modern Synthesis that macroevolutionary patterns can be explained by a simple extension of microevolutionary processes was questioned. Palaeontologists, who led this line of attack, focused on two phenomena that they thought could not be reconciled within this framework: (1) The observation that the fossil record of several clades shows periods of rapid evolutionary change followed by extended periods of stasis led to the hypothesis of punctuated equilibrium, and (2) the observation that most conceivable morphological patterns are not realized, that is, that the morphospace is largely empty and that realized morphologies are clustered in certain domains of the morphospace, seemed to support ideas that morphological change is primarily a consequence of developmental processes, such as heterochrony, rather that strictly determined by (incremental) genetic factors (e.g., Eldredge and Gould 1972, Gould 1977).

Another source of dissatisfaction with the Modern Synthesis was the privileged role of adaptation in dominant theories of evolution (see Gould and Lewontin 1979, as the locus classicus for this critique). The adaptationist paradigm was challenged both by molecular biologists (e.g., with the neutral theory of evolution) as well as by evolutionary developmental biologists, who emphasized the role of internal factors in evolution. The most prominent early concepts in this context were the notion of developmental constraint as a limitation on possible phenotypic variation and therefore on adaptation and Rupert Riedl's idea of "burden," which postulates an internal, in addition to the external and environmentally induced, selection pressure (Riedl 1975, Maynard-Smith et al. 1985). The internal conditions, which act as a second (or rather first) selective environment, are those of the developing system. In the context of Riedl's theory these internal factors, or system conditions, as he calls them, can explain both the evolvability as well as the hierarchical organization of organismal forms (Riedl 1975, Wagner and Laubichler 2004).

By the mid-1980s a consensus had been reached that the prevailing version of evolutionary theory needed to be reformed. However, even though there was general agreement that development needed to be part of evolutionary theory, there was far less agreement about the actual research agendas or about the best strategies to accomplish this desired new synthesis. These centrifugal tendencies only grew stronger after the discovery of the conserved sequences of Homeobox genes provided a major boost to the developmental genetic version of evo-devo.

This plurality of approaches to the problem of ontogeny and phylogeny, which continues to this day, despite growing attempts to establish a genuine new "synthesis" of evo-devo, is also reflected in its current fragmented structure. An integrative framework that would unite all these different evo-devo applications within a genuine new "synthesis" has, as of yet, not emerged. What we can observe, however, is a clustering of research questions around two or three main emphases, such as "evo-devo," "devo-evo," and "developmental evolution" (see Hall 2000, Wagner et al. 2000, Müller 2005).

2.   RESEARCH PROGRAMS WITHIN EVO-DEVO

As we have seen in this brief historical overview, present-day evo-devo has many different roots. Correspondingly, there are also several distinct research programs currently pursued under the banner of evo-devo. Some of those clearly overlap, and collectively these different research programs illustrate the methodological pluralism that is characteristic of evo-devo. This multitude of approaches also helps to understand the current prominence of evo-devo–related research within organismal biology. However, the same diversity that makes evo-devo research so productive in many ways also provides a formidable challenge to any attempts to arrive at a synthesis of different perspectives (more on that later). The main research programs within evo-devo are the following (see also Laubichler 2005, Müller 2005):

1. *The comparative program.* Comparative research is one of the oldest paradigms within biology. It continues to raise important questions and provides the foundation of all further investigations.

Today comparative research covers many different levels of biological organization, from morphology and anatomy to embryology and genomics. In the context of the latter the emphasis has now shifted from simple sequence comparison to comparison of gene expression patterns and gene products and their interactions, thus allowing a comparative study of developmental sequences at high resolution. The comparative program is also important for evo-devo in that it provides us with detailed phylogenies. These are important in assessing any number of genuine evo-devo questions, such as hypotheses about evolutionary transformations or homology.

2. *The experimental program.* The experimental program is a continuation of the venerable tradition of experimental embryology and developmental physiology that dominated organismal biology during the first decades of the twentieth century. The focus of these experimental approaches lies in the elucidation of the mechanisms of development. Even though most experimental research in developmental biology today focuses on the role of genes in development, there is also an increasing awareness of the importance of epigenetic and environmental factors in morphogenesis and evolution. We have already seen that regulation has become one of the central concepts in describing both developmental systems and their evolution. Epigenetic factors, such as differential methylation patterns, genomic imprinting, posttranscriptional control, and RNA editing, as well as biophysical properties of cells and tissues, geometrical patterns of self-organizing systems, and environmental factors such as temperature, all contribute to the regulatory machinery of developmental systems and their evolution. In manipulating the parameters of developing systems the experimental program within evo-devo has produced some interesting results. Among those are the recreation of ancestral morphological patterns in limbs or the elucidation of the rules of digit reduction (Alberch and Gale 1983, Müller 1989). In both cases morphological transformations that closely resemble actual evolutionary transformations were accomplished by means of experimental manipulation of nongenetic parts of the developmental systems, such as constricting the size or manipulating the geometry of limb buds. These approaches contribute greatly to our understanding of the patterns and processes of evolutionary transformations.

3. *The program of evolutionary developmental genetics.* The focus of this research program is on the genetic components of developmental systems and their interactions. In many ways evolutionary developmental genetics is the most visible part of current evo-devo. The discoveries first of *Hox* genes and of other transcription factors that together make up the regulatory gene networks controlling the expression of genes and the differentiation of embryonic anlagen and then of the high degrees of conservation of these very genes have received the most attention. These discoveries also contributed (falsely, as we have seen) to the impression that evo-devo began with the *Hox* story. Among the ideas that emerged in the context of these investigations are the notion of a *genetic toolkit for development* – a set of regulatory elements that are involved in the development of the main features of animal bodies, such as segmental patterning and axis formation – and the proposal to reconstruct a so-called *Urbilateria* as the ancestral condition of all higher animals. The latter combines the insights of the comparative program with the discovery of the genetic toolkit for development (see Carroll et al. 2005 for an overview).

4. *The theoretical and computational program within evo-devo.* Evo-devo has also triggered a lot of theoretical research, especially during the last decades. This part of evo-devo is only becoming more prominent as newly developed formal and mathematical approaches begin to add more rigor to long-standing conceptual ideas. The theoretical program is especially important as it has the potential to provide conceptual unification to otherwise diverse experimental approaches (see later discussion). It also represents a disciplinary counterweight to the program in evolutionary developmental genetics. While the latter is largely dominated by researchers trained in molecular or developmental biology, the former is the domain of evolutionary biologists, who often refer to their approach as developmental evolution or devo-evo. In this context Günter Wagner, a trained evolutionary biologist and editor-in-chief of the *Journal of Experimental Zoology, Part B: Molecular and Developmental Evolution*, one of the leading evo-devo journals, always speaks of developmental evolution. These differences are more than just semantics, as they also correspond to different epistemological convictions and explanatory frameworks. Simply put, developmental biologists tend to be more interested in

structural and typological explanations based on molecular and cellular mechanisms, while evolutionary biologists focus more on dynamic processes on a population and species level (see also Hall 2000 for a discussion about the differences of evo-devo and devo-evo and Amundson 2005 for a detailed account of the different epistemological and explanatory frameworks). The integration of these two approaches is anything but straightforward, although recent theoretical developments that include work on genotype-phenotype maps including questions of how best to characterize phenotype space and morphospace, theoretical and empirical analyses of modularity and robustness, phenotypic plasticity, life history, as well as evolvability, all contribute to a framework that might just prove flexible enough to integrate the different empirical and theoretical traditions (see Callebaut and Raskin 2005, Schlosser and Wagner 2004, West Eberhard 2003, Müller and Newman 2003, Hall and Olson 2003). Theoretical work is also greatly aided by new developments in computational methods and representations. The databases of the various genome projects have been indispensable for identifying developmentally active genes and establishing their evolutionary history. As functional annotations of genes in these databases increase and gene ontologies become more sophisticated, it will soon be possible to extract the kind of information about developmental genes that is necessary for a more detailed under-standing of the evolution of developmental systems. What we have learned for *Hox* and related genes, how their expression domains shift in different species, and how this correlated with morpholog-ical changes, for instance, or what the consequences of certain duplication events have been, will soon be available for a large number of transcription factors, signaling genes, and receptor proteins. In addition, computational reconstructions of gene expression patterns in developing embryos will organize data in a way that will greatly aid mathematical modeling of developmental systems. One example of this approach is Adam Wilkins's analysis of the evolution of genetic pathways in developmental systems, which suggests that certain changes, such as upstream addition of control elements, are more likely than others (Wilkins 2002). To sum up, the theoretical and computational program within evo-devo is about to get a great boost from the increasing success of experimental approaches. Interesting times thus lie ahead.

Analyzing the different research programs within evo-devo is just one way to capture the diversity, excitement, and potential of the field. Another way to help us understand why evo-devo is such a central part of current organismal biology is to investigate some of the concrete research problems that make up the core of present day evo-devo.

3.   RESEARCH QUESTIONS IN EVO-DEVO

There are many specific questions that are currently investigated within the framework of evo-devo so we will have to be selective here. But generally most of the individual problems will be connected to one of six areas of research (Laubichler 2005, Müller 2005).

1. *The origin and evolution of developmental systems.* This question focuses on the evolutionary transformation of the developmental systems themselves. Developmental systems, as any other character of organisms, undergo evolutionary transformations. Research in this area reveals the modular architecture of developmental systems and investigates their robustness and how different developmental modules are combined and regulated.

2. *The problem of homology.* The problem of homology is one of the central questions of all of biology. It is often separated in a historical homology concept, used mainly in phylogenetic analyses, which describes the distribution of homologues, and a biological homology concept, which attempts to explain the existence of homologies in the first place. An evo-devo framework employs both notions of homology and tries to integrate them.

3. *The genotype-phenotype relation.* Mapping the genotype onto the phenotype has emerged as a main problem within evolutionary and quantitative genetics during the last decades. It is also the question that most directly involves developmental mechanisms. For a long time population genetic models assumed that development does not affect the mapping of genotypic onto phenotypic variation in any important way – development was thus treated as a constant. As this position can no longer be upheld, investigations into the formal properties of the genotype-phenotype map have become a major focus within evo-devo.

4. *The patterns of phenotypic variation.* It has long been known that patterns of phenotypic variation are highly clustered and

constrained. Explanations of this phenomenon have always included references to the developmental system, mostly in the form of developmental constraints that limit the possible phenotypic variants. Recently it has also become clear that the evolvability of certain lineages also crucially depends on the existence of developmental constraints. This question is thus also a main concern for evo-devo researchers.

5. *The role of the environment in development and evolution.* More recently environmental factors have also been incorporated into the evo-devo research program. The more we learn about the molecular mechanisms, such as DNA methylation or endocrine disruption, through which the environment can influence the phenotypic expression, the more it becomes obvious that the environment has to become a larger part of explanations within evo-devo.

6. *The origin of evolutionary novelties.* Explaining the origin of novel phenotypic traits has been one of the major challenges of evolutionary biology. As evo-devo offers the first integrated perspective that has the potential to comprehensively address this problem, we will discuss this question in more detail.

4.   THE PROBLEM OF EVOLUTIONARY NOVELTIES

The major challenge for evolutionary biology is to explain the origin of complex novel structures and functions. Darwin already struggled with this problem in the *Origin of Species.* "What good is 5% of an eye?" is an often repeated question and even today the idea of "irreducible complexity" is taken as evidence for "intelligent design" and for the obvious shortcomings of the neo-Darwinian paradigm based on "an unguided, unplanned process of random variation and natural selection," which simply cannot be true. Part of the problem is that proponents of intelligent design are consciously misrepresenting evolutionary biology as neo-Darwinism and are systematically ignoring all the evidence that has accumulated over the last twenty-some years in the context of evo-devo, evidence that helps us understand how something new can actually emerge in the course of evolution. But, as the case of intelligent design also shows, evo-devo and especially what it can teach us about the origin of evolutionary innovations are of more than just academic interest (see also Kirschner and Gerhart 2005).

Since the first draft sequence of the human genome was completed in 2001, some of the simplistic assumptions about the relationship between genotype and phenotype have been challenged by the realization that the actual number of human genes is relatively low, only around 30,000, and that most of these genes (close to 19,000) are already present in tiny nematode worms, such as *C. elegans*. Many of the important regulatory genes and transcription factors, such as the *Hox* genes, are also highly conserved between lineages that have been separated by hundreds of millions of years. Novel features are thus not just a consequence of new genes or even new versions of old genes.

What then accounts for the obvious phenotypic differences between groups of organisms and for the emergence of novel structures in the course of evolution? The short answer to this question is that changes in the developmental systems of these organisms and more specifically changes in the regulatory networks of genes are responsible for these differences. In other words, the same mechanisms that lead to differentiation of cells in the course of individual development (ontogeny) also account for emerging differences in the course of evolution (phylogeny). Intuitively this makes sense. All phenotypic differences, whether they are just variations of a common theme or something radically different, emerge during the development of individual organisms. Developmental processes will thus always be the immediate or proximate causes of phenotypic variation. Still several questions remain: Exactly how do developmental mechanisms contribute to phenotypic changes, and how can we integrate such developmental explanations into the theoretical framework of evolutionary biology?

The key idea that helps us understand these issues is the concept of regulation. Development itself is a highly regulated process. How would it otherwise be possible that despite constant environmental disturbances the outcomes of development are generally predictable such that humans give birth to humans and sea urchin eggs develop into sea urchins? While developmental biology focuses on individual developmental sequences describing the transformation from simple (a fertilized egg) to complex (an adult organism) through the differentiation of cells and the emergence of anatomical and histological structures, evo-devo tries to understand how in the

course of evolution something new can emerge within these regulated developmental sequences. But before we can address potential
explanations of such evolutionary novelties we first have to define
the problem more precisely.

Addressing the problem of evolutionary innovations requires us
first to define clearly what we mean by "novelties" and, second, to
develop a set of causal hypotheses that will allow us to identify the
developmental changes involved in the emergence of an evolutionary novelty. In their seminal analysis of the problem Müller and
Wagner defined a morphological novelty as "a structure that is
neither homologous to any structure in the ancestral species, nor
homonomous to any other structure of the same organism" (Müller
and Wagner 1991, 243). While this rather general definition still
leaves open many details, it does have one practical implication. The
problem of identifying novelties is squarely placed within the
comparative program of evo-devo as their recognition depends on
both a good phylogeny and a detailed assessment of homology.

Setting aside, for the moment, many of the practical problems
connected with recognizing novelties (which are similar to the
problems of assessing homology) we can identify some of the steps
required for establishing a causal hypothesis about the origin of
evolutionary novelties within the context of evo-devo. The first
question that needs to be addressed is, What specific developmental
mechanisms are responsible for a new derived character state that
has been identified as an evolutionary novelty? Answering this
question requires the detailed analysis of the developmental
mechanisms that generate a specific phenotypic character. It is thus
part of the experimental program of evo-devo. The second question
builds on this analysis: Did the developmental mechanisms that are
responsible for the derived character state originate at the same time
as this character state? This is already a difficult question that
requires us to compare the developmental mechanisms of ancestral
and derived character states. In many cases this will not be possible,
as it is impossible to reconstruct the exact ancestral condition of
developmental processes, especially if the transformation in question happened hundreds of millions of years ago. The same problems
also apply to the third and fourth questions: What were the exact
developmental mechanisms responsible for the initial changes in
the character state? Are the observed genetic differences between

these two developmental systems sufficient to account for the observed phenotypic differences? These last two questions focus on the mechanistic details of the changes in the developmental system and the extent to which observable genetic changes provide a complete explanation of evolutionary transformations (Wagner et al. 2000).

We still face many practical difficulties with most cases of evolutionary novelties currently under investigation. Part of the problem can be attributed to the selection of examples and model organisms, which tend to focus on major morphological transformations, such as the fin-limb transition in early vertebrates. However, there are some model systems that do allow us to address questions about the origin of evolutionary novelties experimentally as well as theoretically. For example, social insects display a remarkable diversity in behavior ranging from solitary to eusocial. For many of these species we know their phylogeny, genetics, developmental mechanism, as well as their physiological and behavioral repertoire. And we can manipulate them experimentally in the lab, in several cases actually inducing novel types of social behavior among solitary species. Social insects are therefore an ideal model system for the study of evolutionary novelties. This work is only just beginning, but we can expect that it will lead to many important insights into the problem of evolutionary novelties.

Another issue that needs to be mentioned here is whether genetic differences (including differences in the genetic parts of regulatory networks) alone provide a sufficient explanation for the origin of evolutionary novelties. Our fourth question specifically addresses this issue. By now we have ample evidence that epigenetic and environmental factors, which are part of the developmental system in many organisms, play an important part in the origin of evolutionary novelties, especially during the initial phases of character transformation. On the basis of such observations Müller and Newman have suggested a three-step model for the origin of evolutionary novelties that assumes that the initial emergence of new characters is often caused by epigenetic and environmental mechanisms that are later stabilized by associated genetic changes (Müller and Newman 1999, 2003). While this is still a rather controversial idea, it can be tested, especially with social insects, among which many emergent colony-level traits, such as division of labor and caste distribution, are often a consequence of within-colony interactions, rather than simple mutations.

5.   EVO-DEVO AND THE PROBLEM OF
INTERDISCIPLINARY INTEGRATION

Conceptual integration of rather diverse research paradigms is the main challenge that evo-devo currently faces. Therefore, it can easily be argued that the future of evo-devo as a true inter-disciplinary synthesis will depend more on theoretical advances than on additional experimental data. As we all know, appearances can be deceiving, and the deluge of exciting experimental results over the last two decades has in many ways hidden an underlying conceptual tension that will have to be resolved for evo-devo to succeed. In his recent book Ron Amundson pointed out in great detail how developmental and evolutionary explanations are based on rather different epistemological foundations (Amundson 2005). The former are rooted in what Amundson refers to as a structuralist paradigm that is based on a causal understanding of general molecular and cellular mechanisms, such as molecular gradients, cell-surface interactions, and cell-cell signaling, while the latter are predicated on a mathematical formulation of underlying population-level dynamics, such as the replicator equation, the generalized selection equations, or stochastic processes, such as random genetic drift.

The earlier success of the Modern Synthesis of the 1930s and 1940s, which is often held up as a model for an emerging evo-devo synthesis, was based on matching patterns – of the transmission of genetic information, of speciation and adaptation, and of the fossil record – with dynamical processes within populations (see Mayr and Provine 1980 for an overview of issues related to the Modern Synthesis). Something similar will have to be accomplished for evo-devo. Observed patterns of morphological and behavioral evolution will have to be matched with the possibilities and constraints of developmental systems and the dynamical processes within populations. Such integration will depend on a conceptual framework and associated dynamical models that adequately represent both the actual phenomena to be explained (patterns of morphological and behavioral evolution) as well as the underlying dynamics that generated them (see also Wagner et al. 2000, Wagner's series of editorials in *Molecular and Developmental Evolution* [2000, 2001], Wagner and Larsson 2003, Laubichler 2005).

What form could such integration take? As we have seen in our brief historical overview, developmental mechanisms have always featured prominently in attempts to explain patterns of phenotypic diversity and transformation. Whether these were conceptualized as "laws of variation" (Darwin), "recapitulation" and "terminal addition" (Haeckel), "homeotic mutations" and "hopeful monsters" (Goldschmidt), or "genetic assimilation" (Waddington), the general argument has always been the same: phenotypic changes have to arise during ontogeny; therefore, changes in ontogeny will have to be responsible for observed patterns of phenotypic evolution. The details and specific concerns of these explanations differed, of course, but this variation does not detract form their underlying similarity. Evo-devo falls within the same explanatory paradigm, but it also differs in several important ways. For one, our current understanding of the molecular, cellular, genetic, and epigenetic mechanisms of development is much more detailed and our abilities to manipulate developmental systems experimentally have advanced rather dramatically. This increased understanding of developmental mechanisms has led to a more refined conceptual representation of development that is no longer based on simple mechanical forces, such as the actions of a somewhat mysterious "organizer," or gradients of molecules. Our current understanding of developmental differentiation includes complex causal pathways and interactions between genetic and epigenetic regulatory networks.

We have already indicated several times that regulation has become one of the central concepts in both developmental biology and evo-devo. Indeed, many scientists argue that modifications in the regulatory networks – so-called regulatory evolution – can account for the observed phenotypic transformations (see, for example, Carroll et al. 2005, Davidson 2001, 2006, Wilkins 2002). While changes in the regulatory networks are certainly an important part of the explanation of patterns of phenotypic diversity, our explanatory schema will have to be expanded. First we will have to establish a more adequate conceptual framework for what we want to explain. Evolutionary novelties are a prominent part of the evo-devo explanandum, but related concepts such as facilitated variation, developmental constraints, modularity, robustness, and evolvability, all concepts that have recently been analyzed within the context of evo-devo, will have to be fully integrated into its conceptual framework (see, for instance, Callebaut and Gutman 2005, Kirschner and Gerhart 2005, Maynard Smith et al. 1985, Schlosser and Wagner 2004). These concepts are also tied into the different research programs within evo-devo, A detailed epistemological analysis, of the sort provided by Ron Amundson, of these programs and their underlying assumptions is therefore crucial for the future of evo-devo as a synthetic enterprise.

But before we attempt such a synthesis we should also consider what we mean by synthesis, what exactly should be synthesized, and what we expect from such a synthesis. The Modern Synthesis was successful because, after an initial phase of intense discussions, it emerged with a clear explanatory agenda – how population genetic models can be used to explain adaptation and speciation. And even though these simple principles or their underlying assumptions were almost never "true" for any real case, they nevertheless anchored a very productive experimental and theoretical research enterprise, which ultimately led to our current period of methodological and conceptual pluralism. If we take the historical lesson of the Modern Synthesis to be that the role of a synthesis is to provide a conceptual foundation for different research programs that will ultimately explore the fuzzy edges and areas beyond the core of the synthesis, then we advocate an open conception of synthesis rather than a closed view based on integration of existing paradigms.

Seen that way the role of an evo-devo synthesis would be to provide a set of core concepts and assumptions that allow further research, realizing that this research will eventually transcend the explanatory framework of the current synthesis. How would such an open-ended synthesis look? It is probably too early to tell, but it will have to be based on a conceptual structure that allows the integration of developmental mechanisms into evolutionary explanations at a higher level of resolution than the current ideas about regulatory evolution and the evolution of the genetic toolkit suggest. It will also have to develop a more comprehensive conception of mechanistic causes for both development and evolution that includes genetic, epigenetic, and environmental factors. And it will have to develop a set of paradigmatic model systems that will allow us to study these questions. As we have seen, there are some very promising new model systems currently developed that might just be the right tool for the job. In any case, these are exciting times for both scientists and philosophers to collaborate on some of the most fundamental problems of biology.

19 Molecular and Systems Biology and Bioethics

Molecular biology has set itself the task of looking for the fundamental pieces with which the biological jigsaw is to be put together. Not surprisingly (but with surprising efficacy), it has found many of them, and there are certainly more to come. Once found, these pieces can be arranged on a page next to one another in a reasonable sequence, and ... Behold! An organism! Well, not quite.

Cohen and Rice 1996, 239

The philosophy of molecular biology was, for a time, entirely preoccupied with reduction and reductionism: primarily the reduction of classical genetics to molecular genetics (Kitcher 1984, Waters 1994, Sarkar 1998), but also and more recently the reduction of complex organismal phenotypes to genes (Rosenberg 1997, Sarkar 1998). While these remain of substantial interest, some new areas of interest have also emerged, including philosophical attention to molecular mechanisms (Machamer, Craver, and Darden 2000, Darden and Tabery 2005) and mathematical models (Keller 2002, Sarkar 2005). In-depth focus on the intricate details of the science is increasingly commonplace (e.g., Schaffner 2000, Burian 2004, Sarkar 2005). Molecular biology has also proved to be of philosophical interest not only for its own sake, but also in the service of molecular explanations of evolution (e.g., Burian 2004), disease (e.g., Kitcher 1996), and behavior (e.g., Schaffner 2000), inter alia.

Just as the philosophy of molecular biology has changed in the past few decades, so too has the science, especially with the introduction of new tools and techniques and the novel opportunities afforded by advances in genome sequencing, computer modeling,

361

and bioinformatics. Indeed, some have proclaimed that the 'omics era heralds "the end of molecular biology as we know it" (Laubichler 2000, 287). While reductionism remains of central concern, it is evident that with every successful experiment and new discovery at the molecular level appears a new puzzle at the level of organisms, for how to relate fundamental discoveries to physiological outcomes remains almost as opaque as ever. The sentiment evidenced in the epigraph has been captured as well by many other scientists and commentators, including Eva Neumann-Held: "So far, biology can describe organisms down to the molecular level of genes. However, the interactions of genes [with other genes and] with other, non-genetic components to form an organism is far from being understood.... In the description of organisms (more generally: of systems), biology still has to perform the integrative part" (Neumann-Held 1999, 107; see also Robert 2004).

What to do with genome sequence data is a case in point. When the Human Genome Project was first proposed, its proponents promised dramatic new insights into human beings, in sickness and in health, as well as a steady supply of medical treatments and metaphysical revelations (see, e.g., Cook-Deegan 1994, Nelkin and Lindee 1995). Skeptics pointed out (correctly) that these promises would not be easily fulfilled, even if the seemingly impossible dream of completely sequencing the human genome could be achieved in the first place (see, e.g., Lippman 1992). Remarkably, the latter dream was achieved, with the publication in 2000 of draft sequences and the announcement in 2003 of the finalized draft (Collins et al. 2003). But understanding the human genome sequence (functional genomics), in comparison with those of other organisms (comparative genomics), remains an outstanding task. As several early critics (e.g., Lewontin 1992, Tauber and Sarkar 1992) understood full well, human genome variability would prove crucial, for no single sequenced genome (or composite of genomes) could be meaningfully said to 'represent' the diversity of genetic variation that characterizes any species. Epidemiologists are thus actively sampling DNA from volunteers in the quest for polymorphisms (variants) both between and within groups, and bioinformaticians and others are continually developing new tools for analyzing the vast quantities of genomics data generated through such research.

While making sense of this morass of data remains a daunting task, actually generating treatments for diseases and improving human health outcomes on the basis of the Human Genome Project are more daunting still. On one hand, this is because of the complexity of health (Robert and Smith 2004): individual health and disease are significantly affected by developmental, ecological, and social components that are not reducible to genes or particularly amenable to genetic intervention; at the population level, social, economic, and demographic factors explain health outcomes far better than genetic variations. Accordingly, identifying individual genetic risks may have only minimal effects on health outcomes, whether at individual or population levels. On the other hand, organismal development is far more complicated than glib pronouncements about genetic instructions and blueprints would suggest (Robert 2004): the 'pathway' from gene to phenotype comprises gene-gene, cell-cell, cell-tissue, and environmental interactions; epigenetic effects; and developmental stochasticity; gene effects may be pleiotropic (single genes have multiple, divergent effects); many traits are epistatic (correlated with the activity of many genes); in short, development itself as a dynamic, temporal repertoire of processes determines many aspects of the many-many genotype-phenotype relation. Accordingly, in most cases it is unclear how to proceed from a genome sequence to the identification of genes to a full understanding of the aetiology of a phenotype – and from any of these to determine how to intervene to prevent or promote its manifestation.

That development is complex should come as no surprise to biologists – though some would prefer to ignore developmental complexity in favor of the relative simplicity of developmental genetics and molecular biology (which themselves are far, of course, from simple). Indeed, upon announcing the completion of the Human Genome Project, the National Human Genome Research Institute also announced its elaborate "Vision for the Future of Genome Research" (Collins et al. 2003). The vision includes a series of "grand challenges" for genomics research, intended as "bold, ambitious research targets for the scientific community" (Collins et al. 2003, 2) in basic science, health-related applied science, and policy science and ethics. These grand challenges – like the Human Genome Project on which they build – have as their aim the translation of genomics data into biological understanding and human health outcomes in the genomic (or 'postgenomic') era. Whether these aims will be achieved depends on the development of a more integrative biology, and on biologists taking seriously the charge to understand organismal life in all its complexity.

FROM MOLECULAR BIOLOGY TO SYSTEMS BIOLOGY?

Systems biology has emerged as one 'field' within which biologists and other scientists are attempting to make molecular biology work "in the real world of the organism" (Cohen and Rice 1996, 251). Given the array of molecular data awaiting interpretation, and in the context of a growing awareness of the limits of certain kinds of simplifying strategies in dealing with the complexity of living organisms, molecular biologists have intensified collaborations with computer scientists, engineers, organismal and integrative biologists, and many others under the banner of 'systems biology' (Fujimura 2005). While that term may appear to be a fuzzy-sounding buzzword to some, nonetheless there are now systems biology research centers (such as the Institute for Systems Biology in Seattle), Ph.D. programs (as in the Department of Systems Biology at Harvard Medical School), and journals (including *Nature*'s recent launch of *Molecular Systems Biology*). Indeed, systems biology was recognized by the editors of *Science* (vol. 310, 23 December 2005) as a runner-up for the 2005 Breakthrough of the Year.

But what systems biology *is* is not always clear. A *Nature* editorialist has provided a somewhat macabre though apt response: "What is the difference between a live cat and a dead one? One scientific answer is 'systems biology'. A dead cat is a collection of its component parts. A live cat is the emergent behaviour of the system incorporating those parts" (Anonymous 2005, 1). Generically, then, a systems approach to biology refers to the interdisciplinary study of manifold and complex interactions among DNA, RNA, proteins, cells, and biomodules of various sorts that constitute living entities.

A survey of recent review articles and commentaries reveals that the name 'systems biology' is applied to a number of distinct research programs, each worthy of philosophical scrutiny in its own right (see, e.g., Hartwell et al. 1999, Ideker, Galitski, and Hood 2001, Kitano

2002a, b, Auffray et al. 2003, Grant 2003, Hood et al. 2004, Aderem 2005, Kirschner 2005, Moore, Boczko, and Summar 2005, Sorger 2005, Strange 2005; cf. Newman 2003 and Fujimura 2005). A systems approach has been advocated and pursued in many domains – for instance, in developmental biology (e.g., Gilbert and Sarkar 2000, Robert 2004), drug discovery (Butcher, Berg, and Kunkel 2004), neuroscience (Grant 2003, Wulff and Wisden 2005), physiology (Strange 2005), psychology (Oyama 1985), stem cell biology (Robert, Maienschein, and Laubichler 2006), and toxicology (Waters and Fostel 2004).

The underlying appeal of a systems approach, however realized, is clear, as evidenced in the following depictions of systems biology:

Systems biology is an emerging discipline focused on tackling the enormous intellectual and technical challenges associated with translating genome sequence [data] into a comprehensive understanding of how organisms are built and run.... Systems biology is integrative and seeks to understand and predict the behavior or "emergent" properties of complex, multicomponent biological processes. (Strange 2005, C968)

New technologies have inundated researchers with a deluge of information on genes, proteins, cellular dynamics, and organisms' responses to mutations and the environment. But they haven't explained what makes whole organisms tick. Systems biologists are taking on that challenge, relying heavily on mathematics and statistics to integrate data into a more complete picture of how biological networks from cells to whole organisms function. They are building models and making predictions about how biological systems will behave; the ultimate goal is to understand deep mysteries – such as how cells divide, animals develop, plants flower, and humans breathe. (Pennisi 2003, 1646)

The goal of systems biology is to offer a comprehensive and consistent body of knowledge of biological systems tightly grounded on the molecular level, thus enabling us to fully integrate biological systems into more fundamental principles.... System-level knowledge should be grounded in such a way that the system is composed of molecules; and molecules follow the laws of physics. However, how a system operates can only be described by a set of theories that focuses on system-level behaviors. The point is that such theories must reflect the realities of biological systems and molecules, without abstracting the essential aspects of biology. (Kitano 2002a, 2).

Systems biology is a comprehensive quantitative analysis of the manner in which all the components of a biological system interact functionally over time. Such an analysis is executed by an interdisciplinary team of investigators that is also capable of developing required technologies and computational tools. (Aderem 2005, 511)

The features that distinguish the new field of systems biology are: first, the grounding in molecules, primarily at the level of the gene; second, the goal of systematic, comprehensive and quantitative analysis of all of the components that constitute the system; third, the vertical integration of analysis to develop a structure to the system; fourth, the simulation and computational approaches to modelling the system, and the bioinformatics approaches of data handling. (Grant 2003, 577)

The careful reader will have picked out some common themes across these passages. For instance, one general aim of systems biology is to move beyond analysis of individual and isolated components of a system toward quantification, modeling, and comprehension of the integrated organization and interaction of the components as part of a system. Understanding components of systems will remain critically important in systems biology (as anywhere in biology), but focusing on the structural and interactive properties of systems will help to complete our understanding and improve our predictive abilities.[1] To this end, Strange (2005, C968) stresses that "a systems level characterization of a biological process addresses three main questions. First, what are the parts of the system (i.e., the genes and the proteins they encode)? Second, how do the parts work? Third, how do the parts work together to accomplish a task?"

But the careful reader will also have noticed some important discontinuities in these various characterizations of systems biology, of what it entails, how it proceeds. For instance, these passages suggest *both* that the field has developed primarily as a positive response to the challenge of dealing with all the new data from "'omics" research (genomics, proteomics, metabonomics, etc.), *and* that systems biology has emerged primarily in response to the limits of reductionism in understanding living organisms (Friboulet and Thomas 2005; cf. the essays collected in Hull and Van Regenmortel 2002). On the latter view, however productive reductionism has proved to be, living organisms are not entirely reducible, and so novel approaches are required not only to make sense of the molecular data, but also to integrate data from other levels of analysis.

This is not to say that systems biologists are antireductionists or emergentists – though some are so inclined (e.g., Aderem 2005, Friboulet and Thomas 2005), others maintain a methodological commitment to reductionism within a systems biology framework (e.g., Ideker et al. 2001, Sorger 2005). These tensions signal the need for critical analysis by historians and philosophers of biology, as well as science studies scholars, ideally based in rich and detailed case studies of particular exemplars of systems biology. Of special interest will be historical analysis of 'holistic' understandings of organisms in biology and medicine (Laubichler 2000) and conceptual articulation of what is new (and what is not) in systems biology.

FROM SYSTEMS BIOLOGY TO SYSTEMS BIOMEDICINE?

In general, systems biology involves synthesizing and integrating biological data into mathematical models that simulate the behavior of biological systems (organisms, diseases, etc.) and allow for predictions of future states. This idea that systems biology will contribute to greater predictability underwrites a systems approach to biomedicine, wherein drug discovery and design (Nicholson and Wilson 2003, Aderem 2005, Strange 2005) promise to be 'rationalized' and treatments personalized according to dynamic models of disease manifestation and progression (Hood et al. 2004, Weston and Hood 2004).

These are not new expectations. They are the sorts of claims commonly made about biotechnologies in general and genomics research in particular. The idea of molecular medicine in particular has generated and continues to generate much excitement. A strong motivation for undertaking the Human Genome Project was the promise that genomics will "profoundly alter" medicine, by shifting from treatment of acute disease to "prevention based on the identification of individual risk", for instance, and by tailoring pharmaceuticals to individual genotype (see, e.g., Guyer and Collins 1993 and van Ommen, Bakker, and Dunnen 1999). Is systems biology just the latest in a never-ending string of overhyped biomedical fads? Or is there finally some substance behind the rhetoric?

My cautious response is, a little of both. While there is no doubt that systems biology is among the current pantheon of biological fetishes (systems biology "promises to revolutionize our understanding of complex biological regulatory systems and to provide major new opportunities for practical application of such knowledge" [Kitano 2002b, 1664]), there is also reason for optimism that this interdisciplinary, integrative endeavor will prove biologically and biomedically fruitful.

The source of my cautious optimism is molecular biologists' apparent realization (finally!) that developmental systems, such as living organisms, are dynamic, temporal, and mutable rather than static (cf. Gilbert and Sarkar 2000). And while they are internally complex and heterogeneous, organisms are also embedded in externally complex and heterogeneous environments that can jointly and severally have dramatic impacts on development, physiology, and behavior (cf. van der Weele 1999). To begin to take these considerations seriously in experimental design, model building, and theory elaboration is to begin, perhaps, to realize the ambitions of molecular and systems biomedicine.

Whether these ambitions will be realized depends on a large number of considerations, including the will of funding agencies and universities, and the ingenuity and collaborative abilities of biologists and other scientists. It also depends, in part, on constructive collaborations with a new breed of bioethicist to assess the societal desirability of novel developments, the risks and benefits of particular eventualities, and the appropriate application of new knowledge in the service of human (and humane) ends. Many philosophers of biology have no interest in ethics; many bioethicists have no interest in the details and complexities of biological science; and, to date, much of the bioethical literature on genetics, genomics, and molecular biology has been unfortunately superficial in its treatment of the relevant science. But an emerging cohort of hybrid scholars engaging in both conceptual (biophilosophical) and normative (bioethical, political) analysis may be able to restructure and reinvigorate scientific and public debates about desirable outcomes of science and technology (e.g., Kitcher 2001, Maienschein 2003, Robert and Baylis 2003, Robert et al. 2006).

FROM ELSI TO SYSTEMS BIOETHICS?

When the Human Genome Project (HGP) was initiated in the United States, its proponents dedicated a remarkable 3 percent of the HGP's total annual research budget to fund an Ethical, Legal, and Social Implications program (ELSI). Programs similar to ELSI, though somewhat less well funded, were established in Canada (the Medical, Ethical, Legal, and Social Implications component of the Canadian Genome Analysis and Technology program), the United Kingdom (through the Wellcome Trust, for instance), and elsewhere. In the United States, at least, ELSI had both research and educational mandates. In terms of research, the ELSI program identified several principal ethical, legal, and social issues for initial study: the privacy of genetic information, the provision of safeguards against genetic discrimination, the protection of participants in genetics research, and the secure introduction of genetic tests into clinical medical practice (Juengst 1991).

A great deal of research was generated through the ELSI program, some of it quite good. Of course, ELSI did not fund all of the bio-ethical work undertaken in relation to the HGP, but through the ELSI program, North American philosophers, ethicists, historians, sociologists, and clinicians were able to obtain unprecedented funding to investigate the likely ethical, legal, and social impact of the HGP. Yet after fifteen years of the prolific ELSI program, there is still no end in sight in resolving the ethical problems raised by the HGP. In part, this is because ELSI scholars have tended to focus on issues related to Mendelian genetics, and less so on genomics (or even developmental genetics). Moreover, the focus has been on rare single-gene disorders rather than much more common conditions with multifactorial aetiologies. Finally, many ELSI analyses were atomistic (in the sense of not being integrated with larger issues in the philosophy of biology, philosophy of medicine, health technology assessment, or health policy) and reductionistic (in the sense of being portrayed in simplistic terms, such as false dilemmas and dichotomies).

Like the HGP itself, these ELSI endeavors were necessarily partial at best – a good starting point, perhaps. Just as biologists are now moving toward a more integrative systems biology, perhaps so too should bioethicists move toward a more integrative 'systems bioethics' (Robert et al. 2006). Research in systems bioethics would be emergent from ELSI-style research, as integrated with studies in the history and philosophy of biology and social and political studies of science and technology, and employing a variety of methods from the humanities and social sciences (especially Guston and Sarewitz's 2002 'real-time technology assessment'). It would generate new ways to frame, explore, understand, and alter moral dimensions of scientific research by defining relevant concepts, interests, and values; probing their nature; and establishing an understanding of the dynamic, interactive relations among the disparate components of the particular bioethical 'system' under study. This would then serve as a basis for proactive deliberation about scientific, ethical, and political issues together and interactively, aimed at making bioethics work in the real world of complex, pluralistic civil societies.

Thus, one critical aim of systems bioethics would be to make normative and conceptual analysis actually matter in practice and in policy contexts. This might be achieved through the collaborative interaction of methods from multiple fields of inquiry to integrate key concepts relevant to a joint public and scientific discourse about biological research. But translating bioethics research into policy options is, as bioethicists can attest, no easy task. Social systems, especially in pluralistic civil societies such as democracies characterized by competing interests and value claims, are tremendously complex; to genuinely understand (and alter) social systems requires a coordinated effort across traditional disciplinary boundaries and across multiple sectors and spheres of influence, and not just the coordination of independent efforts, much less the coordination of independent disciplinary efforts.

Alas, like systems biology, this may sound nice in the abstract. Whether and how it might actually be instantiated remain to be seen.

CONCLUSION

Molecular biology has undergone tremendous changes in the past seven decades. While some of the philosophical issues remain constant (e.g., reductionism), new methodological and conceptual considerations have emerged over time, requiring philosophical clarification and scrutiny. In this essay, I have not sought to analyze these issues so much as to sketch in rough outline the rich terrain that comprises the philosophy of molecular biology and, now, of systems biology.

This is fertile ground for historians and philosophers – *What, precisely, is 'systems biology', whether in particular cases or more generally? What, if anything, is new in systems biology, and what are its historical precursors? What are its various concepts, conceptual assumptions, and motivations? What methodological and analytical problems does it raise (and/or solve)? What is the relationship between molecular biology and systems biology, and what is the nature of 'molecular systems biology'?* – as well as for social scientists, including science studies scholars – *How is this new 'field' organized? Does systems biology reflect a dramatically new style of interdisciplinary scholarship in biology, or is it part of a longer tradition (and, if so, which one[s])? What is the 'culture' of systems biology, and what are its emerging traditions?* – and ethicists, too – *What are the determinants of 'translational' systems biology research, and what are the societal implications of advances in systems biology? What are the benefits and opportunity costs of strategic translational research?* These and related questions should preoccupy the ethics, history, philosophy, and social studies of molecular and systems biology in the years to come, ideally in collaborative engagement with scientists and other interested parties. Only then may we determine whether molecular biology can be made to work in the real world – and made to work toward beneficial ends.

NOTES

1. On this view, notice that systems biology is not only the science of systems, but also the science of the 'spaces between' molecules and larger systems – the study of what happens in these spaces, how, and to what end. Cf. Hans Westerhoff, as cited in Henry (2003): "'Systems biology is not the biology of systems,' he emphasizes. Instead, he says, it is the region between the individual components and the system, which is why it's new. 'It's those new properties that arise when you go from the molecule to the system,' he says. 'It's different from physiology or holism, which study the entire system. It's different from reductionist things like molecular biology, which only studies the molecules. It's the in-between.'"

20 Ecology

INTRODUCTION: UNDEAD DOGMAS OF EMPIRICISM

I suspect the demand for evidence about individuals is a bastardized version of an old positivist claim: the claim that theoretical terms must be defined in observational ones, in particular individual sensory experiences.

Kincaid 1996, 182

The philosophy of biology has matured quite a bit over the last two decades. Back in 1988, Ruse noted a conspicuous dearth of work on ecology. But by 1999, Sterelny and Griffiths devoted an entire chapter to it in their introduction to the philosophy of biology. There is still plenty of room, and reason, for more philosophical attention to a science so vital for understanding and addressing environmental concerns. But at least several people now make philosophy of ecology their academic specialty.

In the following I shall very briefly survey recent developments in both ecology and the philosophy thereof. One important aspect of the developments within ecology is an expansion to larger spatio-temporal scales of investigation. This wider focus has often, though not always, resulted in a shift in perspective, from viewing ecological entities as closed systems to treating them as open systems. I will take a closer look at three examples of scale expansion and tease out some of their implications for environmental policy, on one hand, and still-common reductionist philosophies of science, on the other. Finally, I will consider the philosophical implications of the search for mechanisms, when the open nature of the systems under study is acknowledged.

But first, some wider context. In the early twentieth century, the logical positivists espoused a certain kind of ethical subjectivism, a certain kind of reductionism, and a certain kind of instrumentalism. Respectively, these doctrines held that ethical statements are meaningless emotional outbursts, that theoretical predicates are reducible to observational predicates, and that questions about the truth or falsity of statements regarding "unobservable" entities are also meaningless. Most philosophers have since rejected all three of these positions. But other versions of subjectivism, reductionism, and instrumentalism persist within both philosophy and science.

By the end of the twentieth century, scientific realism had arguably eclipsed instrumentalism within philosophy of science. Nevertheless, new versions of instrumentalism still crop up. One might think that a field like ecology should be largely immune to the whole debate. The entities that it deals with are mostly "observable", and hence not subject to the skepticism traditionally leveled by instrumentalists against "unobservable" objects like atoms. Yet Sober has argued for instrumentalism regarding hypotheses about the degree to which the corn plants in two fields differ in average height (Sober 1999). I have shown that even in the new "Akaikean" statistical framework invoked by Sober, theories achieve predictive accuracy in the way that realists say they do. Predictive success results from getting at the underlying truth – not from the kind of "cosmic accident" required by instrumentalist accounts (Mikkelson in press).

It may be that most philosophers of science now consider themselves to be antireductionists. And yet most arguments against reductionism attack only an extreme version of it: the idea that lower-level processes *completely* explain higher-level processes (but not vice versa). I submit that it is time also to question a slightly milder version of reductionism that seems to guide scientific funding policy, as well as many scientists' views about proper methodology. According to this type of reductionism, lower levels "merely" play a far more important role than higher levels, in explanations of most phenomena. For example, scientific funding patterns imply that genetic causes of human disease are far more important than environmental causes. In the following, I shall offer some reason to doubt this kind of position, at least within ecology.

The debate over ethical subjectivism versus ethical realism has not attracted much comment in philosophy of biology (Sober 2000 is an exception). And yet this debate has profound implications for the question of whether science can or should be "value-free" (Putnam 2002). In the following, I shall limit my remarks to noting some of the policy implications of recent research on the causes and consequences of biodiversity.

1.   A VERY BRIEF SKETCH OF RECENT ECOLOGY AND PHILOSOPHY THEREOF

Beatty cited four sets of "interesting foundational and methodological problems" in ecology (Beatty 1998):

1. "[P]roblems of clarifying the differences and causal connections between the various levels of the ecological hierarchy (organism, population, community, ecosystem)"
2. The "issue of how central evolutionary biology is to ecology"
3. "[L]ong-standing issues concerning the extent to which the domain of ecology is more law-governed or more a matter of historical contingency" and
4. The "related question of whether ecologists should rely more on laboratory/manipulative versus field/comparative methods of investigation."

Since Beatty published his overview, ecologists have discovered numerous interlevel causal links and other lawlike generalizations (cf. 1 and 3; see, e.g., Kinsey 2002, Marquet et al. 2005). Some progress has also occurred with regard to at least one aspect of the relationship between ecology and evolution, namely, the extent to which entire ecological communities or ecosystems are targets of natural selection (see 2; Swenson, Wilson, and Elias 2000). And "laboratory/manipulative" and "field/comparative" research programs have both continued to proceed with vigor (4). Nor do I know of any major shifts in emphasis between the two.

Since 1998 philosophers, and scientists offering philosophical commentary, have paid the most attention to the third topic; namely, laws in ecology (Cooper 1998, Colyvan and Ginzburg 2003, Mikkelson 2003). Severe ambiguity about what laws are still plagues this body of writings, just as it dogs more general discussions of laws in philosophy of science. Partly for this reason, I shall herein avoid the question of lawhood in general and focus instead on other aspects of particular ecological "laws". Some work has been done on the first two topics, though. In Sections 2 and 3, I offer new examples and arguments regarding the issue of links between levels of ecological organization. As for the relationship between ecology and evolution, philosophers of biology have tended unfortunately to depict the former as a handmaiden of the latter (Sterelny and Griffiths 1999).

I am not aware of any sustained philosophical discussions since 1998 of Beatty's fourth topic, different empirical approaches in ecology. His dichotomy of "laboratory/manipulative" versus "field/observational" obscures important aspects of this subject, though. For one thing, many of ecology's most important advances have resulted from manipulative field experiments (e.g., those of Paine 1966, Simberloff and Wilson 1969, Likens et al. 1970; see Diamond 1986 for a general discussion, and Section 2c for another example). And yet such experiments straddle Beatty's two categories, rather than fitting neatly into either of them.

For another thing, the relative prominence of different empirical techniques seems to be a less crucial methodological issue than the degree of integration among theory, experiment, and observation. As in any field, these three modes of ecological research are sometimes tightly coupled, and at other times fairly insulated from each other. Intuition as well as a cursory historical survey of science in general and ecology in particular, suggest that better integration yields more substantial scientific progress. The example discussed in Section 2a nicely illustrates the benefits resulting from improved integration of theory, experiment, and observation.

Before proceeding further, I should mention at least a few other cases of recent philosophical work on ecology. Odenbaugh (2001) dealt with philosophical issues stemming from a period of strident reductionism, particularism, and pessimism in ecology (from the late 1970s through the 1980s). De Laplante (2004) has considered relationships between ecology and the social sciences. And environmental ethicists have continued to write about ecological science, as they have for several decades. In some cases, these efforts have yielded astute analyses of ecology resulting from long and deep reflection on the nature of this science and its moral implications (cf. Skipper et al. in press).

2.   GOING MACRO

The common theme uniting the three case studies considered in this section is a shift toward larger scales. Historically, ecologists concentrated on the internal dynamics of local populations, communities, and ecosystems. Now they more often also take into account, or turn their focus toward, the larger wholes that contain local ecological systems. These include metapopulations, metacommunities, landscapes, regions, biotic provinces, and, at the largest scale to date, the entire biosphere or ecosphere. In many cases, such a change in focus has resulted in promising new insights. In this section, I shall assess the philosophical significance of expanded spatiotemporal perspectives on relationships between (a) plants and herbivores, (b) area and number of species, and (c) number of species and the total density of biomass summed across all species.

a.   Responses of Different Trophic Levels to Nutrient Enrichment

Leibold and colleagues noted a striking mismatch between certain manipulative experiments, including field experiments, along with theoretical models, on one hand, and observational surveys of unmanipulated ecological systems, on the other (Leibold et al. 1997). The theory, experiment, and observation in question all focused on changes in plant and herbivore biomass density[1] due to increases in the nutrients that plants need to grow.[2] Most of the "nutrient enrichment" experiments surveyed by Leibold and colleagues fit one of two patterns: as nutrient levels increased, either plants increased proportionally much faster than herbivores, or vice versa.

These experimental results fit nicely with the predictions of some simple theoretical models. According to such models, if the plant species in question are all relatively edible, then increasing the nutrient supply does not increase total plant density very much. Instead, herbivores "chow down" on most of the "extra" plant production, leading to a proportionally much greater increase in their own density. I loosely follow Leibold in calling this scenario a closed "food chain" (Leibold 1996). See the steepest line in Figure 20.1.[3]

Similar models predicted that if some of the plants are relatively inedible, nutrient enrichment should increase their density

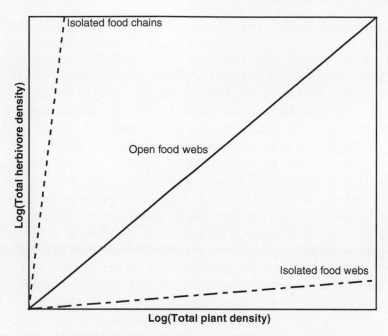

Figure 20.1 Schematic depiction of contrasting relationships between the responses of plants and herbivores to nutrient enrichment

substantially. This is because the inedible plants do not suffer as much chowing down. This relative immunity gives them a competitive advantage against more edible plants. And this, in turn, leaves the herbivores with relatively less to eat. So the herbivores increase little, if at all, in density with increased nutrient levels. This is the "closed food web" scenario[4] (cf. the shallowest line in Figure 20.1).

The nutrient addition experiments and theoretical models described contrast markedly with observations of unmanipulated natural ecosystems. Among these systems, plant and herbivore densities both increase significantly with nutrient levels, as depicted by the middle line in Figure 20.1. A shift in spatial and temporal perspective enabled Leibold and associates to reconcile these observations with theory and experiment. They noted that the theoretical and experimental models at hand both treated ecosystems as if they were isolated from their surrounding landscapes or regions. Consequently, these models did not allow for the colonization of new species from outside a given ecosystem.

Theoretical models that take colonization into account predict significant increases of both plant and herbivore density with nutrient levels (an "open food web" scenario).[5] To understand this, consider that relatively inedible plants generally require greater amounts of nutrients than do more edible species. This is because resistance to herbivory, for example, through the production of noxious chemicals, can be physiologically taxing. Nutrient enrichment therefore allows progressively more inedible species to colonize. Colonization by more inedible species, in turn, increases overall plant density and reduces herbivore density, relative to the food-chain scenario. In Figure 20.1, a tilt downward from the steepest line would represent this outcome.

This raises the question of whether colonization by inedible species should cause herbivore density to decline so dramatically that it results in the same pattern found in the isolated food web scenario. The reasons given by Leibold and associates that this does *not* happen are somewhat obscure. It would suffice, however, for nutrient enrichment to permit the colonization of herbivore species with the ability to overcome the defenses of the "inedible" plants. Overcoming plant defenses can also be physiologically taxing. Therefore, herbivores able to do it stand a better chance in the presence of higher nutrient levels, and thus greater overall plant production. Colonization by this kind of herbivore would result in a tilt upward from the shallowest line in Figure 20.1.

Just as in these theoretical models of open ecosystems, experimental models that allow new species to colonize from outside the ecosystem result in roughly proportional increases in plant and herbivore density with nutrient levels. One methodological take-home message is that models treating local ecosystems as open systems accord with observed patterns in nature. Theories and experiments that treat such ecosystems as though they were closed do not fit these observations.

This take-home message potentially undermines the type of reductionism discussed in Section 3. According to that doctrine, the parts of an entity should play a much more important role than its environment in explanations of that entity's behavior. Since a closed system, by assumption, has *no* environment – no larger system that could exert *any* material influence – treating an entity as a closed system takes this reductionist prescription to an extreme degree. But an open-system approach leaves open the possibility that the environment plays an equally, or even more, critical explanatory role than the parts.

Some reductionists have rationalized their position by claiming that reduction is "the only method of attaining unitary science that appears to be seriously available" (Oppenheim and Putnam 1958). Some antireductionists concede the idea that reduction is the only viable route to the unity of science (e.g., Dupré 1993). Yet tracing causal influences both "upward" from the parts and "downward" from the environment offers more chances to unify our understanding of different levels of organization than would limiting attention to internal dynamics alone.

b.   Species-Area Relations

As in the plant-herbivore studies outlined previously, ecologists have also recently expanded the spatial scale at which they investigate relationships between area and number of species. The "species-area relation", or "law", is one of the most venerable patterns in ecology. At least in part because of practical constraints, ecologists historically began exploring species-area relations among relatively small patches of habitat. These patches generally fell into one of two categories – either contiguous parts of larger habitats or islands. However, Rosenzweig has issued a bold new claim about species-area patterns among entire "biotic provinces" (Rosenzweig 1995).

Strictly speaking, Rosenzweig did not distinguish contiguous patches from islands from provinces on the basis of size (area). Instead, he differentiated them along an axis of immigration. Contiguous patches experience the most immigration. A contiguous patch is so well connected to other patches of the same kind that most of the time, when populations inside the patch decline, they are "rescued" by immigration from outside. An island, as Rosenzweig has defined it, is too isolated for this rescue effect to dominate population dynamics. If a given island population declines, it will go extinct unless births on the island – rather than immigration from outside it – turn the tide. But an island is still connected enough to a mainland for immigration, along with extinction, to dominate diversity dynamics. Most of the species on an island owe their presence there to immigration, rather than to speciation on the

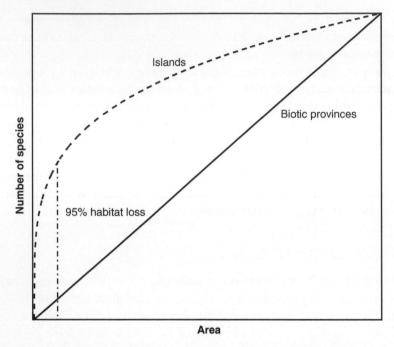

Figure 20.2 Schematic depiction of contrasting relationships between area and number of species

island itself. In contrast, a province is an area so isolated from other, similar habitats that most of its species "arrive" through in situ speciation. For Rosenzweig, Hawaii is isolated enough to count as its own province. But most provinces are much larger.

Rosenzweig's bold claim is that the species-area relationship among provinces is linear, or nearly so. Contiguous patches and islands, in contrast, show a diminishing-returns pattern of species richness with area; Figure 20.2 illustrates the contrast between islands and provinces. If Rosenzweig's claim proves robust, we will have to increase dramatically the estimates of how many species our roadbuilding, agriculture, urbanization, and other avenues of habitat destruction are driving extinct. Importantly, species-area curves provide no information about *how fast* extinctions due to habitat loss will occur. The longer those extinctions take, the more time we will have to restore lost habitat and thereby prevent mass die-offs.

A linear relationship implies that if humans destroy 95 percent of the native habitat in a given province – as we have already done to the Atlantic rain forest of South America, the tallgrass prairie of North America, and other provinces – eventually 95 percent of the species there will go extinct. In contrast, previous estimates of anthropogenic extinction were based on extrapolation from island curves. Those estimates predicted that losing 95 percent of a province would only drive around 50 percent of its species extinct. To see the contrast, imagine starting with the largest area represented in Figure 20.2 – all the way to the right. Now imagine destroying 95 percent of the habitat – that is, moving 95 percent of the way leftward across the diagram. And note the difference between the fractions of the original species count that the island-based curve versus the provincial line predict to remain.

As the example of plant-herbivore density relations does, the species-area example illustrates how new insights are gained by increasing the scale of ecological investigation. In this case, however, the change in scale resulted in a shift from seeing provinces as open systems analogous to near-shore islands to seeing them as relatively closed. The large proportion of species that are endemic to – that is, found nowhere else than in – particular biotic provinces warrants treating provinces as relatively closed with respect to immigration (though of course not isolated from solar energy input, etc.). This case therefore reminds us that in some cases, and in some respects, there may be good reason to treat a given ecological entity as a closed system.

c.   Biodiversity and Ecosystem Function

Besides a trend toward larger scales, another change in the 1990s was a shift from viewing the number of species in a community or ecosystem strictly as a dependent variable – that is, as the *effect* of other ecologically important properties, such as area, as discussed earlier – to investigating its potential role as a *cause* of important "ecosystem functions" as well (Naeem 2002). One prominent strand of this research deals with the influence of plant species richness on total plant density. Plant density, in turn, affects the ability of ecosystems to provide "services" such as carbon storage, flood and drought control, and wildlife habitat. Darwin asserted a positive effect of species richness on total density, so this is not exactly a new topic (Mikkelson 2004). What is new is the large amount of attention – and

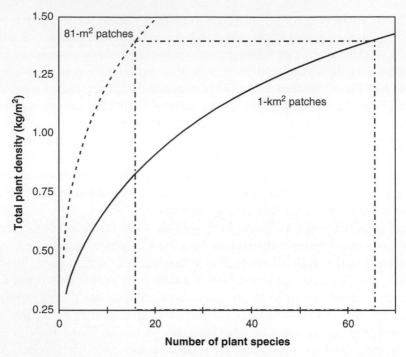

Figure 20.3 Schematic depiction of contrasting relationships between number of plant species and total plant density

controversy – that diversity-density and other diversity-ecosystem-function relationships have recently attracted.

In one experiment, 9-meter by 9-meter (81-m^2) grassland plots sown with more species tended to have substantially greater total density than those planted with fewer species (Tilman et al. 2001). The experimental treatments ranged from 1 to 16 plant species and yielded a range of around 0.4 to 1.3 kilograms/m^2 of total plant density. See the dotted curve in Figure 20.3.[6] For practical reasons, these grassland field experiments and others like it have occurred at scales that are small relative to the size of most farms, let alone larger management units such as national forests or ecoregions.

The species-area law can be used to predict what should happen to the diversity-density relationship at such larger scales (Tilman 1999). At progressively larger scales, it should take more and more species to achieve any given biomass density. For example, we can use a recent estimate of the species-area relation for contiguous patches to extrapolate Tilman and colleagues' results. According to Rosenzweig (2003), roughly $S = cA^{0.15}$, where S is number of species, c is an adjustable parameter, and A is area. This equation entails that for each 81-m^2 plot in a square kilometer of grassland to contain 16 species, the 1-km^2 patch would have to contain 66 species. Thus, to attain the maximum density realized in Tilman and coworkers' experiment throughout a 1-km^2 patch, it would take 66 species, rather than only 16 (cf. the solid curve in Figure 20.3).

One way to understand this contrast between diversity–total-density relations at different scales is to consider that the larger the patch, the more heterogeneous the temperature, moisture, and soil conditions, and so on, are across it. This means that the particular set of species able to attain a certain total density in one part of the larger habitat will differ from the species that are collectively able to attain the same density in other parts. This, in turn, entails that achieving any given density across a large habitat requires more species than are needed to achieve the same density within any smaller part of it.

3.   MECHANISTIC RESEARCH IN OPEN SYSTEMS

In all three areas of research exemplified in the previous section of this paper, ecologists have sought to go beyond the discovery of so-called phenomenological patterns, by uncovering the "mechanisms" responsible for them. In this section, I aim to correct the misconception that mechanistic research is necessarily reductionistic. This misconception seems common among both philosophers and ecologists, and among both advocates and opponents of reductionism. As one antireductionist philosopher of social science put it, "The microexplanation tells us the mechanism by which the macroexplanation operated" (Garfinkel 1981, 58). Darden criticized another philosopher, Thagard, for his "reductive" view of certain genetic mechanisms (Darden 2000). My argument in the following is similar to Darden's, but more fully developed, and framed within the context of ecology. I have elsewhere spelled out a different, complementary reason why mechanistic research often contradicts, rather than fulfilling, reductionist prescriptions (Mikkelson 2004).

a.   Definitions

Before going further, let us consider what is meant by the words "mechanism" and "reductionism". I submit that in general, to provide a mechanism for an ecological relationship is to "fill in the blank". Given that A causes B, to describe a mechanism for that relationship is to show that A causes M, which then causes B. In other words, the mechanism $A \to M \to B$ partly fleshes out the causal pathway $A > B$. This is fairly minimalist compared to the explications discussed by Tabery (2004). I am not convinced that any more is really needed than I provide here, but if more is required, then perhaps my interpretation still works as a partial definition of the term "mechanism", providing necessary, though not sufficient, conditions.

To illustrate this construal of the "mechanism" concept, suppose that an increase in one population causes a second, competing population to decrease. The Lotka-Volterra competition equations – a relatively "phenomenological" approach – can describe this kind of effect with reasonable accuracy. More "mechanistic" models go beyond the Lotka-Volterra equations by specifying *how* populations exert competitive effects on each other. For example, resource-competition equations can represent the following causal chain: An increase in one plant population leads to increased uptake of a certain nutrient by that population, thereby reducing the amount of that resource left in the environment. This decrease in the nutrient supply then causes a second population, dependent on the same nutrient, to decline.

What type(s) of reductionism is (are) at stake in ecology? Schoener seems to have captured at least one important kind. He also assumed without discussion that "mechanistic" means "reductionistic":

[A]dvocacy of a reductionist approach coincides with emphasizing internal, rather than external, factors when simplification is necessary. Thus mechanistic people will stress behavioral and physiological detail at the expense of, say, food-web detail. (Schoener 1986)

Schoener's article focused on explaining population dynamics. His comments therefore imply that mechanistic explanations of such dynamics must involve more detail about the organisms composing the populations than about the communities or ecosystems containing the populations. In other words, he argued that the mechanisms of population dynamics are predominantly micro- or lower-level.

At the extreme, this form of reductionism entails treating populations as completely closed systems that do not interact at all with macro- or higher levels. The density, for example, of a closed population would be explained strictly in terms of the properties of that population, including the individual organisms within it. Such higher-level phenomena as the densities of other populations, or the total number of species, within the same community, would play no role in reductionistic explanations of this kind.

b.   A Higher-Level Mechanism

Let us now consider the implications of a recent theoretical study that instead treats populations as open systems embedded within ecological communities (Kilpatrick and Ives 2003). The authors of this study offered a mechanistic explanation for "Taylor's law". This law relates the variability of a population to its average density: $\sigma^2 = c\mu^z$. σ^2 is the variance of population density, c is a coefficient greater than zero, μ is mean population density, and z is an exponent less than 2. Ecologists have found this pattern – observed among "more than 400 species in taxa ranging from protists to vertebrates" – interesting because it differs from the "null" expectation. If populations experienced constant per capita variability, z in the equation would equal 2. The fact that z is less than 2 for most species means that an increase in population density leads to a decrease in per-capita variability.

Kilpatrick and Ives explained this pattern in terms of populations' "diffuse interactions" with the other species in their communities. Roughly speaking, their mechanism works as follows: Any given "focal" population undergoes some variation due directly to fluctuations in its physical environment. Fluctuations in the populations of its competitors add a second source of variation in the focal population. Suppose that the focal population experiences a permanent increase in its mean density. Other factors being equal, this would force a decrease in the mean densities of at least some of its competitors.[7] This, in turn, would reduce the pressure exerted by those competitors on the focal population. Each member of the focal population would then be less affected by fluctuations in the densities of these competitors. If the direct contribution of environmental "noise" to per-capita variability remains the same, then overall per-capita variation in the focal population should decline.

We thus have an example of a higher-level mechanism for a same-level relationship. This is a different sense of "higher-level mechanism" than the one Glennan attributes to Wimsatt (Glennan 2002). The latter refers to any mechanism that does not "plunge" all the way down to some "fundamental", such as microphysical, level. In contrast, here I mean a mechanism that involves properties at a higher level than the relationship being explained. In this case, a community-level mechanism explains a population-level relationship, Taylor's law. This case therefore contradicts the common assumption, implied by Garfinkel (1981) and expressed by Schoener (1986), that mechanisms entail reductionism. In other words, this case demonstrates that mechanistic research need not emphasize the parts of a focal entity at the expense of its context or environment.

Incidentally, the two examples cited in this section also illustrate the relativity of the phenomenological/mechanistic distinction. The Lotka-Volterra competition equations are phenomenological relative to resource-competition equations. But relative to Taylor's law, the Lotka-Volterra model that Kilpatrick and Ives used to describe *how* mean population density affects population variability is mechanistic.

CONCLUDING SUMMARY

In this essay, I have described a recent shift in perspective from smaller to larger scales in ecology. This shift has revealed that species diversity is more important, and under greater threat, than was previously known. It also motivates greater appreciation of the role played by higher-level causes in nature. I have also illustrated how recognizing the open character of most ecological systems leads naturally to the discovery of higher-level mechanisms. Earlier in the chapter, I situated these points within the context of recent work in the philosophy of ecology. And I framed the discussion in terms of logical positivism's tenacious legacy.

NOTES

1. Living mass per unit area or volume.
2. Many of these surveys actually involved algae, which, strictly speaking, are not plants but play the same ecological role as plants, namely, "photosynthesizer" or "primary producer".
3. Those familiar with food chains know that this scenario presupposes an even number of trophic levels (e.g., just plants and herbivores, or plants plus herbivores plus "primary" carnivores plus "secondary" carnivores").
4. Represented in Leibold's paper by "Model 1".
5. Leibold's Models 2 and 3.
6. Estimated by visual inspection of the Year 2000 data shown in figure 1B of Tilman et al. (2000).
7. Even if the increase in the focal population did not reduce other populations, it would still reduce their *relative* densities – relative, that is, to the density of the focal population.

21 From Ecological Diversity to Biodiversity

1. INTRODUCTION

During the last three decades, biodiversity conservation has emerged as a central focus of environmental concern in many regions of the world (Sarkar 2005). As a result, large-scale efforts are being devoted to devising systematic protocols for conservation, sometimes involving computational efforts unprecedented in ecology (Margules and Pressey 2000). These efforts presume that a sufficiently precise concept of biodiversity is at hand. Some philosophical attention has also been focused on elaborating an adequate normative basis for conservation.[1] These attempted normative justifications for biodiversity conservation also depend on what is meant by biodiversity. Yet, "biodiversity" remains a contested term.

The term "biodiversity" was first used in 1986 as a contraction for "biological diversity" by Walter G. Rosen while planning for a (U.S.) National Forum on Biodiversity (Takacs 1996). Subsequently, temporarily mutated as "BioDiversity," it was used as the title for the proceedings from that meeting (Wilson 1988). No attempt was made to define the term precisely, even as its use spread – the chronology in Table 21.1 includes the most salient episodes.[2] Conservation biology also emerged as a distinct interdisciplinary research area during the 1980s with its central aim the protection of biodiversity (Takacs 1996, Sarkar 2002, 2005). From the very beginning the term "biodiversity" was notable for its polyvalence, meaning different things to different users.[3] Such polyvalence should neither occasion excessive surprise nor, by itself, be taken to indicate some special difficulty with a term. Biological terms such as "species" or "gene" are at least as polyvalent while remaining sufficiently precise for

388

Table 21.1. *Ecological Diversity, Biodiversity, and Conservation Biology: A Brief Chronology*

Year	Development	References
1943	First index of ecological diversity.	Fisher et al. 1943
1948	Shannon publishes his index for the quantity of information in a message. Preston attempts to improve upon Fisher.	Shannon 1948; Preston 1948
1949	Simpson's index introduced.	Simpson 1949
1955	MacArthur uses Shannon index as a measure of stability, makes a diversity-stability claim precise.	MacArthur 1955
1958	Margalef uses Shannon index as a measure of diversity. Elton produces empirical evidence of a diversity-stability relationship.	Margalef 1958; Elton 1958
1959	Hutchinson pays homage to tropical species richness, raising the questions of the determinants of the latitudinal diversity gradient.	Sarkar 2005
1960	Whittaker distinguishes among α-, β-, and γ-diversity.	Whittaker 1960
1967	MacArthur and Wilson publish *The Theory of Island Biogeography*.	MacArthur and Wilson 1967
1968	*Biological Conservation* starts publication.	Sarkar 2004
1969	Pielou provides a unified theoretical analysis of diversity indices in a textbook treatment.	Pielou 1969
1970	Hutcheson produces first dissertation devoted to empirical studies of diversity measures.	Hutcheson 1970
1971	Hurlbert questions the use of diversity indices.	Hurlbert 1971
1973	U.S. Endangered Species Act is passed. May publishes *Stability and Complexity in Model Ecosystems*.	May 1973
1975	Diamond suggests island biogeography as the model for the design of nature reserves. Pielou publishes book-length treatment of ecological diversity measures.	Diamond 1975; Pielou 1975
1976	U.S. National Forest Management Act is passed.	Sarkar 2005
1978	Holling introduces concept of adaptive management.	Sarkar 2005

(*continued*)

Table 21.1 (*continued*)

Year	Development	References
1982	Patil and Taillie publish statistical review of diversity measures. Rao invents the quadratic entropy index capturing distinctiveness. Margule and others effectively criticize the use of island biogeography to design reserves.	Patil and Taillie 1982; Rao 1982; Sarkar 2005
1983	Kirkpatrick uses the first complementarity-based area selection algorithm.	Kirkpatrick and Harwood 1983
1985	Formation of the Society for Conservation Biology (United States). Soulé publishes the manifesto "What Is Conservation Biology?"	Sarkar 2005; Soulé 1985
1986	Publicity material for "National Forum on BioDiversity" includes first use of the term "biodiversity." Janzen publishes the manifesto "The Future of Tropical Ecology." Recognition of stalemate in the SLOSS debate.	Takacs 1996; Janzen 1986
1987	*Conservation Biology* begins publication.	Sarkar 2004
1988	Publication of Wilson, ed., *BioDiversity*, first book with "biodiversity" in its title. First computer algorithm based on complementarity. Last book-length treatment of diversity measures until 2004.	Wilson 1988; Margules et al. 1988; Magurran 1988
1989	U.S. GAP Analysis Program launched to identify features of biodiversity not protected in the conservation areas.	Scott et al. 1993
1991	First issue of *Canadian Biodiversity*, the first journal with "biodiversity" in its title; name changed to *Global Biodiversity* in 1993.	Sarkar 2004
1992	Convention on Biological Diversity (Rio de Janeiro). Weitzman attempts to renew interest in the general statistics of diversity measures.	Takacs 1996; Weitzman 1992
1999	Tilman makes the case for a richness-stability relation.	Tilman 1999
2000	Margules and Pressey publish the consensus framework for systematic conservation planning in *Nature*.	Margules and Pressey 2000
2004	Magurran publishes *Measuring Biological Diversity*.	Magurran 2004

everyday use in biology. The interesting question is whether there is anything special about "biodiversity." It will be argued later, appearances to the contrary, that operational concepts of biodiversity used within contemporary conservation biology suffer from very little ambiguity.

However, the polyvalence of "biodiversity" and the fact that there was almost no attempt to quantify its measures during the first decade of its use do suggest that the concept of biodiversity was introduced de novo, that is, in discontinuity with the decades-old tradition of attempts to define quantitative measures of ecological diversity which preceded it. The new conservation biologists in the United States, convinced that they were involved in the establishment of a discipline with no antecedent, at least inadvertently promoted such a view.[4] They also encouraged the idea that an entirely new conceptual apparatus was being created. This chapter will explore this issue in detail, emphasizing the roots of concepts and measures of biodiversity in earlier quantitative measures of ecological diversity. These connections seem to have been largely ignored in both the biological and the philosophical literature but are important for a philosophical understanding of the conceptual structure of conservation biology and its relation to ecology. The gist of the argument is as follows:

- Explicit scientific discussions of biodiversity, after the contraction was first introduced, implicitly presumed what had been learned from the earlier discussions of diversity within ecology.
- These earlier discussions of ecological diversity had reached an impasse to a large extent because of a stalemate in the diversity-stability debate by the early 1980s. Consequently, it was unclear whether quantitative measures of ecological diversity captured anything important. Conservation biologists seemed to have reacted to this situation by avoiding these measures in their practice.
- Any discussion of biodiversity had to keep part of its focus on the normative justification for its conservation, in contrast to the earlier discussions of ecological diversity. This encouraged a shift of discussion away from the quantitative measures to concepts that are more easily used in contexts of discussions of normativity.

- Moreover, biodiversity conservation always occurred in sociopolitical contexts in which conservationists had to contend with other claimants on land, for instance, those who would convert natural habitats into agricultural or industrial production systems. Discussions of biodiversity were constrained by this context, which required general accessibility to multiple stakeholders and helped move the discussions away from earlier technical treatments of ecological diversity.
- Nevertheless, when attempts were made to quantify biodiversity within conservation biology, those earlier discussions provided many of the conceptual tools that became relevant though the historical antecedents of the new measures were often ignored. In particular, measures of what is known as β-diversity are critically relevant to the operationally quantified measures of biodiversity that must be used in systematic conservation planning. The crucial innovation reponsible for this shift was that planning for conservation emerged as an *algorithmic* process that necessitated operationalization of biodiversity as a quantitative concept.

Philosophers and historians have explored the recent history of ecology and history of conservation biology so sporadically that the historical claims made during this argument must be regarded only as a first attempt to reconstruct these developments, subject to welcome revision in the future.

2.   BACKGROUND: MEASURES OF ECOLOGICAL
       DIVERSITY

The first quantitative index of ecological diversity was proposed by Fisher, Corbet, and Williams (1943) to relate the number of individuals to the number of species in a sample drawn from a natural community. Starting in the late 1940s, Preston (1948, 1962) attempted to extend Fisher's work. All these statistical models were phenomenological, based on the fit with data, rather than on biological principles, though Fisher's index has recently been given a theoretical foundation by Hubbell (2001). However, during the same period, MacArthur (1957) approached the same problem using explicit mechanistic models based on presumed biological interactions. For instance, MacArthur's (1957) broken-stick model, which was supposed to predict diversity, assumed that the species in a community apportioned resources at random.[5] MacArthur's many innovations in the 1950s paved the way for a resurgent theoretical ecology in the 1960s that went beyond population ecology, leading to the theory of island biogeography and, eventually, the diversity-stability question (see Section 3).

However, the most common diversity measures used in ecology emerged during the same period from a different conceptual background. Simpson (1949) introduced an index of "concentration" ($\sum_{i=1}^{n} \pi_i^2$ where π_i is the frequency of the ith type and $\sum_{i=1}^{n} \pi_i = 1$), the complement or inverse of which provides a natural measure of diversity. Margalef (1958) proposed the use of Shannon's information index ($-\sum_{i=1}^{n} \pi_i \log \pi_i$) as a diversity index. The introduction of these two measures led to a variety of others and a large set of empirical studies in the 1960s that measured diversity in the field (reviewed by Hutcheson 1970; see Table 21.2). Nevertheless, the fundamental question remained open: what is the justification of these measures? Hurlbert (1971) posed the question forcefully and no fully satisfactory answer has yet been produced.

Conceptually there are two obvious options for answering this question: (i) it can be shown that a proposed measure captures some unproblematic intuition about diversity by laying down explicit adequacy conditions and, ideally, proving that some proposed measure is the only one satisfying these conditions; or (ii) the use of a measure can be justified by showing it plays an important theoretical role within ecology by being used in relevant models. There are again two ways in which the second option can be carried out: (a) the proposed diversity measure may capture some feature that is a result of ecological processes; or (b) it may be something that plays a determining role in ecological processes. Starting in the 1960s, all three of these strategies of justification have been tried but with limited success.

Proposed adequacy conditions for diversity measures have required that higher diversity must result in an increase of (1) the *richness* or number of species, (2) the *evenness* (or *equitability*) of their presence, (3) their average *abundance rarity* or level of

Table 21.2. *Adequacy Conditions for Ecological α–Diversity*

Adequacy condition	Explanation	Measures incorporating condition
Richness	Number of species at an area	Richness (Patil and Taillie 1982); Simpson (1949) measure; Shannon measure (Margalef 1958)
Evenness/ equitability	Equal proportional occurrence of all species at an area	Simpson (1949) measure; Shannon measure (Margalef 1958); Hill (1973) index; Alatalo (1981) index; Pielou's (1975) indices; redundancy (Patten 1962); standard deviation (Fager 1972); number of moves (Fager 1972); all measures of intrasample variability can be used for this purpose
Average abundance rarity	Low proportional occurrence of a species at an area	Simpson (1949) measure; Shannon measure (Margalef 1958); Hurlbert's (1971) PIE index; Patil and Taillie's (1994) index
Geographical rarity	Low total occurrence of a species in a region	Rarity (Sarkar et al. 2004)
Distinctiveness	Phenotypic/taxonomic uniqueness of a species at an area	Quadratic entropy (Rao 1982); all measures of phylogenetic difference can be used for this purpose
Abundance transfer	Transfer of proportion from a less abundance-rare species to a more abundance-rare species (i.e., the rare species increases in frequency while the common species decreases)	Richness (Patil and Taillie 1982); Simpson (1949) measure; Shannon measure (Margalef 1958); Patil and Taillie's (1994) index

occurrence, (4) the highest *geographical rarity* of species present, (5) their *distinctiveness*, and (6) the effect of *abundance transfer* between the rarest and the most common species. Table 21.2 explains these conditions in more detail and links some common measures of ecological diversity to them; the classification given there is an extension of those provided by Pielou (1975) and Magurran (1988) but probably still not complete; a complete classification would not be entirely unnoticed by ecologists. With the exception of condition (4) all of these measures are "local" in the sense that their assessment does not require access to information outside a system; geographical rarity has global scope, but the rarity of species within a system is still a feature of that system – this point will be important later.

The trouble with this justificatory strategy is that no single measure can satisfy all these adequacy conditions though this result apparently has not been proved in full generality in a unified mathematical framework. Pielou (1969) showed formally how the Shannon and Simpson measures incorporated conditions (1) and (2). Ricotta (2005) shows that condition (5), formalized as Schur-concavity, and condition (6), formalized as quadratic entropy (Rao 1982), together leave only richness as a possible measure of diversity (that is, no other measure can satisfy them). The last result is unsatisfactory because diversity intuitively means more than just the number of species. At the very least it must refer to the distribution of abundances of the species (as, for instance, captured by the evenness condition). Thus, no single measure of diversity can satisfy all five conditions of Table 21.2.

These negative results[6] may be interpreted as showing that there is no concept of diversity at all to be explicated. Alternatively, they may be interpreted as showing that no single measure of diversity will simultaneously optimize all prevailing intuitions. The second interpretation is more interesting: (i) it suggests that these intuitions be examined to determine which are less dispensable than others, and (ii) it leaves open the option to prefer a measure of diversity on the ground that it is connected to ecological processes. Table 21.2 shows that the Simpson and Shannon measures perform better than all other measures with respect to the number of conditions they meet, a point that has often been ignored when new measures are proposed.

Efforts to find connections between diversity measures and ecological processes have also foundered. With respect to the ecological determinants of biodiversity, Patil and Taillie (1976, 1979, 1982) initiated an ambitious program of deriving diversity, interpreted as average abundance rarity, from probabilistic models of interspecific and intraspecific encounters between organisms. The Shannon and Simpson measures then emerge as different ways to compute the average.[7] Most commentators since the 1980s have ignored these efforts. The trouble is that these models have no obvious bearing on any other question in ecology. Consequently, any diversity index emerging from these models may not be relevant in other ecological contexts; in the absence of such connections, diversity will play no significant theoretical role in ecology.

However, if diversity is interpreted as richness, any attempt to derive species-area curves from fundamental principles is a model of diversity. Similarly, the theory of island biogeography (MacArthur and Wilson 1967) is also such a theory of diversity qua richness, as, more explicitly, is Hubbell's (2001) neutral theory of biodiversity. Hutchinson's justly famous query about the source of variation of species richness with latitude also constitutes a call for such a theory of diversity qua richness. But, as noted earlier, there is more to diversity than richness. Diversity, by any measure other than richness, is not easily interpreted as a result of known ecological processes.

3.   STALEMATE: DIVERSITY AND STABILITY

Turning to the putative ecological effects of diversity, a much richer – and still living – tradition attempts to connect diversity to ecological processes, to productivity, and more importantly, to stability (Tilman 1999). The idea that diversity and stability are somehow connected has a long and checkered history (Pimm 1991). MacArthur is probably the first to have made the claim precise, with Elton (1958) and Pimentel (1961) providing initially promising empirical support. Connell and Orias (1964) analyzed patterns of diversity and MacArthur (1975) produced theoretical arguments in support of a relationship; however, Whittaker (1975) gave grounds for caution. Meanwhile, also in the 1960s, the appropriate definition of ecological stability became as controversial as the definition of diversity. Lewontin (1969a) introduced a variety of influential exact definitions, none of which captured all ecological intuitions. Table 21.3 provides a taxonomy of common definitions of stability.

A result of these multiple definitions is that any presumed diversity-stability relationship can be interpreted in a large variety of ways. There are at least 16 different measures of α-diversity in Table 21.2, and 7 different measures of stability in Table 21.3, giving at least 112 possible paired relations. These are low estimates: Pimm

(1991) identified 5 alternative uses of "stability" but these were embedded in a hierarchy of 3 levels of complexity and 3 levels of organization, which result in 45 measures of stability. Only a tiny fraction of the possible relationships have ever been explored experimentally or theoretically. The last point does not seem to be generally recognized in the ecological literature though it has a significant consequence: any negative assessment about the possibility of a diversity-stability relation cannot be more than tentative, subject to future revision as more potential relations are explored.

The theoretical models of the diversity-stability relation that have been explored often produced results that seemed to depend on modeling strategies, when models could be mathematically analyzed at all. In particular, May (1973) analyzed a large class of models in which increased diversity (in the form of what he called complexity) did not lead to increased stability and often delimited the set of conditions in which stability could be maintained. Many theoretical models remain practically impossible to test in the field. In the 1980s, as conservation biology was emerging as a new recognizable discipline, the diversity-stability debate had reached a stalemate. The situation is not very different some twenty years later.

In the last few decades, much empirical work has increasingly questioned support for a diversity-stability relation (Pimm 1991, Mueller and Joshi 2000). Recently, however, interesting positive results were reported by Tilman and collaborators (Lehman and Tilman 2000), with diversity interpreted as richness and stability interpreted as constancy. Yet, these reports have been followed by equally compelling negative ones with richness found to be inversely correlated with stability, now interpreted as resilience and resistance (Pfisterer and Schmidt 2002).[8] Just as in the case of the stalemate of the early 1980s, the debate over the diversity-stability relation appears once more to be at an impasse though experimental work continues on several fronts.

Yet, in the context of biodiversity conservation, a putative diversity-stability relation has immense normative force and continues to be used in this fashion.[9] If diversity is at least partly defined by species richness, such a relationship may provide a rationale for the rivet argument (Sarkar 2005): the disappearance of each species individually may appear to be insignificant. But, like the final rivet that decides a plane's fate, one final extinction may lead to the

Table 21.3. *Definitions of Stability Modified from Sarkar (2005)*

	Category	Definition	Measure
Perturbation-based categories	Local stability	Probability of return to a reference state or dynamic after a change in the value of a system's state variables	Measured by the same parameters as the state variables
	Resilience	Rate at which a system returns to a reference state or dynamic after a perturbation	The inverse of the time taken for the effects of a perturbation (e.g., of species' abundances or densities) or to decay relative to the initial size
	Resistance	Inverse of the magnitude of the change in a system, relative to a reference state or dynamic after a perturbation	(1) Inverse of the change of species' densities or abundances relative to the original state; (2) change of species composition relative to the original composition
	Perturbation tolerance/ domain of attraction	Size of the perturbation a system can sustain and return to a reference state or dynamic (irrespective of time taken)	Perturbation size measured in natural units (perturbation may be biotic or abiotic)

Perturbation-independent categories	Persistence	Ability of the system to continue in a reference state or dynamic	(1) The time for which a system sustains specified minimum population levels, e.g., nonextinction of a proportion of its species; (2) The time a system will sustain specified species compositions
	Constancy	Inverse of the variability of a system (community or population)	The inverse of the size of fluctuations of some parameter of the system such as species richness, size, or biomass abundance
	Reliability (Lehman and Tilman 2000)	Probability that a system (community or population) will continue "functioning"	Measured in terms of how faithfully and efficiently a system processes energy and materials and engages in other biogeochemical activities

collapse of the biosphere. Some version of this argument lies behind many consequentialist normative rationales for biodiversity conservation. As Goodman (1975, 261) puts it:

The diversity-stability hypothesis has been trotted out time and time again as an argument for various preservationist and environmentalist policies. It has seemed to offer an easy way to refute the charge that these policies represent nothing more than the subjective preferences of some minority constituencies.... From a practical standpoint, the diversity-stability hypothesis is not really necessary; even if the hypothesis is completely false it remains logically possible – and, on the best available evidence, very likely – that disruption of the patterns of evolved interaction in natural communities will have untoward, and occasionally catastrophic, consequences. In other words, though the hypothesis may be false, the policies it promotes are prudent.

Nevertheless, cogent normative justifications of biodiversity conservation must rely on other arguments.

These negative developments seem to suggest that there may be no veridical concept of ecological diversity and, consequently, no such concept of biodiversity. But this would be an unwise interpretation: a conclusion about biodiversity does not follow from the premise about ecological stability unless the two concepts are relevantly related. In the context of biodiversity and its conservation, the concepts of diversity that were explored earlier may well have been inappropriate *inventory*-based concepts, referring only to what occurs within systems localized at areas, rather than *difference*-based concepts, comparing the differences in the inventories at two or more areas. Whittaker (1960) elaborated the relevant distinctions long ago, in 1960. He distinguished among α-diversity, the diversity within an area, β-diversity, that between areas; and γ-diversity, or the total diversity of a region, including both α- and β-diversity. The latter two are what are relevant to biodiversity, as the next section will emphasize. The measures and adequacy conditions we have been considering are all related to α-diversity; even geographical rarity of species at an area, as was noted earlier, refers directly only to what occurs at that area.

4.   BIODIVERSITY

On the basis of a sociological analysis, Takacs (1996) and others have argued that the term "biodiversity" was introduced to move public interest away from the conservation of single charismatic species (megafauna such as large colorful birds or large mammals) to all species. This is supposed to have been particularly important for conservation in the neotropics, which are lacking in charismatic species but were under special threat due to accelerating habitat destruction since the early 1960s (Janzen 1986).

This view is partly supported by Takacs's catalog (1996, 46–50) and definitions such as that used in the Convention on Biodiversity (Article II): " 'Biological diversity' means the variability among living organisms from all sources including, *inter alia*, terrestrial, marine and other aquatic ecosystems and the ecological complexes of which they are part; this includes diversity within species, between species and of ecosystems." More recently, Wilson (1997) argues: "Biodiversity is defined as all hereditarily based variation at all levels of organization, from genes within a single population or species, to the species composing all or part of a local community, and finally to the communities themselves that compose the living parts of the multifarious ecosystems of the world." The gratuitous reference to heredity aside, both definitions are essentially the same. Definitions such as these are still inventory based, relying on the contents of individual systems like the definitions of ecological diversity discussed in an earlier section; however, many of Takacs's respondents explicitly appeal instead to difference-based definitions, which look at differences between entities within systems and, especially, between systems. From both points of view, earlier discussions of diversity in the ecological literature seem to be at best marginally relevant to the question of formulating a definition of biodiversity adequate for conservation biology.

However, at least three sets of arguments raise questions about the accuracy of the last thesis, that is, the denial of the relevance of the earlier work. *First*, none of the concepts invoked in the adequacy conditions for ecological diversity measures (see Table 21.2) depends on charismatic species. Further, from the beginning, it was recognized that defining diversity measures was a formal statistical problem with the measures applicable to any set of entities, not just species. This is clear in the discussions of (among others) Fisher and associates (1943), Preston (1948), Patil and Taillie (1982), Rao (1982), and Weitzman (1992). Consequently, the generalization of concern from charismatic species to all features of biological diversity did not preclude the use of the earlier measures and analyses.

Second, even if some of the early discussions of biodiversity ignored the work on ecological diversity, starting in the 1990s, quantitative measures of biodiversity began to be proposed within conservation biology though, initially, these were oriented toward taxonomic diversity.[10] Later work showed that many of the non-taxonomic measures can be formally related to ecological diversity measures.[11] Recent quantitative work on biodiversity also draws heavily on these earlier discussions (see, e.g., Hubbell 2001).

Third, and most important, once β- and γ-diversity are included (besides α-diversity), the available measures of diversity capture even the new difference-based definitions of biodiversity (leaving out only the diversity of processes; see Section 6 for elaboration of this point). For instance, some of the recent work on defining bio-diversity for use within conservation biology has focused on quantitative operational concepts, especially complementarity (Vane-Wright, Humphries, and Williams 1991, Justus and Sarkar 2002), which can be explicitly related to β-diversity (discussed later). These operational quantified measures of biodiversity are important because systematic conservation planning emerged as an explicitly *algorithmic* process that must have precise goals by which the relative success of a plan can be assessed. The major goal of the algorithms is to select the best area for conservation using various criteria, given a list of areas and at least lists of biodiversity "surrogates" present in them (with these surrogates being features that are supposed to measure the relevant aspects of biodiversity for conservation planning). However, this raises the interesting unresolved philosophical question about the relations between these measures and the intuitive concepts used by those who provided definitions for Takacs's (1996) catalog (see the next paragraph).

The quantitative complementarity value of an area, relative to a set of areas already selected for conservation, is the number of new, or as yet only inadequately protected, biodiversity surrogates (such as species, species assemblages, habitat types) that this area contains.[12] Areas with higher complementarity value are given preference for protection over areas with lower values. Notice that this definition assumes that the problem of what surrogates should be used has already been solved. But the definitions cataloged by

Takacs (1996) are intended to specify what should count as surrogates. Operational definitions of biodiversity assume that the appropriate surrogates are given and focus on their quantification for conservation planning. The ambiguity – and freedom – in the choice of surrogates is unavoidable (Sarkar 2002, 2005), but measures of ecological diversity are equally subject to this problem. In ecology, α-, β-, and γ-diversity are traditionally measured using species (as, also, was complementarity), but no biological reason precludes their being measured by using other surrogates – recall that these measures are all ultimately statistical concepts.

Complementarity is a critical concept in the design of conservation area networks as part of biodiversity conservation planning because it enables the construction of such networks to maximize biodiversity representation while minimizing the area included. This is important when budgetary constraints limit the amount of land that can be dedicated to conservation.[13] The algorithmic (usually iterative) selection of areas using measures such as complementarity is known by various names, including "conservation area network selection," "place prioritization," "site selection," and "reserve selection" in conservation biology.

Now, starting with Whittaker (1960), quantitative measures for β-diversity have been proposed and debated though not to the same extent as α-diversity. Traditional statistical measures such as the Jaccard index and the Hamming distance have been used to measure β-diversity. Many others have been suggested.[14] It is clear that the measures of complementarity and β-diversity are formally related. Perhaps because complementarity was not explicitly introduced as a concept of β-diversity, the connection has only recently received an overdue explicit recognition in the ecological literature.[15] As Magurran (2003, 172) puts it: "Complementarity is ... β-diversity by another name – the more complementary two areas are, the higher their β-diversity." This assessment is basically correct but requires an important modification in the context of conservation planning.

Because complementarity is defined with respect to an existing selected set of areas, unlike traditional measures of β-diversity, it is not a symmetric concept except in the simplest (or mathematically degenerate) case in which the existing selected set consists of exactly one area. If areas are selected iteratively, as in many algorithms used in conservation planning, the complementarity value of an area changes with each iteration. In general, distances based on complementarity measures are nonmetric: they do not satisfy the triangle inequality.[16] In any case, the conceptual relation between complementarity and β-diversity reiterates the point that there are more connections between the older work on ecological diversity and concepts of biodiversity than has been recognized.

Moreover, complementarity is not the only concept by which biodiversity can be operationally quantified for use in planning. Several studies have advocated the use of geographical rarity along with complementarity. Sarkar and colleagues (2004) even used a mixture of complementarity and Shannon's and Simpson's indices for α-diversity though these generally did not perform as well in practice as a mixture of rarity and complementarity in representing biodiversity as economically as possible in a selected set of areas. Thus, even the older discussion of α-diversity reenters biodiversity in the practice of contemporary conservation biology. When multiple concepts are used to prioritize areas, Sarkar (2002) has argued that the corresponding algorithm used implicitly defines biodiversity in that context. There are many other ways to quantify biodiversity operationally, and γ-diversity does not enter this discussion only because it is yet to be systematically used.

There are thus two sources of the polyvalence of biodiversity with which this chapter started: (i) the choice of surrogates and (ii) the choice of an operational definition for biodiversity. On this issue, biodiversity is similar to the earlier ecological concepts of diversity. Once biodiversity surrogates and operational definitions are selected, no further ambiguity of the term remains. In this respect, biodiversity is much better situated than definitions of species or genes. For the last two concepts, unlike the case of biodiversity, disambiguation problems remain even after a particular definition is selected.

5.   HISTORICAL SPECULATIONS

It remains to explain why, in the context of the 1980s, as conservation biology emerged as a recognizable science, distinct from ecology but drawing heavily from it, biodiversity as a concept was not explicitly connected to the widely available older measures of ecological diversity. In the absence of sufficiently detailed historical reconstructions of the emergence of conservation biology, any explanation must be partly speculative.

One likely possibility is that the stalemate in the diversity-stability debate generated skepticism about whether traditional concepts of ecological diversity can be salvaged for any purpose at all. Hurlbert (1971) argued that diversity is a "non-concept" as early as 1970 and such concerns were only exacerbated by continued failures to resolve the debate. As Table 21.1 indicates, by the mid-1980s, interest in devising and studying measures of ecological diversity largely evaporated. Even when there was a revival of interest in the diversity-stability question in the 1990s, richness was the only measure of diversity to be tracked and, as noted earlier, richness is not a good measure of diversity. This contrasts sharply with stability; in that case, all the measures in Table 21.3 have been used in recent years (see McCann 2000). As noted in the last section, the question whether richness correlates with any measure of stability remains open. How other measures of diversity would perform remains insufficiently explored. Given all this uncertainty and lack of scientific success, in the 1980s it would probably have been at least practically imprudent to base political arguments for conservation on the existing ecological diversity measures.

It is also likely that not much attention was paid to the earlier work because it was not envisioned that conservation biology would require a quantitative concept of biodiversity or even that it would be useful. The sociopolitical context of conservation required conservation proposals to be understood by nonspecialist stakeholders. Popular scientific books rather than technical monographs were perhaps inevitably the primary means of communicating with these stakeholders, and quantitative measures of diversity (which were believed to be of doubtful utility, in any case) had little role to play in such a context.[17] From the beginning, a central theoretical and practical problem has been the design of networks of reserves and other conservation areas. However, early efforts concentrated on the use of island biogeography theory (Diamond 1975); even though that theory was quantitative, its use to suggest heuristic rules for reserve design remained nonquantitative. It spawned the "single-large-or-several-small" (SLOSS) debate, which, after a decade of acrimony, also ended in a stalemate. Meanwhile the use of island biogeography theory for this purpose was severely criticized by several groups on the ground that the analogy between oceanic islands and conservation areas was faulty and empirically ungrounded (see Sarkar 2005 for a history). All these discussions remained nonquantitative.

Quantitative planning-oriented efforts at defining biodiversity occurred in a changed context after systematic conservation planning (Margules and Pressey 2000) had emerged as an *algorithmic* enterprise within conservation biology. While these algorithmic approaches date back to Kirkpatrick's work in Tasmania in 1983 (Kirkpatrick and Harwood 1983), with Margules, Nicholls, and Pressey (1988) already producing software implementing complementarity-based algorithms in 1988, they remained controversial even in the late 1990s (Sarkar 2005). Even the GAP Analysis Program, launched in the United States to identify gaps in what is protected in conservation areas nationally, was slow to adopt quantitative methods for the selection of new conservation areas (Scott et al. 1993). It is, therefore, not surprising that there was little concern for quantitative measures of biodiversity during the first decade of conservation biology, 1986–96. As seen earlier in this section, though attempts to connect biodiversity to earlier measures of ecological diversity began in the early 1990s, almost all such work dates from the late 1990s in the changed context in which quantitative measures of biodiversity were being devised and explored because they were necessary for algorithmic selection of reserves and other conservation areas.

6.   FINAL REMARKS

The discussions of the earlier sections should have made clear that there is a deep connection between biodiversity, when operationally quantified through measures such as complementarity, and the traditional concept of β-diversity. Once operational quantification begins to use other measures, including geographical rarity, even α-diversity becomes important. It is only because biodiversity often functioned as a normative concept in the 1980s, and because conservation biology was initially not conceived as requiring quantitative assessments of biodiversity, that these connections have not previously been explored. The polyvalence of "biodiversity" is also relatively easily resolved.

However, an important caveat is necessary: the concepts of diversity (ecological diversity or biodiversity) considered refer only to the structural and taxonomic hierarchies of biological organization and not to questions of functional or behavioral organization. Yet, it is clear that some behaviors, such as the migration of North American monarch butterflies (*Danaus plexippus*), constitute endangered biological phenomena (Brower and Malcolm 1991) that should form part of the concept of biodiversity. Sarkar (2002, 2005) has pointed out that not only do proposed definitions of biodiversity fail to capture such behavioral diversity but attempts to operationalize these definitions are yet to show signs of plausibility, let alone success.

Two final points, one about the normative context of biodiversity conservation, and the other about cultural differences that must be faced in practical contexts, will end this chapter. From its very inception, conservation biology was conceived of as not only being a descriptive science, but also as having a normative goal that makes it more analogous to medicine than biology (Soulé 1985). There have been many attempts to view biodiversity as a normative concept. Meanwhile, there has been much effort within conservation biology to connect biodiversity with ecosystem function, and these attempts are clearly analogous to the attempts to connect diversity with stability. However, there is a critical difference: in the case of biodiversity, the appeal to ecosystem function typically forms part of a normative (consequentialist) argument to defend conservation (Sarkar 2005). In the case of ecological diversity, the putative connection to stability (or productivity, etc.) does not necessarily have that normative role though the motivation for the search for a diversity-stability relationship may well lie in the hope for finding such a normative conclusion (Goodman 1975). Moreover, biodiversity conservation does not occur in a sociopolitical vacuum. Rather, conservation must compete with other potential uses of land, including biological and mineral resource extraction, recreation, conversion for agricultural and industrial development, and wilderness preservation. If biodiversity conservation must trump these other claims, there must be a convincing normative rationale for its greater value. A question that bears further investigation is whether – and, if so, to what extent – such a normative role for biodiversity has influenced the way it is supposed to be defined.

Varying social and cultural contexts also impose different constraints on the goals and value of conservation planning. Local and global stakeholders may differ in what constitutes the biodiversity that deserves protection. Critics from the South have long argued that Northern attitudes to the natural world are irrelevant to the exigencies of those materially struggling for survival (Sarkar 2005). However, this debate has been more about the value of wilderness than biodiversity though there is also disagreement about whether traditional cultures value biodiversity in general or only that part of it that should be conserved as resources. The discussions in this chapter simply ignore this kind of question about the meaning of biodiversity. It assumes that cultural concerns about what constitutes biodiversity can be incorporated into the choice of surrogates.

Ultimately it may well be that the normativity of conservation biology and the necessity of incorporating sociopolitical considerations in different cultural contexts are what distinguish biodiversity from earlier concepts of ecological diversity.

ACKNOWLEDGMENTS

Thanks are due to Justin Garson and James Justus for comments on an earlier draft of this paper.

NOTES

1. See, for example, Norton (1987) and Sarkar (2005).
2. The history relevant for this chapter is relegated to a table for lack of space.
3. Gaston (1996) and Takacs (1996) provide catalogs of variant definitions defended by conservation biologists.
4. Sarkar (2005) gives a detailed history and analyzes the rhetoric of the early years of conservation biology.
5. MacArthur also had two other models, but, in these discussions, the broken-stick attained iconic status as a null model while the others were largely ignored – these others will also be ignored here.
6. There are a large number of other similar results – see, e.g., Hill (1973), Alatalo (1981), Rousseau et al. (1999). They have also been reviewed by Peet (1974), Poole (1974), Pielou (1975), Magurran (1988), Smith and Wilson (1996), and, most recently, Ricotta (2005).

7.   Even earlier, Hurlbert (1971) had made related suggestions.

8.   For the state of play, see the commentary by Naeem (2002).

9.   See McCann (2000) for a recent example.

10. See, for example, Vane-Wright et al. (1991).

11. Ricotta (2005) provides a critical discussion.

12. Justus and Sarkar (2002) give the history of the use of complementarity in conservation planning.

13. Sarkar et al. (2004) review what can and cannot be achieved by complementarity-based algorithms, especially in comparison to optimal algorithms to solve the same formalized problems. Earlier literature is condensed in this piece.

14. Koleff et al. (2003) provide a fairly comprehensive survey, with the latter exploring twenty-four measures applicable to binary species presence-absence data.

15. See, however, Hooper (1998) for some scattered remarks that indicate some recognition.

16. In the literature on algorithm design, the use of complementarity is known as the "greedy algorithm."

17. Some of these issues are documented in Takacs (1996), but much more historical analysis remains to be done.

22 Biology and Religion

INTRODUCTION

The historical, conceptual, and cultural interplay between biology and religion involves a complex and philosophically fascinating set of relationships. Certainly the simple view that religion has uniformly been a hindrance to biological research is an unfair caricature. Religion has sometimes had a positive effect, often indirectly stimulating or even directly encouraging scientific research of the biological world. Similarly, biology has had a profound effect on religion, sometimes offering challenges that require believers to reassess basic theological assumptions. Scholars have examined both directions of influence, finding both expected and unexpected connections (e.g., Cantor 2005, Rolston 1999, Russell, Stoeger, and Ayala 1999). This chapter will give a necessarily selective account of some of the mutual influences between biology and religion. Its major purpose will be to highlight a deep pattern underlying their interplay, namely, the pervasive effect of the religious idea of the divinely created normativity of nature.

How this pattern is exemplified in multiple ways in the ongoing creationist controversy will be a recurring example, but we will also examine how it may be seen in a wide range of issues from questions about what it means to be human; to religious warnings against "playing God"; to views about gender roles and sexual morality, environmentalism, personhood, and the status of the fetus; and ultimately to metaphysics and the question of ultimate priority in creation. We begin by looking at biological evolution, a critical topic that intertwines with many of these other cases.

RELIGION AND EVOLUTION

Certainly one of the best cases by which to examine the interplay between science and religion is the ongoing religious controversy over evolution and creation. Although this controversy is often stereotyped as little more than a simple attack upon biology by naïve biblical literalists, this superficial view misses the way the battle distills the essence of the deep conceptual divisions in the ways that people conceptualize and deal with scientific and religious worldviews and their implications. As we will see, the issue touches not only upon the gross features of the so-called culture war and the broad struggle between science and religion, but is connected to a wide range of the philosophically interesting topics in the relationship of biology and religion, including evidential, ethical, metaphysical, and even existential questions.

Being Human

The effect that the discovery of evolution had and continues to have upon the religious, especially the Christian, worldview may be even more profound than the earlier scientific revolutions that displaced human beings from the physical center of creation. If we were not specially created, what does it mean to be human? Does humanness begin and end with the biological notion of the species *Homo sapiens*? The traditional Christian understanding of human dignity took it to be based in the Genesis notion that humans are created in God's image. What can be retained of human uniqueness and dignity if we evolved from apes and are "just" one more branch on the evolutionary tree? And what happens now that genetic technology puts our future evolution in our own hands?

Through its discoveries and the questions that these lead to, biology has challenged religious thinkers to reassess old notions of Imago Dei. Though a literal reading of "God's image" would suggest that human beings resemble God in visual shape, surely it is an odd notion in more ways than one that human beings get their dignity because their body looks like God's. Evolution gives further impetus to separate that notion from the body itself and to focus on more abstract candidates for the human essence. It is common now for religious thinkers to identify instead traits such as human freedom and ethics. Some like the philosopher of religion and biology Patricia A. Williams place this explicitly in an evolutionary framework and see human freedom as emerging as hominids evolved to be able to recognize and make choices. Williams has no problem accepting evolution and argues that God's love was first manifested in the universe when it could be expressed through the evolution of creatures with symbolic language. The scientific discovery of our evolutionary connections to all living things, she says, helps us understand that the commandment to love one's neighbor should embrace all of the living world (Williams 2001).

While Williams's particular account may not yet be widely shared, it demonstrates how religious views are absorbing and processing discoveries from biology. Similar processing and accommodation have happened numerous times in the past. To give just one further example, consider historical changes in views regarding the possibility of living beings on other worlds, an idea that evolutionary biology makes one consider more seriously.

Extraterrestrial Life

Discussions of the possibility of life on other planets go back well before Darwin. In de Fontenelle's 1686 classic *Conversations on the Plurality of Worlds*, a philosopher and his hostess discuss the implications of the new Copernican cosmology and find the idea of people on the Moon and planets to be a light-hearted and entrancing possibility. But William Whewell's 1853 treatment of the issue was more sober. In *Of the Plurality of Worlds*, Whewell argued that no life existed anywhere else in the universe (Whewell and Ruse 2001). His argument bore the stamp of a theological worry: God's relationship to humans is supposed to be personal and unique, with the idea of His appearing on Earth as a man, the savior Jesus, being so fundamental to the Christian view that it would be impossible and equally repugnant to imagine either other worlds with analogous saviors or bereft of salvation. Admitting the possibility of living beings on other worlds would threaten our special relationship with God, not to mention opening the door to supporters of evolution. Indeed, Thomas Paine used just this kind of an argument to criticize Christian beliefs, citing the absurdity of the idea of God's traveling from world to world in an endless succession of death to redeem the progeny of alien Adams and Eves. But few contemporary Christians seem bothered by the possibility of extraterrestrial life and the Catholic philosopher of science Ernan McMullin (2000) has explained how Christian theology can be formulated to accommodate such notions.

Rethinking Religion in Light of Evolution

It is important to recognize the extent to which most mainstream religions have already accommodated evolutionary biology. Even Christian theology, for which evolution might be thought to pose the greatest challenges, has for the most part made peace with the findings of Darwin and the evolutionary biologists who have followed. Mainstream Christianity has done this in much the same way that it eventually came to terms with the earlier challenges from physics and astronomy.

Rather than following a simplistic reading of Scripture, religious thinkers have followed Galileo's advice that truth should not contradict truth and that believers should allow God to speak through the book of the world and not just the book of the word. Rather than insisting that God must have used a direct form of miraculous creation to bring about biological complexity, they have taken a more broad-minded and generous view of God's powers of creation. They allow, or even insist, that God created the world indirectly, by endowing it with a complete set of laws that did not require periodic intervention and adjustment. Rather than tying Christian theology to metaphors of God's "design" or "plan", theologians like John Haught (2000) suggest that these may be misleading, and that notions of God's "vision" or "dream" for the universe might be better. Rather than getting hung up over the evolution of our material body, they say, remember that what is essential to Christian doctrine is the immaterial soul.

For these and other reasons, the Catholic Church and other Christian denominations say that evolution should not be seen as in conflict with Christian faith. Nor are these only recent accommodations. Even in Darwin's day many evangelical Christians were among the earliest defenders of the theory, as documented in David Livingstone's (1987) revealing history of the encounter between evangelical theology and evolutionary thought.

Creationism

However, evolution continues to be viewed as theologically anathema to most Christian fundamentalists, especially in the United States. With about a third of all Americans identifying themselves as fundamentalist according to polls, and many evangelicals holding similar views, there is a huge receptive audience for various forms of creationism.

The general concept of creationism is the rejection of the scientific account of evolution in favor of supernatural special creation, but there are many variations of this idea that reflect different theological assumptions. By far the dominant form of creationism is the variety that holds that the Earth is 6,000 to 10,000 years old. These "young Earth" creationists calculate from the days and the generations listed in the Bible to get this figure. However, other creationists read the days of creation in Genesis as being long ages of time, since a day from God's point of view need not correspond to our solar day, and so they accept the standard scientific chronology. Other "old Earth" creationists interpret Scripture in other ways that allow this. Some creationists hold that the major global geographical features were caused by a catastrophic worldwide flood, while others believe that Noah's flood may have been local or "tranquil". And so on. Moreover, there are also non-Christian creationists who reject evolution in favor of the creation stories of their own religions (Pennock 1999).

The public controversy about creationism as we experience it today can be traced to the *Epperson v. Arkansas* decision of the United States Supreme Court in 1968, which ruled that laws that banned the teaching of evolution were unconstitutional. When the courts also ruled that teaching biblical creationism was unconstitutional, young Earth creationists tried a new approach. Dropping the overt references to Scripture, they claimed to promote merely a scientific view, which they called "creation science". They had some success at getting legislatures to pass "balanced treatment" acts, but again the courts found such laws to be unconstitutional. The testimony of the philosopher of biology Michael Ruse on the nature of science provided a critical element of the decision. In 1987, the Supreme Court ruled in *Edwards v. Aguillard* that creation science was not science but was disguised religion, and thus that teaching it in the public schools violated the Establishment Clause of the Constitution. Quickly adapting to this loss, creationists changed their terminology. For instance, manuscripts of a major creationist biology textbook – *Of Pandas and People* – that was in preparation dropped the term "creation-science" in 1987 immediately after the *Edwards* decision and replaced it with the term "intelligent design". For obvious reasons, intelligent design (ID) advocates now deny that they are creationists, but their history and their substantive views show otherwise.

The intelligent design movement is most characterized by what it called "the Wedge", a strategy devised by Phillip Johnson, a law professor who is credited with negotiating a truce between young Earth and old Earth creationists to improve their chance of success by uniting around a banner of "mere creation". So long as they stayed focused on their common view that evolutionary thinking is profoundly anti-Christian, and that God, not natural processes, created the world and human beings, they could agree temporarily to drop their battles about the flood and the age of the Earth. Such issues could be introduced later after the sharp edge of the wedge had penetrated the constitutional barrier. ID is often misidentified as a form of old Earth creationism; in fact, the young Earthers in the group have simply agreed to hold off pressing their case until ID reaches the classroom. The Wedge was also a metaphor they used to speak of how they would split apart the materialist, naturalist worldview of science so they could replace it with their theistic science. When lobbying for their view to be taught in the public schools, they continued the old creation science claim that ID was based entirely in science and was not religious. When speaking with supporters, however, they revealed the hidden agenda. Here is a representative example from Phillip Johnson:

My colleagues and I speak of "theistic realism" – or sometimes, "mere creation" – as the defining concept of our movement. This means that we affirm that God is objectively real as Creator, and that the reality of God is tangibly recorded in evidence accessible to science, particularly in biology. We avoid the tangled arguments about how or whether to reconcile the Biblical account with the present state of scientific knowledge, because we think these issues can be much more constructively engaged when we have a scientific picture that is not distorted by naturalistic prejudice. If life is not simply matter evolving by natural selection, but is something that had to be designed by a creator who is *real*, then the nature of that creator, and the possibility of revelation, will become a matter of widespread interest among thoughtful people who are currently being taught that evolutionary science has shown God to be a product of the human imagination. (Johnson 1996, emphasis added)

The ultimate goal of ID creationists is to reintroduce supernatural explanations into biology and into science generally. They claim that biological complexity in particular, but also other functional complexity they believe they can identify in the universe, can be explained only by the purposeful action of a transcendent intelligence.

RELIGIOUS EXPLANATION IN BIOLOGY

Of course, creationists are not alone in offering mystical explanations. Nor is it only in trying to explain the creation of life and its myriad complexities that religious believers have appealed to the supernatural. We can here only briefly touch upon a few representative cases of the many examples of purported religious explanations in biology.

An important set of examples involves how to understand illness and disease. Early inklings of the transition from a religious to a scientific explanatory framework are often illustrated in the Hippocratic writings on epilepsy. Rather than thinking of epileptic seizures as a "sacred disease" involving some sort of divine possession, Hippocrates recommended that medical doctors understand it in natural terms. The idea that sickness is the result of possession, perhaps by evil spirits, is common across many religious traditions, not just animistic ones. Renaissance Christians offered similar explanations for the dancing mania, what became known as St. Vitus dance. On this kind of religious view of disease, cures necessarily will involve an appeasement of or struggle with immaterial spirits. The materialistic explanations of medical science may be viewed as irrelevant or even as suspect.

A related religious concept of disease is that it is the result of sin. In some cases the disease is taken to be a punishment for sin, sometimes even directly inflicted by God. A classic example of this view of disease as divine punishment view was seen in the plagues of the Middle Ages, which led to self-flagellation and other displays of public penance as believers attempted to atone for some unknown offense to God. Paintings from the period display Christ's throwing bolts of plague from heaven to punish the sinners.

Nor are such religious "explanations" in terms of immaterial spirits and divine punishment confined to the dark ages of history. Among the many recent examples one particularly salient one was the view expressed by many fundamentalist Christians that the disease that eventually was identified as AIDS but that had initially been called the "gay disease" was God's punishment for the sin of homosexuality. Even after the human immunodeficiency virus (HIV) was identified as the cause, many continued to view the disease as a modern plague sent to punish sinners and questioned whether HIV was indeed the cause.[1]

Such examples of religious explanations of disease are illustrative of the way in which this influential religious worldview mixes empirical and moral issues. What we are seeing in such cases is the playing out of the religious view of the normativity of nature.

NORMATIVITY OF NATURE

The idea that nature has a built-in moral structure is common to many religions, but here we will focus on the way this is seen from a Christian perspective. The basic idea is simple. God designed the world with a plan in mind. He set it up and saw to it that the world was good. Leaving aside the complexities associated with the Fall, the general idea is that God created the world with an innate moral structure. Persons who overlook this divinely created inherent normative structure cannot possibly have ethical relationships with the world and with God.

Creation and Morality

The intelligent design leader Phillip Johnson has explained what he and what all creationists take to be the significance of the evolution/creation debate.

If you have a biblical creation story, then getting the right relationship with God and getting to heaven are the most important things. If you throw that overboard and you have a naturalistic creation story, those things become unimportant and what becomes important is how we apply scientific knowledge to make a heaven here on earth. That's a dream of various kinds of reform programs – socialism, for example. (Quoted in Goode 1999)

This view that the Christian creation story informs us of how to have the right relationship with God is common among many Christian believers. Fundamentalists in particular look to the biblical stories of creation for guidance about the plan that God has for human beings. For instance, the story of how God created one man, Adam, and then later one women, Eve, is seen to be informative of the proper biological order of creation. It defines the proper core components of the family – one man and one woman. It defines what are supposed to be the appropriate gender roles: the women is to be the helpmate of her husband. It explains why women should suffer the pains of childbirth: they are a punishment for Eve's sin. And so on. Again, the world is seen to have a built-in normativity, much of it related to sexual morality.

This view is by no means peculiar to creationism, but it is pervasive in creationist writings. A Freudian could easily analyze creationism as a sublimation and displacement of a repressed sexual obsession.

ID AND SEX

Creation science writings are rife with warnings about how evolutionist thinking is to blame for sexual promiscuity, divorce, pornography, abortion, and even bestiality. Intelligent design creationists make reference to exactly the same list. Philip Johnson and others regularly illustrate what they take to be at stake in the battle between the naturalistic worldview of evolution and the theistic worldview of intelligent design using examples involving sex. Premarital sex, adultery, divorce, and flexible gender roles, all purportedly the fruits of the former, are put up against chastity, faithful, stable marriages, and "proper" gender roles that are supposedly the fruits of the latter. In a single book Johnson twice mentions sex-education classes in which girls practice unrolling condoms over cucumbers as an example of the sorry effects of the evolutionist, naturalist worldview.

But creationists typically reserve their greatest ire for what they, as do other Christian fundamentalists, see as the worst of the sexual sins against nature. In arguing against evolution, creationists regularly cite the lines from Romans 1 that says that ever since the creation of the world God's invisible attributes of eternal power and divinity have been able to be understood and perceived in what he has made. Those who refuse to see this are fools without excuse. Johnson says that the self-deceptive thinking of evolutionists further affirms the correctness of that chapter of the biblical worldview. And what else does Romans 1 say? That when they exchange the truth of creation for the lie of evolution, God hands these sinners over to the unnatural passions of homosexuality.

HOMOSEXUALITY

Christian fundamentalists and evangelicals take homosexuality to be one of the major problems of the day. The culture war against what they see as the immorality of homosexuality goes hand in glove with the war against evolution. Both arise in part from the religious idea of normativity built into nature. This is another example in which biology as a science both affects and is affected by religion.

Science in general and biology in particular potentially have a lot to say about homosexuality. The pioneering Kinsey report made people question long-held assumptions about the prevalence of homosexual behavior. Could it be true that 10 percent of people are homosexuals? Studies by ethologists revealed cases of homosexual behavior among animals in the wild. Are there really lesbian seagulls? Studies by geneticists suggested the possibility of a genetic basis to homosexuality. Is there really a gay gene? Such information is extremely salient for religious believers who think that God built normativity into creation.

Religious conservatives take sexual orientation to be a matter of choice and believe that homosexuality is a sinful choice that goes against nature as God intended it. But if homosexuality has a biological basis, then it is not so easy to dismiss it as "unnatural". If people are "born gay", then can they be blamed for what they are? If sexual orientation is fixed biologically, then does it make sense to say that counseling can cure it? Studies have shown that people are less likely to view homosexuality as inherently immoral if they believe it is biologically determined.

This reaction to information from biology reveals a tension in the basis of the religious objection to homosexuality. Most Christian objections to homosexuality stem from taking it to be morally condemned in the Bible. But if Creation was designed with an intrinsic normative structure, then biological information may necessitate a reassessment. Indeed, this kind of argument from biology is sufficiently compelling that some Christians modify their view to say that homosexual orientation is not itself immoral but that homosexual behavior is. Others challenge the science, arguing that studies suggesting a biological basis of sexual orientation are flawed and that gays can indeed be cured of what really is a pathology.

In such an atmosphere, it is no wonder that biological research on homosexuality such as that of Dean Hamer and Simon LeVay, who claimed to find evidence of a genetic influence in male homosexuality, becomes highly contentious and politicized. The debate is made even more complicated by some philosophers who are dismissive of biology and argue that homosexuality is simply a social construction. And there is the more general criticism that much of this argument on both sides is based on a fundamental mistake in ethical reasoning involving the naturalistic fallacy. Moreover, there are also reasonable, independent objections to genetic reductionism and the idea that there could be a gay gene. Philosophers like Michael Ruse (1990) have stepped in to help sort out these and other issues.

ENVIRONMENTALISM

It is worth giving one more important example to illustrate the way in which biology and religion become entangled because of assumptions about the normativity of nature. Religious assumptions may have a profound effect on the way that people view the value of the environment.

In an influential article "The Historical Roots of Our Ecologic Crisis", Lynn White Jr. argued that Judeo-Christian religious assumptions that permeate Western culture are largely the source of the attitudes that he blamed for environmental degradation. Among these attitudes, he claimed, is a faith in perpetual progress that is indefensible apart from a particular kind of teleology. It arises from a story of creation in which God makes the world and all living things for the express benefit of man. (Woman is an afterthought and also created for man's benefit, to prevent him from being lonely.) According to this religious view no item in physical creation, says White, has any purpose but to serve man's purposes. This is a religious philosophy that sees man as the master of nature and as having dominion over all, second only to God Himself.

Psalm 8, for example, speaks of God as giving man dominion over creation, putting all living things under his feet. And of course the justification reaches back to Genesis 1, where God says to Adam and Eve, "Be fruitful, multiply, fill the earth and conquer it. Be masters of the fish of the sea, the birds of heaven and all living animals on the earth." White says that in this sense Christianity is the most anthropocentric religion the world has ever seen. He warned that "we shall continue to have a worsening ecologic crisis until we reject the Christian axiom that nature has no reason for existence save to serve man" (White 1967, 54).

White recognized that Christianity is complex and looked to see whether it had theological resources that might mitigate this basic problem. He suggested the model of Saint Francis of Assisi, who emphasized humility of man individually and as a species, and attempted to promote what White thought was a more democratic vision of God's creation, exemplified in his notion of Brother Ant and Sister Fire. This was probably too simplistic a solution. Although Saint Francis did talk with the animals as brothers, when they spoke back to him they repeated the same problematic biblical teleology, saying that they existed "for your sake, o man".

However, Christian theologians who had felt rebuked by White's charges subsequently tried to find an alternative scriptural basis for an environmental ethic. Shifting emphasis from passages in which the dominion model is rooted, they drew upon passages like that of God's covenant with Noah and all of creation and upon the parable of the good steward in the Gospel of Luke. The rainbow was a sign of God's covenant not just with man but with the Earth and a promise never to destroy it again. And Jesus's parable of the good steward reinterprets the idea of dominion to include a responsibility of stewardship – "When a man has had a great deal given to him on trust, even more will be expected of him" (Luke 12: 48–49). As the steward cares for the household, so should man care for the Earth

and its creatures (Wright 1989). While liberal Christian denominations may not have needed this kind of justification, its articulation
has had a documentable effect on some conservative believers. The
nascent environmental movement among evangelical Christians
who have adopted this perspective is a positive development.

Unfortunately, there is another religious impediment to solving
our environmental problems, this one especially found in the beliefs
of certain fundamentalist Christian sects in the United States who
hold that we are already living in what they call the "end times".
They hold that biblical prophesies, particularly in the Book of
Revelation, that discuss the signs and events that will herald the end
of the world indicate that this will likely happen in our lifetime.
Why worry about loss of biodiversity, global warming, or other
environmental problems if the world is about to end and the true
believers are to be taken to heaven to sit by God? This attitude was
exemplified by James Watt, secretary of the interior under President
Reagan, who in public testimony before Congress said that is was
unimportant to protect natural resources because of the imminent
return of Jesus – "[A]fter the last tree is felled, Christ will come
back." Environmental destruction may not only be safely ignored,
but actually welcomed or even hastened for these believers as taking
us more rapidly toward the apocalypse and "the rapture". Given that
polls show that 59 percent of Americans believe that the prophecies
of Revelation are going to come true, this religious impediment
to environmental care may be a larger problem than even the
Christian stewardship model of the good Saint Francis can overcome
(Moyers 2004).

BIOETHICS AND "PLAYING GOD"

This same kind of view of the normativity of nature is behind religious warnings against "playing God" that are common in religious
moral assessments in bioethics. We should not play God and modify
the world, some hold, because God already set it up and saw to it that
the world was good. God has a plan for the world and for each person –
even in birth and death – so it would be wrong to interfere in such
matters. Much of the philosophical literature in bioethics, and
certainly most of what is discussed in philosophy courses on biomedical ethics, involves responses to or defenses of religious moral

beliefs of this sort. We have space here only to enumerate briefly a few of these.

Beginning with bioethical issues involving the end of life, one could examine at length how this kind of religious view plays out in the literature on the definition of death, on euthanasia and physician-assisted suicide. The biological notion of brain death, for instance, continues to meet resistance from religious conservatives, as does the practice of removing life-sustaining technology or hastening death of the terminally ill by lethal injection. Indeed, there were even early objections to what was called the "God squad", a hospital committee that would decide how to allocate scarce kidney dialysis machines when these were first developed. Only God should decide who should live and die and when a life is over. Similarly, for many believers, only God should decide when a life may begin. Moral objections against contraception, in vitro fertilization (IVF), and reproductive cloning in large measure derive from this kind of religious view that human beings may not usurp God's powers to create new life. There continues to be a vigorous public debate about these issues. Bioethicists have struggled with this kind of religiously based objection to biomedical technologies, most recently as articulated in the works of Leon Kass, appointed by the second President Bush to head the President's Council on Bioethics in large part because of Kass's support of these sorts of religious views.

One finds the same kind of assumption behind religious objections to other sorts of biotechnology. Although there are other kinds of arguments used as well, a common reason for opposition to genetic engineering is that nature is put together as it is for a good reason and that it is hubris to presume to improve upon it. It is no wonder that the Frankenstein story, which draws upon the myth of Prometheus's theft of fire from the gods – essentially a creation story of the origin of our control of fire – is the dominant trope used against bioengineering.

Of course one should not ignore other important religious assumptions that underlie contemporary bioethical controversies. The controversy over stem cell research, for instance, exemplifies how many different religious assumptions can come into play. Thinkers writing from a Jewish perspective, for instance, have held that their religious teachings would support stem cell research. Judaism places a high value upon human life and holds that one has a religious duty to care for one's body. The sick should accept all medical treatments, and it is our duty to do everything reasonably in our power to combat the ravages of the body. Catholic religious thinkers, on the other hand, as well as conservative Protestants, view this particular issue more in light of their religious views about the status of the fetus. If the fetus is a person even in the earliest moment after conception, then how can it be used to supply stem cells, even for a good purpose?

METAPHYSICS

This last point takes us to issues of religious metaphysics. The way in which such believers judge the moral status of the fetus is clearly a function of their religious beliefs about the true nature of person-hood. It goes without saying that this is the main source of religious opposition to abortion. Information from biology is used by all sides in the abortion controversy, but no biological data can resolve what at base depends on the metaphysical assumption that God infuses the conceptus with a human soul and that that is what really determines personhood.

Such metaphysical religious beliefs about the soul are equally in play in debates about the nature of mind and of the possibility of free will. Those who hold that it is the immaterial soul that makes human beings unique among the animals will be unlikely to accept biological explanations of human action.

This takes us back to the ID creationists, who hold just this kind of a view. Their basic assumption is that neither evolution, nor any biological process, nor even any natural physical process is capable in principle of accounting for intelligent action. The creation of information, they say, necessarily requires a designing intelligence and this intelligence cannot be merely natural; it must transcend the material world. This includes human beings, who are "embodied" intelligences, which is their way of speaking of the soul. This is the key to their argument against evolution, against scientific nat-uralism, and for the existence of the transcendent master intellect who is responsible for the complexity of the universe.

It is for this reason that ID creationists see Christian metaphysics as at stake in the evolution debate. Philip Johnson explained the significance of this issue for IDCs when he was asked why he focused upon Darwinism.

I wanted to know whether the fundamentals of the Christian worldview were fact or fantasy. Darwinism is a logical place to begin because, if Darwinism is true, Christian metaphysics is fantasy. (Quoted in Anonymous 2002)

But this is asking too much. Understanding the limits of scientific methodology ought to help us here. To the degree that such spiritual possibilities are understood as truly supernatural, they will forever remain outside science. Solving such ultimate religious metaphysical questions is more than one should ask of biology.

PRIORITY CLAIMS: WHICH EXPLAINS WHICH?

At the end of the day, much of what is taken to be at stake turns on the question of whether people think that religion explains biology or biology explains religion. As we have seen, creationists think that if evolution is true, then all of their religious beliefs must be false, so it is no wonder that they oppose it so vigorously. The justification of a moral code, writes Philip Johnson, depends upon getting the creation story right. The Christian story is one in which God created human beings, whose sins separate them from God and who must be saved from sin to become whole. The Enlightenment story is of human beings whose mastery of science enables them to escape from superstition and eventually realize that their ancestors created God rather than the reverse. For many religious believers, that priority dispute is the ultimate philosophical question for understanding the relationship between biology and religion. Did God create human beings, or did human beings evolve and then make up the idea of God?

Psychological accounts of the idea of God as a projection of infantile images of the father and mother are well known, and some biologists have given similar deflationary accounts of religion. The philosopher Daniel Dennett (2006) recently applied the biologist Richard Dawkins's concept of the meme, a cultural analog to the biological concept of the gene, to religious ideas, analyzing religion as a natural phenomenon. Others have tried this in even more reductionist evolutionary terms. In the book *The Biology of Religion*, for instance, Vernon Reynolds and Ralph Tanner investigated how individuals' religious faith or membership in a religious group affected their chances of survival and their reproductive success.

That is to say, they attempted to give a sociobiological account of religious belief and religious practices. Marshaling historical and contemporary cross-cultural data about a wide range of religions, they looked at the biological effects of religious beliefs and practices involving conception and contraception, abortion and infanticide, birth and childhood, marriage, death, disease, and more. They concluded that religious practices were biologically adaptive, arising from past survival strategies and continuing to enhance reproductive fitness. Religions, as they rather audaciously stated it, are "culturally phrased biological messages ... a primary set of 'reproductive rules'" (Reynolds and Tanner 1983, 294).

It is notable that Reynolds and Tanner took to heart the critique of sociobiology and what the philosopher of biology Philip Kitcher called its "vaulting ambition" to explain such cultural complexities. They dropped the theoretical framework of sociobiology almost entirely when they revised the book, retitling it *The Social Ecology of Religion* and coining the term "socioecology" for their more eclectic (and ultimately unsatisfying) theoretical perspective. They also found empirical weaknesses in their earlier view, recognizing that religious practices can exacerbate as well as reduce the risk of disease, a fact that should not have been the "surprising discovery" they claimed it to be (Reynolds and Tanner 1995, 17). Their major revised conclusion is the weaker one that religions "evolved to provide legitimating of 'safe' ways of dealing with those events in life that bring human beings into a state of danger or fear or anxiety or just an overwhelming feeling of pointlessness" (1995, 42) and "to endow ... life events with meaning" (1995, 308). However, there is little biological content to the notion of "evolved" here anymore. They believe that biology still has something to say about why religions exist, but their analysis is not aimed as explaining away religious belief.

The God Module

However, biology may yet try to do this. Biology has added fuel to the debate recently in a different way, in what has become known as the field of neurotheology. Some biologists have claimed to find evidence of a "God module" in the brain. In particular, they find an association between epileptic seizures in the left temporal lobe and feelings of ecstasy sometimes described as experiences of the presence of God. This work takes off from a fact that has long been known, namely, that some subjects affected by temporal lobe epilepsy report having intense spiritual experiences during their seizures; some claim that God spoke to them directly. Such patients would often become preoccupied with spiritual issues even during seizure-free periods. Experiments using transcranial magnetic stimulators showed that one could produce these kinds of effects in subjects with no history of temporal lobe seizures. One researcher who stimulated his own temporal lobes reported being amazed at having the experience of God for the first time in his life. As may be expected, while some people have taken this brain area to be the seat of a special human faculty for experiencing the divine, others see it as confirmation that such religious experiences are delusions caused by electrical disturbances in the brain. The neurologist V. S. Ramachandran discusses his research on the neural basis of religious experience in the same way as he does his work with people who feel phantom limbs or who see cartoon characters in a visual blind spot. This rephrases the earlier question so that we may now ask, Did God create the brain, or did the brain create God?

In recent work, the psychiatrist Eugene D'Aguili and the radiologist Andrew Newberg used high-tech imaging devices to observe the brains of Buddhists and nuns during meditation. When these subjects reported subjective feelings of oneness with the universe or of the presence of God while in focused meditation, the researchers observed decreased activity in the brain's "object association areas" that purportedly process and mediate the boundary between self and themselves.

The data show, claim Newberg and D'Aguili (2002), that mystical experience is not a mere fabrication or a simple result of wishful thinking, but rather has a real, neurological basis. Moreover, they say that these experiences occur as part of normal, healthy neurophysiology and should not be dismissed as random, pathological events. Mystical experience, they say, is biologically, observably, and scientifically real. They argue that humans seek God because our brains are biologically programmed to do so, hypothesizing that spiritual experience is intimately interwoven with human biology. Continuing research is investigating questions such as whether religious ritual can create its own neurological environment, and whether there is a connection between religious ecstasy and sexual orgasm. Belief in God will not go away, they conclude, because the religious impulse is hard-wired in the biology of the human brain. Some suggest that this is the common biological origin of all religions.

Theological responses to this kind of work range from taking it to be suggestive that God is both real and reachable, to criticizing it as a form of scientism. Critics question the appropriateness of trying to measure mystical experience, suggesting that it is a mistake to think that theological notions of the transcendent could correspond to empirical observations.

Did God create nature, or did nature create God? Suffice to say, neither biology nor religion is yet in a position to claim the final answer.

NOTE

1.   Although he did not speak of it in these terms, it is an interesting coincidence that at the same time that Phillip Johnson was writing his initial articles and book on intelligent design, he also began writing in support of Peter Duesberg's dissident view that HIV does not cause AIDS but rather that it is the result of the homosexual lifestyle, including long-term consumption of recreational drugs and/or the AIDS drug AZT. Johnson recently wrote of a "racket" of "AIDS careerists" who may be covering up a "ghastly mistake" (2004). He and other ID creationists make the same kind of intimations of a conspiracy among scientists to cover up the purported false and fraudulent evidence for evolution.

23 The Moral Grammar of Narratives in History of Biology

The Case of Haeckel and Nazi Biology

INTRODUCTION: SCIENTIFIC HISTORY

In his inaugural lecture at Cambridge as Regius Professor of Modern History in 1895, Lord John Acton urged that the historian deliver moral judgments on the figures of his research. Acton declaimed:

> I exhort you never to debase the moral currency or to lower the standard of rectitude, but to try others by the final maxim that governs your own lives and to suffer no man and no cause to escape the undying penalty which history has the power to inflict on wrong. (Acton 1906, 234)

In 1902, the year after Acton died, the president of the American Historical Association, Henry Lea, in dubious celebration of his British colleague, responded to the exordium with a contrary claim about the historian's obligation, namely, objectively to render the facts of history without subjective moralizing. Referring to Acton's lecture, Lea declared:

> I must confess that to me all this seems to be based on false premises and to lead to unfortunate conclusions as to the objects and purposes of history, however much it may serve to give point and piquancy to a narrative, to stimulate the interests of the causal reader by heightening lights and deepening shadows, and to subserve the purpose of propagating the opinions of the writer. (Lea 1904, 234)

As Peter Novick has detailed in his account of the American historical profession, by the turn of the century historians in the United States had begun their quest for scientific status, which for

429

most seemed to preclude the leakage of moral opinion into the objective recovery of the past – at least in an overt way. Novick also catalogues the stumbling failures of this noble dream, when political partisanship and rampant nationalism sullied the ideal (Novick 1988).

Historians in our own time continue to be wary of rendering explicit moral pronouncements, thinking it a derogation of their obligations. On occasion, some historians have been moved to embrace the opposite attitude, especially when considering the horrendous events of the twentieth century – the Holocaust, for instance. It would seem inhumane to describe such events in morally neutral terms. Yet even about occurrences of this kind, most historians assume that any moral judgments ought to be delivered as obiter dicta, not really part of the objective account of these events. Lea thought a clean depiction of despicable individuals and actions would naturally provoke readers into making their own moral judgments about the past, without the historian's coercing their opinions.

This attitude of studied neutrality has become codified in the commandments handed down by the National Center for History in the Schools, whose committee has recently proclaimed: "Teachers should not use historical events to hammer home their own favorite moral lesson" (Nash 1996). Presumably this goes as well for the historian teacher, whose texts the students study. And one might suppose that when the narrative describes episodes in the history of science, occasion for intrusive moral assessment would be quite limited.

I believe that these demands that historians disavow moral assessment neglect a crucial aspect of the writing of history, whether it be political history or history of science: the deep grammar of narrative history requires that moral judgments be rendered. And that is the thesis I will argue in this essay, namely, that all historical narratives must make moral assessments. I will be especially concerned with an evaluation that might be called that of "historical responsibility."

The role of moral judgment about past historical characters has, despite causal assumption to the contrary, been brought to eruptive boil recently in one area of history of biology – that of nineteenth- and early twentieth-century evolutionary thought in Germany. The

individual about whom considerable historical and moral controversy swirls is Ernst Haeckel (1839–1919). I will say more about
Haeckel in a moment. He offers a test case for my thesis. Now I will
simply point out that Haeckel, more than any other individual, was
responsible for the warfare that broke out in the second half of the
nineteenth century between evolutionary theorists and religiously
minded thinkers, a warfare that continues unabated in the contemporary cultural struggle between advocates of intelligent design
and those defending real biological science.

My motivation for considering the moral structure of narratives is
encapsulated in the main title of a book that was published not
long ago: *From Darwin to Hitler*. The pivotal character in this historical descent, according to the author, is Ernst Haeckel. He and
Darwin are implicitly charged with historical responsibility for acts
that occurred after they themselves died. I do not think judgments
of this kind, those attributing moral responsibility across decades,
are unwarranted in principle. The warrant lies in the grammar
of historical narrative. Whether this particular condemnation of
Darwin and Haeckel is appropriate remains quite another matter.

THE TEMPORAL AND CAUSAL GRAMMAR OF NARRATIVE HISTORY

Let me focus, for a moment, on two features of narrative history as a
prelude to my argument and as an illustration of what I mean by the
grammar of narrative. This concerns the ways time and causality are
represented in narrative histories. Each seeps into narratives in at
least four different ways (see Richards 1992, 19–53). Let me first
consider, quite briefly, the temporal dimensions of narrative.

Initially, we might distinguish what might be called the *time of
events*. Embedded in the deep structure of narrative is the time
during which events occur; that sort of time flows equitably on into
the future, with each unit having equal duration. Narratives project
events as occurring in a Newtonian time. This kind of time allows
the historian to place events in a chronology, to compare the duration of events, and to locate them in respect to one another as
antecedent, simultaneous, or successive.

But the structuring of these events in a narrative also exhibits what
might be called *narrative time*, and this is a different sort of temporal

modality. Consider, for instance, Harold Pinter's play *Betrayal*. The first scene is set temporally toward the end of the Newtonian sequence dramatized, and the next scene, going in the right direction, occurs a few days later. But the third scene falls back to two years before, and the fourth a year before that, with subsequent scenes taking us back finally to a period six years before the final days with which the play begins. The audience, however, never loses its temporal bearings or believes that time staggers along, weaving back and forth like an undergraduate leaving the local college pub.

The historian might structure his or her narrative in a roughly comparable way. He or she might relate one aspect of the history and then return to an earlier time to follow out another thread of the story. Or the historian might have the narrative jump into the future to highlight the significance of the precipitating antecedent event. Again, when this is done with modest dexterity, the reader is never confused about the Newtonian flow of time.

The *time of narration* is a less familiar device by which historians restructure real time as well as narrative time. One of the several modes by which historians construct this kind of time is through contraction and expansion of sentence duration. Let me illustrate what I mean by reference to a history with which most readers will be familiar – Thucydides's *History of the Peloponnesian War*. At the beginning of his history, Thucydides – a founder, along with Herodotus, of the genre of narrative history – expends a few paragraphs on events occurring in the earliest period of Cretan hegemony through the time of the Trojan War to just before the outbreak of the war between the two great powers of ancient Greece, Athens and Sparta. The period he so economically describes in a few paragraphs extends, in Newtonian time, for about two thousand years. But Thucydides then devotes several hundred pages to the relatively brief twenty-year period of the war, at least that part of the war he recorded. Sentence duration is an indication of the importance the historian places on the events mentioned. Sentence expansion or contraction, however, may have other sustaining causes.

Simply the pacing or rhythm of the historian's prose might be one. The great French scientist and historian Bernard de Fontenelle said that if the cadences of his sentences demanded it, the Thirty Years War would have turned out differently. Some historians will linger over an episode, not because it fills in a sequence vital to the tale, but because the characters involved are intrinsically interesting. Maybe some humorous event is inserted in the story simply to keep the reader turning pages. In histories, centuries may be contracted into the space of a sentence, while moments may be expanded through dozens of paragraphs.

A fourth temporal dimension of narrative is the *time of narrative construction*. This is a temporal feature especially relevant to considerations of the moral structure of histories. A narrative will be temporally layered by reason of its construction, displaying, as it were, both temporal depth and a temporal horizon. The temporal horizon is more pertinent for my concerns, so let me speak of that. Thucydides wrote the first part of his history toward the end of the war that he described, when the awful later events allowed him to pick out those earlier, antecedent events of explanatory relevance – earlier events that would be epistemologically tinged with Athenian folly yet to come. Only the benefits of hindsight, for example, could have allowed him to put into the mouth of the Spartan messenger Melesipus, who was sent on a last desperate peace mission just before the first engagement of the war, the prophetic regret "This day will be the beginning of great misfortunes to Hellas." By the horizontal ordering of time, the historian can describe events in ways that the actors participating in the events could not: Melesipus's prophecy was possible only because Thucydides had already lived through it. This temporal perspective is crucial for the historian. Only from the vantage of the future can the historian pick out from an infinity of antecedent events just those deemed necessary for the explanation of the consequent events of interest.

Different causal structures of narratives correspond to their temporal modalities. I will not detail all of their aspects, but let me quickly rehearse their dominant modes. The most fundamental causal feature of narratives is the *causality of events*. This is simply the causality ascribed to events about which the historian writes. Typically the historian will arrange events so as to indicate their causal sequence, a sequence in which the main antecedent causes are indicated so as to explain subsequent events, ultimately the central events which the history was designed to explain.

Events in a narrative, however, display a different causal grammar from events in nature. We may thus speak of the *causality of narrated events*. When in 433 B.C., the Athenians of Thucydides' history interfered in an internal affair of Corinth, a Spartan ally, they could not have predicted that war would result, though they might have suspected; they certainly could not have predicted their ignominious defeat in the Sicilian campaign twenty years later. From inside the scene that Thucydides has set, the future appears open; all things are possible, or at least unforeseeable. Yet each of Thucydides' scenes moves inevitably and inexorably to that climax, namely, to the destruction of the fleet at Syracuse, the central event of his history. The historian, by reason of his or her temporal horizon, arranges antecedent events to make their outcome, the central event of interest, something the reader, can expect – something, in the ideal case, that would be regarded as inevitable given the antecedent events – all the while keeping his actors in the dark until the last minute.

This is a view about the grammar of narration that some historians would not share. Some try assiduously to avoid surface terms redolent of causality in their narratives. But I think this is to be unaware of the deeper grammar of narrative. The antecedent events are chosen by the historian to make, as far as he or she is able, the consequent events a causal inevitability. That is what it is to explain events historically. To the degree this kind of causal structure is missing, to that degree the history will fail to explain how it is that the subsequent events of interests occurred or took the shape they did. Without a tight causal grammar the narrative will loosen to mere chronicle.

This grammatical feature of narrative has bearing on any moral characterization of the actions of the individuals about whom the historian writes. And this in two ways. First, we do think that when we morally evaluate actions, we assume the individuals could have chosen otherwise. There will thus be a tension between the actors represented as regarding the future as open, as full of possibilities, and the historian's knowledge that the future of the actors is closed. They did what they did because of the narrated events, events carrying those individuals to their appointed destiny.

The second way the causality of narrated events bears on moral assessment has to do with the construction of the sequence of events and their causal connections. The historian will also be making a moral evaluation of the actions of a character – implicitly at least – and will arrange that sequence in which the character's actions are placed so as either morally to indict the individual, or morally to exculpate the individual, or, what is more frequently the case, to locate the individual's actions in a morally neutral ground. I will say more about this feature of the grammar of narrative in a moment.

A third causal modality deployed by historians may be called *the causality of narration*. This aspect of causality has several features, but I will mention only one: this is the location in a narrative of various scenes. So, for example, Thucydides will place one scene before another to indicate what he presumes is an important condition or cause for a subsequent scene, even though the scenes may be at some real temporal distance. A speech made to motivate an action might be placed immediately before the scene in which the action is described, even though the two events may be separated by a fair amount of real time. Such juxtaposition can have a conditioning effect as well. Immediately after Thucydides relates Pericles' great funeral oration, which extols the virtues of Athenian democracy and the glories of its laws, he shoves in a dramatic description of the Athenian plague, when citizens ignored the laws and each sought his own pleasure, thinking it might be his last. Yet the oration and the plague were separated by many months. This kind of causality effectively conditions the reader's response to the realities of Athenian society.

Finally, there is the *causality of narrative construction*. There are two quite different causal features that would fall under this rubric. First, one might discriminate the final cause in narrative construction. Most histories aim to explain some central event – the outbreak of the American Civil War, Darwin's discovery and construction of his theory of natural selection, or the racial attitudes of Hitler. The antecedent events in the history provide the causal explanation of the central event, which latter might be thought of as the final cause, that is, the goal of the construction. Historians in their research use this final cause as the beacon in light of which they select out from an infinity of antecedent events just those that might explain the central event. No historian begins, as it were, at the beginning, rather at the end. Without the final cause as guide – a guide that might alter, of course, during the research – the historian could not even start to lay out those antecedent causes that he or she will finally regard as the explanation for the conclusion of the historical sequence.

A related feature of the causality of narrative construction concerns the motives guiding the historian, of which there may be several. The proclaimed and standard motivation of the great nineteenth-century historian Leopold von Ranke was to describe an event "wie es eigentlich gewesen," how it actually was; and, insofar as how it was becomes in specific instances the central event that needs explanation, that event – the final cause – becomes the motive for constructing the history. Ranke's general standard must be that of every historian. Good historians will want to weigh purported causes of events and emphasize the most important, while reducing narrative time spent on the less important. Yet often other motivations, perhaps hardly conscious even to the historian, may give structure to his or her work. In his suspicious little book *What Is History?* E. H. Carr urged that "when we take up a work of history, our first concern should be not with the facts which it contains but with the historian who wrote it" (Carr 1961, 24). If the reader knows in advance that the historian is of a certain doctrinal persuasion, then a judicious skepticism may well be in order. After all, a historian may select events that have real but minor causal connections with central events of concern, while ignoring even more important antecedent causes. The history would then have a certain verisimilitude, yet be a changeling. Motivations of authors are often revealed by the moral grammar of narratives, another structural feature that lies at the syntactic depths of historical accounts.

THE MORAL GRAMMAR OF NARRATIVE HISTORY

I am going to now turn specifically to the features of the moral grammar of histories, and then illustrate some of the ways that structure characterizes Ernst Haeckel's story. If narratives have these grammatical structures, then it would be well for historians to be reflectively conscious of this and to formulate their reconstructions in light of a set of principles that I believe should be operative. And in a moment I will suggest what those principles ought to be by which we morally judge the behavior of individuals who lived in the past and by which we assess their culpability for the future actions of others.

But let me first pose the question, Do historians make normative judgments in their histories and should they? I will argue that not only should they, they must by reason of narrative grammar. At one level, it is obvious that historians, of necessity, do make normative judgments. Historical narratives are constructed on the basis of evidence: written documents – letters, diaries, published works; also artifacts, such as archaeological findings; and high-tech instruments, such as DNA analysis; or low-tech oral interviews. And historians attribute modes of behavior to actors on the basis of inference from evidence and in recognition of certain standards. Even when doing something apparently as innocuous as selecting a verb to characterize a proposition attributed to an actor, the historian must employ a norm or standard. For example, Thucydides could have had Melesipus *think* that disaster was in the offing, *believe* that disaster was in the offing, *be convinced* that disaster was in the offing, *suspect* that disaster was in the offing, *assume* that disaster was in the offing, or *prophesy* that disaster was in the offing. Whatever verb the historian selects, he or she will do so because the actor's behavior, as suggested by the evidence, has met a certain standard for such and such modal description – say, being in a state of firm conviction as opposed to vague supposition. All descriptions require measurement against standards or norms – which is not to say that in a given instance, the standard and consequent description would be the most appropriate. The better the historian, the more appropriate the norms employed in rendering descriptions.

Virtually all of the historian's choices of descriptive terms must be normative in this sense. But must some of these norms also be moral norms? I believe they must. The argument is fairly straightforward – at least as straightforward as arguments of this sort ever get. Human history is about *res gestae*, things done by human beings, human actions. Actions are not mere physiological behaviors, but behaviors that are intended and motivated. Inevitably these actions impinge on other individuals immediate or remote. But intentional behaviors' impinging on others is precisely the moral context. The historian, therefore, in order to assign motives and intentions to individuals whose behaviors affect others and to describe those motives and intentions adequately – that historian must employ norms governing such intentional behaviors, that is, behaviors in the moral context.

Certainly the assessment of motives and intentions may yield only morally neutral descriptions. But even deciding that an intended behavior is morally neutral is, implicitly at least, also to judge it against standards of positive or negative moral valence and to decide that it conforms to neither or that it lies somewhere in between. Even a morally neutral assessment is a moral assessment. There is no claim here, of course, that such evaluations are generally self-consciously performed by historians. Mostly these evaluations occur quite reflexively, instead of reflectively. And they usually exist not explicitly on the surface of the narrative, but in the interstices.

Let me offer one example more concretely of what I am arguing, and this from a historian whom no one would accuse of cheap moralizing – his moralizing is anything but cheap. His descriptions reveal a rainbow of shaded moral evaluations, which range subtly between the polar categories of shining virtue and darkling vice. Byron called him the Lord of Irony, and it is often through that trope that he makes his moral assessments. I am speaking, of course, of Edward Gibbon.

Let me quote just a short passage from the *Decline and Fall*, where Gibbon is describing what might have been the motives of Julian, as his soldiers were clamoring for his elevation to emperor, even while Constantius was still on the throne. Julian protested he could not take the diadem, even as he reluctantly and sadly accepted it. Gibbon writes:

The grief of Julian could proceed only from his innocence; but his innocence must appear extremely doubtful in the eyes of those who have learned to suspect the motives and the professions of princes. His lively and active mind was susceptible of the various impressions of hope and fear, of gratitude and revenge, of duty and of ambition, of the love of fame and of the fear of reproach. But it is impossible for us to ascertain the principles of action which might escape the observation, while they guided, or rather impelled, the steps of Julian himself... He solemnly declares, in the presence of Jupiter, of the Sun, of Mars, of Minerva, and of all the other deities, that till the close of the evening which preceded his elevation he was utterly ignorant of the designs of the soldiers; and it may seem ungenerous to distrust the honour of a hero, and the truth of a philosopher. Yet the superstitious confidence that Constantius was the enemy, and that he himself was the favourite, of the gods, might prompt him to desire, to solicit, and even to hasten the auspicious moment of his reign, which was predestined to restore the ancient religion of mankind. (Gibbon 1777–88, 2: 319–20)

In the cascade of rhetorical devices at play – zeugma, antithesis, irony – Gibbon explicitly refuses to attribute morally demeaning motives to Julian, and, of course, at the same time implicitly does precisely that. There is another element of judgment that Gibbon evinces here, which is also an important feature of the moral grammar of historical narrative.

Narratives explain action by allowing us to understand character, in this case Julian's character. Gibbon, however, has led us to comprehend Julian's action not only by cognitively suggesting what the motives of a prince might be but also by shaping our emotional response to Julian's character and thus producing in us a feeling about Julian's action. We morally evaluate individuals, partly at least, through feelings about them. The historian can orchestrate outrage – as some dealing with Haeckel have – by cutting quotations from an actor into certain vicious shapes, selecting those that appear damning while neglecting those that might be exculpating. Or, as Gibbon does, the historian can evoke feelings of moral disdain with little more than the magical mist of antithetic possibilities. As a result, readers will have, as it were, a sensible, an olfactory understanding: the invisible air of the narrative will carry the sweet smell of virtue, the acrid stench of turpitude, or simply the bittersweet of irony. These feelings will become part of the delicate moral assessment rendered by the artistry of the historian.

This is just one small example of the way moral judgment exists in the interstitial spaces of a narrative, instead of lying right on the surface. But sometimes such judgments do lie closer to the skin of the history. Let me now focus precisely on a case of this and consider the principles that, I believe, should be operative in making moral judgments of historical figures. This is in the instance of Ernst Haeckel.

ERNST HAECKEL, DARWIN'S CHAMPION IN GERMANY

Haeckel was Darwin's great champion of evolutionary theory in Germany; he was a principal in the theory's introduction there and a forceful defender of it from the mid-1860s until 1919, when he died

(Richards 2004, 2005). Haeckel's work on evolution reached far beyond the borders of the German lands. His popular accounts of evolutionary theory were translated into all the known and unknown languages – at least unknown to the West – including Armenian, Chinese, Hebrew, Sanskrit, and Esperanto. More people learned of evolutionary theory through Haeckel's voluminous writings during this period than from any other source, including Darwin's own work.

Haeckel achieved many popular successes, and, as well, produced more than twenty large, technical monographs on various aspects of systematic biology and evolutionary theory. In these works he described many hitherto unknown species, established the science of ecology, gave currency to the idea of the missing link – which one of his protégés (Eugene Dubois) actually found – and promulgated the biogenetic law that ontogeny recapitulates phylogeny. Most of the promising young biologists of the next generation studied with him at Jena. His artistic ability was considerable and, at the beginning of the twentieth century, he influenced the movement in art called Jugendstil by his book *Kunstformen der Natur* (Art forms of nature). Haeckel became a greatly celebrated intellectual figure, often mentioned for a Nobel Prize. He was also the scourge of religonists, smiting the preachers at every turn with the jawbone of evolutionary doctrine. He advocated what he called a "monistic religion" as a substitute for the traditional orthodoxies, a religion based on science.

As a young student, trying to find a subject for his habilitation, Haeckel roamed along the coasts of Italy and Sicily in some despair. He thought of giving up biology for the life of a Bohemian, spending his time in painting and poetizing with other German expatriates on the island of Ischia. But he felt that he had to accomplish something in biology, so that he could become a professor and marry the woman he had fallen deeply in love with – his love letters sent back to his fiancée in Berlin are something delicious to read. He finally hit upon a topic: a systematic description of a little known creature that populated the seas, the one-celled protist called a radiolarian. It was while writing his habilitation on these creatures in 1861 that he happened to read Darwin's *Origin of Species* and became a convert. Haeckel produced a magnificent two-volume tome on the radiolaria, which he himself illustrated with extraordinary artistic and scientific acumen (Haeckel 1862). The radiolarian monograph's most

immediate and significant effect was to secure Haeckel a professor-
ship at Jena, thus allowing him to marry his beloved cousin, Anna
Sethe.

On his thirtieth birthday in 1863, Haeckel learned he had won a
prestigious prize for his radiolarian work. And on that same day, a day
that should have been of great celebration, his wife of eighteen
months tragically died. Haeckel was crushed. His family feared he
might commit suicide. As he related to his parents, this heart-searing
blow led him to reject all religion and replace it with something more
substantial, something that promised a kind of progressive transcendence, namely, Darwinian theory.

In the years following this upheaval, Haeckel became a zealous
missionary for his new faith, and his own volatile and combative
personality made him a crusader whose demeanor was in striking
contrast to that of the modest and retiring English master whom he
would serve. This outsized personality has continued to irritate
historians of smaller imagination.

THE MORAL INDICTMENT OF HAECKEL

In 1868, Haeckel produced a popular work on the new theory of
evolution, entitled *Natürliche Schöpfungsgeschichte* (Natural history of creation). It would go through twelve editions up to the time
of his death in 1919 and prove to be the most successful work of
popular science in the nineteenth and early twentieth centuries.
There are two features of that work that incited some of the fiercest
intellectual battles of the last part of the nineteenth century and
have led some historians and others to comparably fierce judgments
of Haeckel's moral probity.

The first has to do with what became the cardinal principle of his
evolutionary demonstrations, namely, the biogenetic law that
ontogeny recapitulates phylogeny. This principle holds that the
embryo of a developing organism goes through the same morphological stages that the phylum went through in its evolutionary history:
so, for example, the human embryo begins as a one-celled creature,
just as we presume life began on this Earth in a one-celled form; it
then goes through a stage of gastrulation, and Haeckel believed that
in the far distant past, our primitive ancestors plied the seas in that

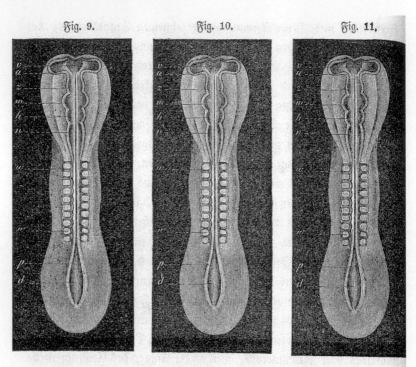

Fig. 9. Embryo des Hundes. Fig. 10. Embryo des Huhns. Fig. 11. Embryo der Schildkröte. Alle drei Embryonen sind genau aus demselben Entwicke-lungsstadium genommen, in dem soeben die fünf Hirnblasen angelegt sind. Die Buchstaben bedeuten in allen drei Figuren dasselbe: v Vorderhirn. z Zwischenhirn. m Mittelhirn. h Hinterhirn. n Nachhirn. p Rückenmark. a Augenblasen. w Urwir-bel. d Rückenstrang oder Chorda.

Figure 23.1   Embryos of dog, chicken, and turtle at the "sandal" stage (from Ernst Haeckel's *Natürliche Schöpfungsgeschichte*, 1868)

cuplike form; then the embryo takes on the morphology of an archaic fish, with gillarches; then of a primate; then a specific human being.

The corollary to the law is that closely related creatures – vertebrates, for example – will go through early embryological stages that are quite similar to one another. Some of Haeckel's enemies charged that he had exaggerated the tail of the human embryo to make it more animal-like – a controversy that became known as *Die Schwanzfrage*. But the deeper, more damaging fight arose with Haeckel's illustration of quite early embryos at the sandal stage, a stage when they look like the sole of a sandal. In the accompanying text to his illustration (Figure 23.1), Haeckel remarks: "If you compare the young embryos of the dog, chicken, and turtle..., you won't be in a position to perceive a difference" (Haeckel 1868, 249).

One of the very first reviewers of Haeckel's book, an embryologist who became a sworn enemy, pointed out that one certainly would not be able to distinguish these embryos, since Haeckel had used the same woodcut three times. He had, in the words of Ludwig Rütimeyer, the reviewer, committed a grave sin against science and the public's trust in science (Rütimeyer 1868).

In the second edition of his book, Haeckel retained only one illustration of an embryo at the sandal stage and remarked in the text: It might as well be the embryo of a dog, chicken, or turtle, since you can not tell the difference. The damage, however, had been inflicted, and the indictment of fraud haunted Haeckel for the rest of his life. The charge has been used by creationists in our own day as part of a brief, not only against Haeckel, but against evolutionary theory generally. Yet not only creationists, but several historians have employed it in their own moral evaluation of Haeckel and his science.

The second feature of Haeckel's work on which I would like to focus really did not create a stir in his own time but has become a central moral issue in ours. This has to do with the assumption of progress in evolution, an assumption that Haeckel certainly made. That assumption is forcefully displayed in the tree diagram appended to his *Natürliche Schöpfungsgeschichte*. The diagram (Figure 23.2) displays the various species of humankind, with height on the vertical axis meant to represent more advanced types. Here the Caucasian group leads the pack (seen in the upper right branch of the tree), arching above the descending orders of the "lower species" – all rooted in the Urmensch or Affenmench, the ape-man. A salient feature of the diagram should catch our attention: among the varieties of the Caucasian species, the Berbers and Jews were thought by Haeckel to be as advanced as the Germans and Southern Europeans. This classification should have had bearing on Haeckel's assignment by some historians to the ranks of the proto-Nazis.[1]

NAZI RACE HYGIENISTS AND THEIR USE OF HAECKELIAN IDEAS

That several Nazi race hygienists appealed to Haeckel to justify their views is clear. One pertinent example is Heinz Brücher's

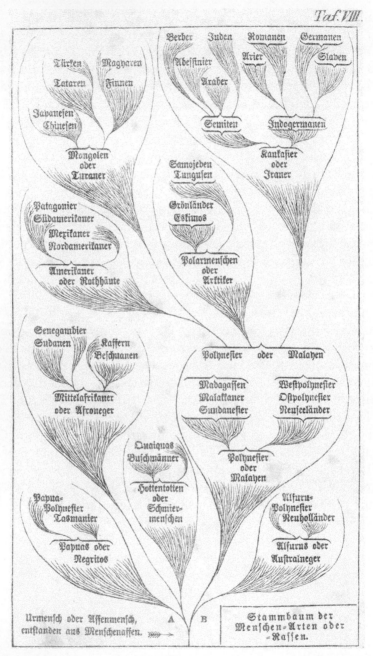

Figure 23.2   The stem tree of the human species (from Ernst Haeckel's *Natürliche Schöpfungeschichte*, 1868)

Ernst Haeckels Bluts- und Geistes-Erbe (Ernst Haeckel's racial and spiritual legacy), published in 1936. Not only did the author look to Haeckel's views of racial hierarchy as support for policies of National Socialism, he first gave full account of Haeckel's own impeccable pedigree. Included with the book was a five-foot chart laying out Haeckel's family tree. The aim of Brücher's racial hygienic analysis was both to demonstrate a new method of showing the worth of an intellectual position and to use that method to justify Haeckel's own doctrine. That is, only the best blood flowed through Haeckel's veins, and therefore we may trust his ideas.

To make the favorable connection between Haeckel and Hitler, Brücher focused on a passage from Haeckel's *Natürliche Schöpfungsgeschichte*, which reads: "The difference in rationality between a Goethe, a Kant, a Lamarck, a Darwin and that of the lower natural men – a Veda, a Kaffer, an Australian and a Papuan is much greater than the graduated difference between the rationality of these latter and that of the intelligent vertebrates, for instance, the higher apes." Brücher then cites a quite similar remark by Hitler in his Nuremberg speech of 1933 (Brücher 1936, 90–91, Hossfeld 2005, 312–16). In this way he has made Haeckel historically responsible, at least in part, for Hitler's racial attitudes.

THE JUDGMENT OF 'HISTORICAL RESPONSIBILITY'

Brücher's attribution of moral responsibility to Haeckel is of a type commonly found in history, though the structure of these kinds of judgments usually is unnoticed, lying as it does in the deep grammar of historiography. For example, historians will often credit, say, Copernicus, in the fifteenth century, with the courage to have broken through the rigidity of Ptolemaic assumption and thus, by unshackling men's minds, to have initiated the scientific revolution of the sixteenth and seventeenth centuries. This, too, is a moral appraisal of historical responsibility, though, needless to say, Copernicus himself never uttered: "I now intended to free men's minds and initiate the scientific revolution." Yet, historians do assign him credit for that, moral credit for giving successors the ability to think differently and productively.

The epistemological and historical justification for this type of judgment is simply that the meaning and value of an idea or set of ideas can be realized only in actions that themselves may take some long time to develop – this signals the ineluctable teleological feature of history. While this type of judgment derives from the moral grammar of history, this does not mean, of course, that every particular judgment of this sort is justified.

THE REACTION OF CONTEMPORARY HISTORIANS

How has Haeckel gone down with contemporary historians? Not well. His ideas, mixed with notions about his aggressive and combative personality, have lodged in the arteries feeding the critical faculties of many historians, causing sputtering convulsions. Daniel Gasman has argued that Haeckel's "social Darwinism became one of the most important formative causes for the rise of the Nazi movement" (Gasman 1971, xxii, and 1998). Stephen Jay Gould and many others concur that Haeckel's biological theories, supported, as Gould contends, by an "irrational mysticism" and a penchant for casting all into inevitable laws, "contributed to the rise of Nazism" (Gould 1977, 77–81). And more recently, in *From Darwin to Hitler*, Richard Weikart traces the metastatic line his title describes, with the midcenter of that line encircling Ernst Haeckel.

Weikart offers his book as a disinterested historical analysis. In that objective fashion that bespeaks the scientific historian, he declares, "I will leave it the reader to decide how straight or twisted the path is from Darwinism to Hitler after reading my account" (Weikart 2004, x). Well, after reading his account, there can be little doubt not only of the direct causal path from Charles Darwin through Ernst Haeckel to Adolf Hitler but of Darwin's and Haeckel's complicity in the atrocities committed by Hitler and his party. These evolutionists bear historical responsibility.

Taking E. H. Carr's advice to heart, we might initially be suspicious of Weikart's declaration of objectivity, proceeding as it does from a member of an organization having strong fundamentalist motivation – the Discovery Institute. Nonetheless, other historians have made similar suggestions – Gould and Gasman, for instance.

It is yet disingenuous, I believe, for Weikart to pretend that most readers might draw their own conclusions despite the moral grammar of his history. Weikart, as well as Gasman, Gould, and many other historians, have created a historical narrative implicitly following – they could not do otherwise – the principles of narrative grammar: they have conceptualized an end point – Hitler's behavior (the final cause) regarded here as ethically horrendous – and have traced back causal lines to antecedent sources that might have given rise to those attitudes of Hitler, tainting those sources along the way. It is like a spreading oil slick carried on an indifferent current and polluting everything it touches.

Now one can cavil, as I certainly would, about many deficiencies in the performance of these historians. They have not, for instance, properly weighed the significance of the many other causal lines that led to Hitler's behavior – the social, political, cultural, and psychological strands that many other historians have in fact emphasized. And thus that they have produced a monocausal analysis that quite distorts the historical picture.

While responsibility assigned Darwin and Haeckel might be mitigated by a more realistic weighing of causal trajectories, some culpability might, nonetheless, remain. Yet is there any consideration that might make us sever, not the causal chain – that is, the chain linking Darwin's writing, to Haeckel's, to Brücher's, to memos of high-ranking Nazis, and finally to Hitler's speeches – but the chain of moral responsibility? After all, Haeckel, and of course, Darwin, had been dead decades before the rise of the Nazis. And as Monty Python might have put it, they are still dead.

Let me summarize at this juncture the different modal structures of moral judgment in historical narratives that I have tried to identify. First is the explicit appraisal of the historian, rendered when the historian overtly applies the language of moral assessment to some decision or action taken by a historical figure. This both is rare and runs against the grain of the cooler sensibilities of most historians, Lord Acton excepted.

Second is the appraisal of contemporaries (or later individuals). Part of the historian's task will often be to describe the judgments made on an actor by his or her own associates or subsequent individuals. In the case of Haeckel, there were those who condemned him of malfeasance, as well as colleagues who defended him against the charge. This mode of moral attribution may be for the historian evidentiary, but hardly decisive.

Third is the appraisal by causal connection. This occurs when the historian joins the decisions of an actor with consequential behavior of moral import. The behavior itself might be that of the actor or behavior displaced at some temporal remove from the actor's overt intentions – Hitler's actions, for example, as supposedly promoted by Haeckel's conceptions. This latter is what I have called "historical responsibility." The causal trajectory moves from past to future, but the moral responsibility flows along the causal tracks from future back to past. And it is the guiding hand of the historian – fueled by a complex of motives – that pushes this historical responsibility back along the causal rails to the past. And it is here that a minor causal relationship can be mistaken for a major moral relationship. I will, in just a moment, indicate how I believe the historian ought reflectively to modulate the flow of responsibility.

Finally, there is appraisal by aesthetic charge. This occurs when the historian through artful design evokes a feeling of positive or negative regard for the actor. In the treatment of Haeckel by Gould, Gasman, and Weikart, the odor of suspicion has been diffused over his character, so that his scent lingers over Hitler's actions.

PRINCIPLES OF MORAL JUDGMENT

This takes me to the final part of my argument, namely, the principles that ought to govern our moral judgments about historical figures, especially for actions that were at some temporal distance from their own historical positions. I believe that the same general principles ought to serve as standards for our moral assessments of historical figures as serve for the assessments of our contemporaries, including ourselves. But much will depend on how those principles are specified when judging historically remote individuals.

First, there is the supreme principle of evaluation: it might be the golden rule, the greatest happiness, altruism, or the categorical imperative. Likely in the cases I have in mind any of these presumptive first principles will yield a similar assessment of moral motives, since they express, I believe, the same moral core. Second there is the intention of the actor: what did he or she attempt to do? What action did the individual desire to execute, to be distinguished, of course, from mere accidental behavior? Third is the motive for acting – that is, the ground for the intention to act in a certain fashion. The motive will determine moral valence. Finally, in assessing moral behavior, we must examine the beliefs of the individual actor and try to determine whether they were reasonable beliefs – and this is the special provenance of the historian. Let me give an example.

When the Hippocratic physicians, during the great Athenian plague that Thucydides so dramatically described, purged and bled the afflicted, their treatment actually hastened the deaths of their patients. But we certainly do not think the physicians malign or malfeasant, since they had a reasonable belief in the curative power of their practice. Their intention was to apply the best therapeutic techniques. And their motive, we may presume, was altruistic, since they risked their own lives to care for the sick. One should judge them, I believe, moral heroes, even though the consequence of their behavior was injury, and even the death of their patients.

The case of Ernst Haeckel is decidedly more problematic. In assessing the moral probity of his replication of woodcuts, the historian would have to examine his intentions and motivations. Did he claim his woodcuts were evidence of his biogenetic law? If so, he must have been motivated to deceive, and we may be thus entitled to suspect his character. Or did he merely intend to provide an illustration of the law for a general audience, maintaining as he did that at an early stage vertebrate embryos can not be morphologically distinguished? (Indeed, the historian does recognize that with the techniques then available the embryos could not be distinguished.) And thus at best, through a false economy, Haeckel may have only committed a very minor infraction, one that would not rise to the level of fraud and moral condemnation.

Concerning Haeckel's conception of a racial hierarchy, the historian has the task of exploring two questions in particular: what did he intend to accomplish by his theory? And how reasonable were the beliefs he harbored about races? To take the first question: Could Haeckel's actions be reasonably interpreted as intending to set in motion something like the crimes of the Nazis? Or minimally, did he exhibit a careless disregard for the truth of his views about races, so that some malfeasant act could, at least, have been vaguely anticipated. It is in answering this question that the grammar of narrative must be carefully observed. The historian may lay down the scenes of his or her history so as to lead causally to a central event, such as Hitler's racial beliefs and their results in the Holocaust, but the historian needs to keep the actors in the dark – insofar as it is reasonable to do so – about those future consequences, in this case, to keep Haeckel oblivious to the future use of his work. The historian may easily slip, since he or she knows the future outcome of the actor's decisions. It is easy to assume the actor also knew or could have anticipated those outcomes, at least in some vague way. More likely, though, the historian might simply fail to reflect on the crucial difference between his or her firm knowledge of the past and the actor's dim knowledge of the future.

In addition to carefully assessing intentions and motives, the historian must also consider the set of beliefs harbored by the actor. For example, was it reasonable for someone like Darwin or Haeckel to believe that evolutionary theory led to a hierarchy of species within a genus or races within a species? Or did they hold these ideas in reckless disregard for the truth? To assess reasonableness of belief in this instance, the historian would have to know what the scientific consensus happened to be in the second half of the nineteenth century. And in this case, a modestly diligent historian would discover that the community of evolutionary theorists – as well as other biologists – did understand the human races to stand in a hierarchy, just as did other animals, which also displayed scalable traits. In the human case, the traits included those of intelligence, moral character, and beauty. Nineteenth-century evolutionary theory implied that races stood in a hierarchy, and all of the available evidence supported it. We might recognize from our perspective certain social factors constraining the judgments of those biologists, but it is safe to say they did not.

Then the historian can further ask, In this particular instance, what does categorizing peoples as branches of a racial hierarchy mean for the treatment of those so classified? This question does not allow for a universal answer, but will depend more particularly on the individual scientist. Weikart, for instance, indicts Darwin for acceding to a belief in a racial hierarchy, but neglects to mention that Darwin did not think any action should be taken to reduce the welfare of those lower in the scale. Haeckel's own attitudes about how one should treat those lower in the hierarchy are less clear, but there is hardly room for moral condemnation, given the obscurity of his views about practical action.

CONCLUSION

It can only be a tendentious and dogmatically driven assessment that would condemn Darwin for the crimes of the Nazis. I will confess, though, that I have not yet made up my own mind about the historical responsibility of Haeckel, with whom I have considerable sympathy.

A historian can not write an extended account of the life of an individual without some measure of identification. If one is going to recover the past with anything like verisimilitude, one must, as R. G. Collingwood has maintained, relive the ideas of the past, which is not only to unearth long interred intellectual structures but also to feel again the pulse of their vitality, to sense their urgency, to admire their originality, and thus to empathize with their authors. And yet one has to do all of this while retaining a reflective awareness of the moral structure in which actors conceived those ideas and perceived their import.

NOTE

1.   There is direct evidence for Haeckel's attitude about Jews beyond his placement of them among the advanced races. In the early 1890s, he discussed the phenomenon of anti-Semitism with the Austrian novelist and journalist Hermann Bahr (1863–1934), who collected almost forty interviews with European notables on the issue, such individuals as August Bebel, Theodor Mommsen, James Arthur Balfour, and Henrik Ibsen. In his discussion with Bahr, Haeckel did acknowledge that lower-class Russian Jews would be regarded as an offense to the high standards of German culture – but then he regarded lower-class Italians – especially the Neapolitans – in the same way. Yet of educated German Jews, he remarked: "I hold these refined and noble Jews to be important elements in German culture. One should not forget that they have always stood bravely for enlightenment and freedom against the forces of reaction, inexhaustible opponents, as often as needed, against the obscurantists [Dunkelmänner]. And now in the dangers of these perilous times, when Papism again rears up mightily everywhere, we cannot do without their tried and true courage" (Bahr 1894, 69). By "dark men" Haeckel likely meant the Jesuits. There is simply no reason to believe Haeckel to be racially anti-Semitic, as Gasman and Weikart do.

REFERENCE LIST

Acton, J. 1906. On the study of history. *Lectures on Modern History*, 5–26. London: Macmillan.

Adcock, G. J., E. S. Dennis, S. Easteal, G. A. Huttley, L. S. Jermiin, and W. J. Peacock. 2001. Mitochondrial DNA sequences in ancient Australians: Implications for modern human origins. *Proceedings of the National Academy of Sciences* **98**: 537–42.

Aderem, A. 2005. Systems biology: Its practices and challenges. *Cell* **121**: 511–13.

Alatalo, R. V. 1981. Problems in the measurement of evennes in ecology. *Oikos* **37**: 199–204.

Alberch, P., and E. A. Gale. 1983. Size dependence during the development of the amphibian foot: Colchicine-induced digital loss and reduction. *Jounal of Embryology and Experimental Morphology* **76**: 177–97.

Alexander, J. M. 2000. Evolutionary explanations of distributive justice. *Philosophy of Science* **67**: 490–516.

Alroy, J. 2000. Understanding the dynamics of trends within evolving lineages. *Paleobiology* **26**, no. 3: 319–29.

Amundson, R. 2005. *The Changing Role of the Embryo in Evolutionary Biology: Structure and Synthesis*. New York: Cambridge University Press.

——— 1996. Historical development of the concept of adaptation. *Adaptation*, 11–53. New York: Academic Press.

Ankeny, R. A. 2000. Fashioning descriptive models in biology: Of worms and wiring diagrams. *Philosophy of Science* **67**: 260–72.

Anonymous. 2002. Berkeley's radical: An interview with Phillip E. Johnson. *Touchstone: A Journal of Mere Christianity*.

——— 2005. In pursuit of systems. *Nature* **435**: 1.

Antonovics, J., A. D. Bradshaw, and R. G. Turner. 1971. Heavy metal tolerance in plants. *Advances in Ecological Research* **7**: 1–85.

Ariew, A. 2002. Platonic and Aristotelian roots of teleological arguments. *Functions: New Readings in the Philosophy of Psychology and Biology*. Editors A. Ariew, M. Perlman, and R. Cummins. New York: Oxford University Press.

Aristotle. 1979. *De Generatione Animalium as Generation of Animals*. Cambridge, MA: Harvard University Press.

1984. Physics. *The Complete Works of Aristotle*. Editor J. Barnes. Princeton, NJ: Princeton University Press.

Arnold, A. J., and K. Fristrup. 1982. The theory of evolution by natural selection: A hierarchical expansion. *Paleobiology* **8**: 113–29.

Arthur, W. 2004. *Biased Embryos and Evolution*. Cambridge: Cambridge University Press.

Auffray, C., S. Imbeaud, M. Roux-Rouquie, and L. Hood. 2003. From functional genomics to systems biology: Concepts and practices. *Comptes Rendus Biologies* **326**: 879–92.

Axelrod, Robert. 1984. *The Evolution of Cooperation*. New York: Basic Books.

1997. The evolution of strategies in the iterated prisoner's dilemma. *The Dynamics of Norms*. Editors C. Bicchieri, R. Jeffrey, and B. Skyrms, 199–220. Cambridge: Cambridge University Press.

Baarsma, E. A., and H. Collewfin. 1975. Changes in compensatory eye movement after unilateral labyrinthectomy in the rabbit. *Archives of Otorhinolaryngology* **211**: 219–30.

Bahr, H. 1894. Ernst Haeckel. *Der Antisenitismus*, 62–69. Berlin: S. Fischer.

Bailey, J. M., and S. Agyei, Y. Cladue, and B. A. Gaulin. 1994. Effects of gender and sexual orientation on evolutionary relevant aspects of human mating psychology. *Journal of Personality and Social Psychology* **66**: 1081–93.

Bailey, J. M., and R. C. Pillard. 1995. Genetics of human sexual orientation. *Annual Review of Sex Research* **5**: 126–50.

Bailey, J. M., R. C. Dawood, K. Miller, and M. B. Pillard. 1999. A family history study of male sexual orientation using three independent samples. *Behavior Genetics* **29**: 79–86.

Bailey, J. M., and K. J. Zucker. 1995. Childhood sex-typed behavior and sexual orientation: A conceptual analysis and quantitative review. *Developmental Psychology* **31**: 43–55.

Bamshad, M., and S. P. Wooding. 2003. Signatures of natural selection in the human genome. *National Review of Genetics* **4**: 99–111.

Barinaga, M. 1995. Remapping the motor cortex. *Science* **268**: 1696–98.

Beatty, J. 1984. Chance and Natural Selection. *Philosophy of Science* **51**: 183–211.

1998. "Ecology." *Routledge Encylopedia of Philosophy*. Editor F. Craig. London: Routledge. Accessed October 8th, 2005, from www.rep. routledge.com/article/Q030.

1995. The evolutionary contingency thesis. *Concepts, Theories, and Rationality in the Biological Sciences: The Second Pittsburgh-Konstanz Colloquium in the Philosophy of Science*. Editors J. Lennox and G. Wolters, 45–81. Pittsburgh: University of Pittsburgh Press.

1981. What's wrong with the received view of evolutionary theory? *PSA 1980*. Vol. 2, 397–426. East Lansing, MI: Philosophy of Science Association.

Beatty, J., and S. Finsen. 1989. Rethinking the propensity interpretation: A peek inside Pandora's Box. *What the Philosophy of Biology Is: Essays Dedicated to David Hull*. Editor Michael Ruse, 17–30. Dordrecht: Kluwer Academic.

Bechtel, W. 2000. From imaging to believing: Epistemic issues in generating biological data. *Epistemology and Biology*. Editors R. Creath and J. Maienschein, 138–63. Cambridge: Cambridge University Press.

Bechtel, W., and A. Abrahamsen. 2005. Explanation: A Mechanist Alternative. *Studies in History and Philosophy of Biological and Biomedical Sciences. Special Issue: Mechanisms in Biology* 36: 421–41.

Bechtel, W., P. Mandik, J. Mundale, and R. Stufflebeam, Editors. 2001. *Philosophy and the Neurosciences: A Reader*. New York: Blackwell.

Bechtel, W., and R.C. Richardson. 1993. *Discovering Complexity: Decomposition and Localization as Strategies in Scientific Research*. Princeton, NJ: Princeton University Press.

Beckwith, J. 1987. The Operon: An historical account. *Escherichia coli and Salmonella typhimurium: Cellular and Molecular Biology*. Editor F. C. Neidhart et al. Vol. 2, 1439–43. Washington, DC: American Society for Microbiology.

Behe, M. J. 1996. *Darwin's Black Box*. Toronto: Free Press.

Benton, M. J. Forthcoming. The history of life. *The Harvard Companion to Evolution*. Editors M. Ruse and J. Travis. Cambridge, MA: Harvard University Press.

Bergsrom, T. C., and Oded Stark. 1993. How altruism can prevail in an evolutionary environment. *American Economic Review* 83, no. 2: 149–55.

Bernheim, D. 1984. Rationalizable strategic behavior. *Econometrica* 52, no. 4: 1007–28.

Berthoz, A. 1988. The role of gaze in compensation of vestibular dysfunction: The gaze substitution hypothesis. *Progress in Brain Research* 7: 411–20.

Binmore, K. 1998. *Just Playing*. Cambridge, MA: MIT Press.

Blair, J.E., and S.B Hedges. 2005. Molecular clocks do not support the Cambrian explosion. *Molecular Biology and Evolution* **22**, no. 3: 387–90.

Blanchard, R., and A.F. Bogart. 1996. Homosexuality in men and number of older brothers. *American Journal of Psychiatry* **153**: 27–31.

Blanchard, R., and P. Klassen. 1997. H-Y antigen and homosexuality in men. *Journal of Theoretical Biology* **185**: 373–78.

Bobrow, D., and J.M. Bailey. 2001. Is male homosexuality maintained via kin selection? *Evolution and Human Behavior* **22**: 361–68.

Bogen, J. 2005. Regularities and Causality: Generalizations and Causal Explanations. *Studies in History and Philosophy of Biological and Biomedical Sciences. Special Issue: Mechanisms in Biology* **36**: 351–67.

Bonner, J.T. 1998. Origins of multicellularity. *Integrative Biology* **1**, no. 1: 27–36.

Boorman, S.A. 1978. Mathematical theory of group selection: Structure of group selection in founder populations determined from convexity of the extinction operator. *Proceedings of the National Academy of Sciences* **69**: 1909–13.

Borgerhoff Mulder, M. 1991. Human behavioural ecology. *Behovioural Ecology: An Evolutionary Approach*. Editors J.R. Krebs and N.B. Davies, 69–98. Oxford: Blackwell.

Bottger, D., et al. 2000. The Cambrian substrate revolution. *GSA Today* **10**, no. 9: 1–7.

Bowler, P. 1971. Preformation and pre-existence in the seventeenth century: A brief analysis. *Journal of the History of Biology* **4**: 221–44.

Boyd, R., 1999. Homeostasis, species, and higher taxa. *Species: New Interdisciplinary Essays*. Editor R.A. Wilson, 141–85. Cambridge, MA: MIT Press.

    1991. Realism, anti-foundationalism, and the enthusiasm for natural kinds. *Philosophical Studies* **61**: 127–48.

Boyd, R., and P.J. Richerson. 1985. *Culture and the Evolutionary Process.* Chicago: University of Chicago Press.

Brady, R. 1985. On the independence of systematics. *Cladistics* **1**: 113–26.

Brandon, R., 1990. *Adaptation and Environment*. Princeton, NJ: Princeton University Press.

    1978. Adaptation and Evolutionary Theory. *Studies in History and Philosophy of Science* **9**: 181–206.

    2005. The difference between selection and drift: A reply to Millstein. *Biology and Philosophy* **20**: 153–70.

    1999. Introduction. *Biology and Philosophy* **14**: 1–7.

    1982. The levels of selection. *Proceedings of the Philosophy of Science Association* **1**: 315–23.

1994. Theory and experiment in evolutionary biology. *Synthese* **99**, no. 1: 59–73.

Brandon, R., and S. Carson. 1996. The indeterministic character of evolutionary theory: No "no hidden variables proof" but no room for determinism either. *Philosophy of Science* **63**: 315–37.

Brandon, R., and F. N. Nijhout. Forthcoming. The empirical non-equivalence of genic and genotypic models of selection: A (decisive) refutation of genic selectionism and pluralistic genic selectionism. *Philosophy of Science*.

Bromham, L. 2003. What can DNA tell us about the Cambrian explosion? *Integrative and Comparative Biology* **43**: 148–56.

Bromham, L., et al. 1998. Testing the Cambrian explosion hypothesis by using a molecular dating technique. *Proceedings for the National Academy of Sciences* **95**, no. 11: 12386–89.

Brower, L. P., and S. B. Malcom. 1991. Animal migrations: Endangered phenomena. *American Zoologist* **31**: 265–76.

Brücher, H. 1936. *Ernst Haeckels Bluts- und Geistes-Erbe: Eine kulturbiologische Monographie*. Munich: Lehmanns Verlag.

Brunet, M., et al. 2002. A new hominid from the Upper Miocene of Central Africa. *Nature* **418**: 145–51.

Budd, G., and S. Jensen. 2000. A critical reappraisal of the fossil record of the bilaterian phyla. *Biological Reviews of the Cambridge Philosophical Society* **75**: 253–95.

Buller, D. J. 2005. *Adapting Minds: Evolutionary Psychology and the Persistent Quest for Human Nature*. Cambridge, MA: MIT Press.

Burian, R. M. 2004a. *The Epistemology of Development, Evolution, and Genetics*. New York: Cambridge University Press.

1993. How the choice of experimental organism matters: Epistemological reflections on an aspect of biological practice. *Journal of the History of Biology* **26**: 351–67.

2004b. Molecular epigenesis, molecular pleiotropy, and molecular gene definitions. *History and Philosophy of the Life Sciences* **26**, no. 1: 59–80.

Buss, D. M. 1995. Evolutionary psychology: A new paradigm for psychological science. *Psychological Inquiry* **6**: 1–30.

Buss, L. W. 1987. *The Evolution of Individuality*. Princeton, NJ: Princeton University Press.

Butcher, E. C., E. L. Berg, and E. J. Kunkel. 2004. Systems biology in drug discovery. *Nature Biotechnology* **10**: 1253–59.

Cabeza, R., and L. Nyberg. 1997. Imaging cognition: An empirical review of PET studies with normal subjects. *Journal of Cognitive Neuroscience* **9**: 1–26.

2000. Imaging cognition II: An empirical review of 275 PET and fMRI Studies. *Journal of Cognitive Neuroscience* **12**: 1–47.

Callebaut, W., and D. Rasskin-Gutman. 2005. *Modularity: Understanding the Development and Evolution of Natural Complex Systems.* Cambridge, MA: MIT Press.

Camerer, C., and R. Thaler. 1995. Anomalies: Ultimatums, dictators, and manners. *Journal of Economic Perspectives* **9**, no. 2: 209–19.

Canguilhem, G. 1979. *Wissenschaftsgeschichte und Epistemologie.* Frankfurt/Main: Suhrkamp.

Cantino, P. D., and K. de Queiroz. 2003. "PhyloCode: A phylogenetic code of biological nomenclature." Available at www.ohiou.edu/phylocode/.

Cantor, G. 2005. *Quakers, Jews, and Science: Religious Responses to Modernity and the Sciences in Britain, 1650–1900.* Oxford: Oxford University Press.

Carr, E. 1961. *What Is History?* New York: Vintage Books.

Carroll, S. B. 2001. Chance and necessity: The evolution of morphological complexity and diversity. *Nature* **409**: 1102–09.

2005. *Endless Forms Most Beautiful: The New Science of Evo Devo.* New York: W. W. Norton.

Carroll, S. B., et al. 2005. *From DNA to Diversity: Molecular Genetics and the Evolution of Animal Design.* Malden, MA: Blackwell.

Carroll, S. B., J. K. Grenier, and S. D. Weatherbee. 2001. *From DNA to Diversity: Molecular Genetics and the Evolution of Animal Design.* Oxford: Blackwell.

Cassidy, J. 1978. Philosophical aspects of the group selection controversy. *Philosophy of Science* **45**: 575–94.

Cavalli-Sforza, L. L., and M. W. Feldman. 1981. *Cultural Transmission and Evolution. A Quanitative Approach.* Vol. 16. Princeton, NJ: Princeton University Press.

Cela Conde, C. J., and F. J. Ayala. 2001. *Senderos de la Evolucion Humana.* Madrid: Alianza Editorial.

Chapdelaine, Y., and L. Bonen. 1991. The wheat mitochondrial gene for subunit I of the NADH dehydrogenase complex: A trans-splicing model for this gene-in-pieces. *Cell* **65**, no. 3: 465–72.

Charles, D. 1995. Teleological causation in the physics. *Aristotle's "Physics": A Collection of Essays.* Editor J. Lindsay. Oxford: Clarendon Press.

Cheney, D. 1986. Interactions and relationships between groups. *Primate Societies.* Editors B. Smuts, D. Cheney, R. Seyfarth, R. Wrangham, and T. Struhsaker, 267–81. Chicago: University of Chicago Press.

1983. Proximate and ultimate factors related to the distribution of male migration. *Primate Social Relationships: An Integrated Approach.* Editor Robert Hinde, 241–49. Oxford: Blackwell Scientific.

Cohen, J., and S. H. Rice. 1996. Where do biochemical pathways lead? *Integrative Approaches to Molecular Biology*. Editors J. Collado-Vides, B. Magasanik, and T. F. Smith, 239–51. Cambridge, MA: MIT Press.

Collins, F. S., E. D. Green, A. E. Guttmacher, and M. S. Guyer. 2003. A vision for the future of genomics research. *Nature* **422**: 835–47.

Collins, F., et al. 1998. New goals for the U.S. Human Genome Project: 1998–2003. *Science* **282**: 682–89.

Colvin, J. 1983. Rank influences rhesus male peer relationships. *Primate Social Relationships: An Integrated Approach*. Editor R. Hinde, 57–64. Oxford: Blackwell Scientific.

Colyvan, M., and L. R. Ginzburg. 2003. Laws of nature and laws of ecology. *Oikos* **101**: 649–53.

Connell, J., and E. Orias. 1974. The ecological regulation of species diversity. *American Naturalist* **903**: 399–413.

Connors, B. W., and M. J. Gutnick. 1990. Intrinsic firing patterns of diverse neocortical neurons. *Trends in Neuroscience* **13**: 99–140.

Cook-Deegan, R. 1994. *The Gene Wars*. New York: W. W. Norton.

Cooper, G. 1998. Generalizations in ecology: A philosophical taxonomy. *Biology and Philosophy* **13**: 555–86.

Cooper, J. 1987. Hypothetical necessity and natural teleology. *Philosophical Issues in Aristotle's Biology*. Editors A. Gotthelf and J. Lennox. Cambridge: Cambridge University Press.

Cosmides, L., and J. Tooby. 1997. The modular nature of human intelligence. *The Origin and Evolution of Intelligence*. Editors A. B. Scheibel and J. W. Schopf, 71–101. Sudbury, MA: Jones & Bartlett.

Cracraft, J. 1990. The origin of evolutionary novelties: Pattern and process at different hierarchical levels. *Evolutionary Innovations*. Editor M. Nitecki, 21–46. Chicago: University of Chicago Press.

1983. Species concepts and speciation analsis. *Current Ornithology* **1**: 159–87.

Craver, C. F. Beyond reduction: Mechanisms, mulitfield integration, and the unity of neuroscience. *Studies in History and Philosophy of Biological and Biomedical Sciences. Special Issue: Mechanisms in Biology*. Editors C. F. Craver and L. Darden. **36**: 373–97.

2002b. Interlevel experiments, mulitlevel mechanisms in the neuroscience of memory. *Philosophy of Science (Supplement)* **69**: S83–S97.

2001. Role functions, mechanisms, and hierarchy. *Philosophy of Science* **68**: 53–74.

2002a. Structures of scientific theories. *Blackwell Guide to the Philosophy of Science*. Editors P. K. Machamer and M. Silberstein, 55–79. Oxford: Blackwell.

Crick, F. H. C. 1958. On protein synthesis. *Symposium of the Society for Experimental Biology* **12**: 136–63.

1959. The present position of the coding problem. *Structure and Function of Genetic Elements: Brookhaven Symposia in Biology* **12**: 35–39.

Cummins, Robert. 2002. Neo-teleology. *Functions: New Readings in the Philosophy of Psychology and Biology*. Editors A. Ariew, M. Perlman, and R. Cummins. Oxford: Oxford University Press.

Damuth, J., and I. L. Heisler. 1988. Alternative formulations of multilevel selection. *Biology and Philosophy* **3**: 407–30.

Darden, L. 2000. Review of Paul Thagard's *How Scientists Explain Disease*. *Philosophy of Science* **67**: 352–54.

2006. *Reasoning in Biological Discoveries: Mechanisms, Interfield Relations, and Anomaly Resolution*. New York: Cambridge University Press.

Darden, L., and J. Tabery. 2005. "Molecular biology." *The Stanford Encyclopedia of Philosophy*. Editor E. N. Zalta. Available at http://plato.stanford.edu/entries/molecular-biology/.

Darwin, C. 1859. *On the Origin of Species*. London: John Murray.

1964. *On the Origin of Species: A Facsimile of the First Edition*. Cambridge, MA: Harvard University Press.

Davidson, E. H. 2001. *Genomic Regulatory Systems: Development and Evolution*. San Diego, CA: Academic Press.

2006. *The Regulatory Genome: Gene Regulatory Networks in Development and Evolution*. Burlington, MA, and San Diego: Academic Press.

Dawkins, R. 1986. *The Blind Watchmaker*. New York: W. W. Norton.

1996. *Climbing Mount Improbable*. New York: W. W. Norton.

1989. The evolution of evolvability. *Artificial Life*. Editor C. G. Langton, 201–20. Redwood City, CA: Addison-Wesley.

1982b. *The Extended Phenotype: The Gene as the Unit of Selection*. Oxford: W. H. Freeman.

1982a. Replicators and vehicles. *Current Problems in Sociobiology*. Cambridge: Cambridge University Press.

1978. Replicator selection and the extended phenotype. *Zeitschrift Fur Tierpsychologie* **47**: 61–76.

1976. *The Selfish Gene*. Oxford: Oxford University Press.

1989. *The Selfish Gene*. 2nd ed. Oxford: Oxford University Press.

1983. Universal Darwinism. *Molecules to Men*. Editor D. S. Bendall. Cambridge: Cambridge University Press.

De Gelder, B. 2000. More to seeing than meets the eye. *Science* **289**: 1148–49.

de Laplante, K. 2004. Toward a more expansive conception of ecological science. *Biology and Philosophy* **19**: 263–81.

de Queiroz, K. 2005. Ernst Mayr and the modern concept of species. *Proceedings of the National Academy of Sciences* **102**: 6600–07.

1998. The general lineage concept of species, species criteria, and the process of speciation: A conceptual unification and terminological recommendations. *Endless Forms: Species and Speciation*. Editors D. J. Howard and S. H. Berlocher, 57–75. Oxford: Oxford University Press.

1992. Phylogenetic definitions and taxonomic philosophy. *Biology and Philosophy* **7**: 295–313.

1988. Systematics and the Darwinian revolution. *Philosophy of Science* **55**: 238–59.

de Queiroz, K., and J. A. Gauthier. 1990. Phylogeny as a central principle in taxonomy: Phylogenetic definitions of taxon names. *Systematic Biology* **39**: 27–31.

1994. Toward a phylogenetic system of biological nomenclature. *Trends in Ecology and Evolution* **9**: 27–31.

de Queiroz, K., and S. Poe. 2001. Philosophy and phylogenetic inference: A comparison of likelihood and parsimony methods in the context of Karl Popper's writings on corroboration. *Systematic Biology* **50**: 305–21.

Dembski, W. A. 1998. *The Design Inference: Eliminating Chance through Small Probabilities*. Cambridge: Cambridge University Press.

1997. Intelligent design as a theory of information. *Perspectives on Science and Christian Faith* **3**: 180–90.

Dennett, D. C. 2006. *Breaking the Spell: Religion as a Natural Phenomenon*. New York: Viking.

1995. *Darwin's Dangerous Idea*. New York: Simon & Schuster.

DeYoe, E. A., and D. C. Van Essen. 1988. Concurrent processing streams in monkey visual cortex. *Trends in Neuroscience* **11**: 219–26.

Diamond, J. M. 1975. The island dilemma: Lessons of modern biogeographic studies for the design of natural reserves. *Biological Conservation* **7**: 129–46.

1986. Overview: Laboratory experiments, field experiments, and natural experiments. *Community Ecology*. Editors J. Diamond and T. J. Case, 3–22. New York: Harper & Row.

Dick, D. M., R. J. Viken, R. J. Kaprio, J. Koskenvou, and M. Rose. 2001. Exploring gene-environment interactions: Socioregional moderation of alcohol use. *Journal of Abnormal Psychology* **110**: 625–32.

Dickemann, M. 1979. Female infanticide and the reproductive strategies of stratified human societies. *Evolutionary Societies and Human Social Behavior*. Editors N. A. Chagnon and W. Irons, 321–67. North Scituate, MA: Duxbury.

1995. Wilson's Panchreston: The inclusive fitness hypothesis of sociobiology re-examined. *Journal of Homosexuality* **28**: 147–83.

Dietrich, Michael. 2000a. Form hopeful monsters to homeotic effects: Richard Goldschmidt's integration of development, evolution and genetics. *American Zoologist* **40**: 738–47.

2000b. The problem of the gene. *Comptes Rendus De L'Academie des Sciences-Series III-Sciences de la Vie* **323**, no. 12: 1139–46.

Dillon, N. 2003. Positions, please … . *Nature* **425**: 457.

Dobzhansky, Th. 1937. *Genetics and the Origin of Species*. New York: Columbia University Press.

1962. *Mankind Evolving*. New Haven, CT: Yale University Press.

1956. What is an adaptive trait? *American Naturalist* **40**, no. 855: 337–47.

Dohrn, G. 1875. *Der Ursprung der Wirbelthiere und das Princip des Functionswechsels*. Leipzig.

Donoghue, M. 1985. A critique of the biological species concept and recommendations for a phylogenetic alternative. *Bryologist* **88**: 172–81.

Dretske, F. 1981. *Knowledge and the Flow of Information*. Cambridge, MA: MIT Press.

Dugatkin, L. A., and M. Alfieri. 1991a. Guppies and the tit-for-tat strategy: Preference based on past interaction. *Behavioral Ecology and Sociobiology* **28**: 243–46.

1991b. Tit-for-tat in guppies (poecilia reticulata): The relative nature of cooperation and defection during predator inspection. *Evolutionary Ecology* **5**: 300–09.

Dupre, J. 1993. *The Disorder of Things: Metaphysical Foundations of the Disunity of Science*. Cambridge, MA: Harvard University Press.

Dupuis, C. 1984. Willi Hennig's impact on taxonomic thought. *Annual Review of Ecology and Systematics* **15**: 1–24.

Edwards, A. W. F. 1972. *Likelihood*. Cambridge: Cambridge University Press.

Ehrlich, P. R. 2000. *Human Natures: Genes, Cultures, and the Human Prospect*. Washington, D. C., and Covelo, CA: Island Press/Shearwater Books.

Eldredge, N. 1995. *Reinventing Darwin*. New York: John Wiley.

1985. *Unfinished Synthesis: Biological Hierarchies and Modern Evolutionary Thought*. New York: Oxford University Press.

Eldredge, N., and J. Cracraft. 1980. *Phylogenetic Patterns and the Evolutionary Process: Method and Theory in Comparative Biology*. New York: Columbia University Press.

Eldredge, N., and S. J. Gould. 1972. Punctuated equilibria: An alternative to phyletic gradualism. *Models in Paleobiology*. Editor T. J. M. Schopf, 82–115. San Francisco: W. H. Freeman.

Elliott, K. 2004. Error as means to discovery. *Philosophy of Science* **71**: 174–97.

Ellis, L., M. A. Ames, and D. Burke. 1987. Sexual orientation as a continuous variable: A comparison between sexes. *Archives of Sexual Behavior* **16**: 526–29.

Elton, C. S. 1958. *The Ecology of Invasions by Animals and Plants*. London: Methuen.

Enc, B., and F. Adams. 1992. Functions and goal-directedness. *Philosophy of Science* **59**: 635–54.

Endler, J. 1986. *Natural Selection in the Wild*. Princeton, NJ: Princeton University Press.

Endler, J., and T. McClellan. 1988. The process of evolution: Towards a newer synthesis. *Annual Review of Ecology and Systematics* **19**: 395–421.

Ernst, Z. 2001. Explaining the social contract. *British Journal for the Philosophy of Science* **52**, no. 1: 1–24.

Erwin, D., and E. Davidson. 2002. The last common bilaterian ancestor. *Development* **129**: 3021–32.

Essock-Vitale, S., and R. Seyfarth. 1986. Intelligence and social cognition. *Primate Societies*. Editors B. Smuts, D. Cheney, R. Seyfarth, R. Wrangham, and T. Struhsaker, 452–61. Chicago: University of Chicago Press.

Fager, E. W. 1972. Diversity: A sampling study. *American Naturalist* **106**: 293–310.

Faith, D. P., and J. W. H. Trueman. 2001. Towards an inclusive philosophy for phylogenetic inference. *Systematic Biology* **50**: 331–50.

Falk, R. 1991. The dominance of traits in genetic analysis. *Journal of the History of Biology* **24**: 457–84.

———. 2000. The gene: A concept in tension. *The Concept of the Gene in Development and Evolution*. Editors P. R. Falk and H. J. Rheinberger Buerton, 317–48. Cambridge: Cambridge University Press.

———. Genetic analysis. In press. *International Handbook of the Philosophy of Biology*. Editors M. Matthen and C. Stephens. New York and Amsterdam: Elsevier.

———. 2005. Genetics. *The Philosophy of Science: An Encyclopedia*. Editors J. Pheiffer and S. Sarkar, 330–39. New York: Routledge Reference.

———. 2001. The rise and fall of dominance. *Biology and Philosophy* **16**, no. 3: 285–323.

———. 1995. The struggle of genetics for independence. *Journal of the History of Biology* **28**: 219–46.

———. 1986. What is a gene? *Studies in the History and Philosophy of Science* **17**: 133.

Farris, J. S. 1979. The information content of the phylogenetic system. *Systematic Zoology* **28**: 483–519.

   1983. The logical basis of phylogenetic analysis. *Advances in Cladistics.* Editors N. I. Platnick, and V. A. Funk, 7–36. New York: Columbia University Press.

Farris, J. S., A. G. Kluge, and J. M. Carpenter. 2001. Popper and likelihood versus "Popper*." *Systematic Biology* **50**: 438–43.

Felsenstein, J. 1978. Cases in which parsimony or compatibility methods will be positively misleading. *Systematic Zoology* **27**: 401–10.

Finta, C., and P. G. Zaphiropoulos. 2001. A statistical view of genome transcription. *Journal of Molecular Evolution* **53**: 160–62.

Fisch, U. 1973. The vestibular response following unilateral vestibular compensation. *Acta Otlaryngology* **76**: 229–38.

Fisher, D. C. 1985. Evolutionary morphology: Beyond the analogous, the anecdotal, and the ad hoc. *Paleobiology* **11**: 120–38.

Fisher, R. A. 1930. *The Genetical Theory of Natural Selection.* Oxford: Clarendon Press.

   1925. *Statistical Methods for Research Workers.* Edinburgh: Oliver and Boyd.

Fisher, R. A., A. S. Corbet, and C. B. Williams. 1943. The relation between the number of species and the number of individuals in a random sample of an animal population. *Journal of Animal Ecology* **12**: 42–58.

Fisher, R. A., and E. B. Ford. 1947. The spread of a gene in natural conditions in a colony of the moth Panaxia dominula. *Heredity* **1**: 143–74.

Fitelson, B., E. Stephens, and C. Sober. 1999. How not to detect design – a review of William Dembski's *The Design Inference. Philosophy of Science* **66**: 472–88.

Flohr, H., J. Bienfold, W. Abeln, and I. Macskovics. 1981. Concepts of vestibular compensation. *Lesion-Induced Neuronal Plasticity in Sensorimotor Systems.* Editors H. Flohr and W. Precht, 153–72. Amsterdam: Springer.

Fogle, T. 2001. The dissolution of protein coding genese in molecular biology. *The Concept of the Gene in Development and Evolution.* Editors R. Falk, H. J. Rheinberger, and P. Beurton, 3–25. Cambridge: Cambridge University Press.

Fox-Keller, E. 2000. *The Century of the Gene.* Cambridge, MA: Harvard University Press.

Francis, R. 2003. *Why Men Won't Ask for Directions: The Seductions of Sociobiology.* Princeton, NJ: Princeton University Press.

Frey, B. S., and I. Bohnet. 1980. Institutions affect fairness: Experimental investigations. *Behavior* **75**: 262–300.

Friboulet, A., and D. Thomas. 2005. Systems biology – an interdisciplinary approach. *Biosensors & Bioelectronics* **20**: 2404–07.

Fujimura, J. H. 2005. Postgenomic futures: Translations across the machine-nature border in systems biology. *New Genetics and Society* **24**: 195–225.

Futuyma, D. 1979. *Evolutionary Biology*. Sunderland, MA: Sinauer.

1986. *Evolutionary Biology*. 2nd ed. Sunderland, MA: Sinauer.

Garfinkel, A. 1981. *Forms of Explanation*. New Haven, CT: Yale University Press.

Gasman, D. 1998. *Haeckel's Monism and the Birth of Fascist Ideology*. Frankfurt: Peter Lang.

1971. *The Scientific Origins of National Socialism*. New York: Science History.

Gaston, K. J. 1996. What is biodiversity? *Biodiversity: A Biology of Numbers and Difference*. Editor K. J. Gaston, 1–9. London: Blackwell.

Ghiselin, M. T. 1997. *Metaphysics and the Origin of Species*. Albany, NY: SUNY Press.

1974. A radical solution to the species problem. *Systematic Zoology* **23**: 536–44.

Gibbon, E. 1777–88. *The History of the Decline and Fall of the Roman Empire*. 6 Vols. London: Strahan and Cadell.

Giere, R. 1988. *Explaining Science: A Cognitive Approach*. Chicago: University of Chicago Press.

Gilbert, S. F., and S. Sarkar. 2000. Embracing complexity: Organicism for the 21st century. *Developmental Dynamics* **219**: 1–9.

Gillespie, J. H. 1977. Natural selection for variances in offspring number: A new evolutionary principle. *American Naturalist* **111**: 1010–14.

1974. Natural selection for within-generation variance in offspring number. *Genetics* **76**: 601–06.

1973. Polymorphism in random environments. *Theoretical Population Biology* **4**: 193–95.

Glennan, S. S. 1996. Mechanisms and the nature of causation. *Erkenntnis* **44**: 49–71.

2005. Modeling mechanisms. *Studies in History and Philosophy of Biological and Biomedical Sciences. Special Issue: Mechanisms in Biology*. Editors C. F. Craver and L. Darden. **36**: 443–64.

2002. Rethinking mechanistic explanation. *Philosophy of Science (Supplement)* **69**: S342–S353.

Godfrey-Smith, P. 1992. Additivity and the units of selection. *PSA: Proceedings of the Biennial Meeting of the Philosophy of Science Association, 1992* **1**: 315–28.

1999. Genes and codes: Lessons from the philosophy of mind? *Where Biology Meets Psychology: Philosophical Essays.* Editor V. Hardcastle. Cambridge, MA: MIT Press.

2001. Information and the argument from design. *Intelligent Design Creationism and its Critics: Philosophical, Theological and Scientific Perspectives.* Editor R. Pennock, 575–96. Cambridge, MA: MIT Press.

2000. On the Theoretical Role of 'Genetic Coding'. *Philosophy of Science* **67**: 26–44.

Forthcoming. The strategy of model-based science. *Biology and Philosophy.*

2001. Three kinds of adaptationism. *Adaptationism and Optimality.* Editors S.H. Orzack and E. Sober, 335–57. Cambridge: Cambridge University Press.

Goldschimdt, R. 1940. *The Material Basis of Evolution.* New Haven, CT: Yale University Press.

Goode, S. 1999. Johnson challenges advocates of evolution. *Insight*, October 25.

Goodman, D. 1975. The theory of diversity-stability relationships in ecology. *Quarterly Review of Biology* **50**: 237–66.

Goodrich, E. S. 1912. *The Evolution of Living Organisms.* London: T. C. and E. C. Jack.

Gould, S.J. 1983. *Hen's Teeth and Horse's Toes.* New York: W. W. Norton.

1977. *Ontogeny and Phylogeny.* Cambridge, MA: Belknap Press.

1980. *The Panda's Thumb.* New York: W. W. Norton.

2002. *The Structure of Evolutionary Theory.* Cambridge, MA: Harvard University Press.

1989. *Wonderful Life: The Burgess Shale and the Nature of History.* New York: W. W. Norton.

Gould, S.J., and R. C. Lewontin. 1979. The spandrels of San Marco and the Panglossian paradigm: A critique of the adaptationist program. *Proceedings of the Royal Society of London, Series B: Biological Sciences* **205**: 581–98.

Gould, S.J., and E. A. Lloyd. 1999. Individuality and adaptation across levels of selection: How shall we name and generalize the unit of Darwinism? *Proceedings of the National Academy of Sciences USA* **96**: 11904–09.

Gould, S.J., and E. S. Vrba. 1982. Exaptation – a missing term in the science of form. *Paleobiology* **8**: 4–15.

Goulson, D., and D. Owen. 1997. Long-term studies of the *medionigra* polymorphism in the moth *Panaxia domincula*: A critique. *Oikos* **80**: 613–17.

Grafen, A. 1991. Modelling in behavioural ecology. *Behavioural Ecology: An Evolutionary Approach.* Editors J. R. Krebs and N. B. Davies, 5–31. Oxford: Blackwell.

Grant, S. G. N. 2003. Systems biology in neuroscience: Bridging genes to cognition. *Current Opinion in Neurobiology* **13**: 577–82.

Greenberg, A. S., and J. M. Bailey. 1993. Do biological explanations of homosexuality have moral, legal, or policy implications? *Journal of Sex Research* **30**: 245–51.

Grene, M. 1990. Evolution, typology, and population thinking. *American Philosophical Quarterly* **27**: 237–44.

——— 2002. Reply to David Hull. *The Philosophy of Marjorie Grene.* Editors R. E. Auxier and L. E. Hahn, 279–83. La Salle, IL: Open Court.

Griesemer, J. R. 2000. Development, culture, and the units of inheritence. *Philosophy of Science* **67**: 348–68.

——— 2005. The informational gene and the substantial body: On the generalization of evolutionary theory by abstraction. *Idealization XII: Correcting the Model: Idealization and Abstraction in the Sciences.* Editors M. Jones and N. Cartwright. Amsterdam/New York: Rodopi.

——— 2004. Three-dimensional models in philosophical perspective. *Models: The Third Dimension of Science.* Editors Soraya de Chadarevian and Nick Hopwood, 433–42. Stanford, CA: Stanford University Press.

Griesemer, J. R., and M. Wade. 1988. Laboratory models, causal explanations and group selection. *Biology and Philosophy* **3**: 67–96.

Griffiths, G. C. D. 1974. On the foundations of biological systematics. *Acta Biotheoretica* **13**: 85–131.

Griffiths, P. E. 2001. Genetic information: A metaphor in search of a theory. *Philosophy of Science* **68**: 394–412.

——— 1996. The historical turn in the study of adaptation. *British Journal for the Philosophy of Science* **47**: 511–32.

——— 1999. Squaring the circle: Natural kinds with historical essences. *Species: New Interdisciplinary Essays.* Editor R. A. Wilson, 209–28. Cambridge: Cambridge University Press.

Griffiths, P. E., and R. D. Gray. 1994. Developmental systems and evolutionary explanation. *Journal of Philosophy* **91**, no. 6: 277–305.

——— 1997. Replicator II: Judgment day. *Biology and Philosophy* **12**, no. 4: 471–92.

Griffiths, P., and E. Neumann-Held. 1999. The many faces of the gene. *Biosciences* **49**: 656–64.

Guston, D. H., and D. Sarewitz. 2002. Real-time technology assessment. *Techonology in Society* **24**: 93–109.

Güth, W., R. Schmittberger, and B. Schwarz. 1982. An experimental analysis of ultimatum bargaining. *Journal of Economic Behavior and Organization* **3**: 367–88.

Guyer, M. S., and F. S. Collins. 1993. The human genome project and the future of medicine. *American Journal of Diseases of Children* **147**: 1145–52.

Haeckel, E. 1862. *Die Radiolarien (Rhizopoda radiaria). Eine Monographie.* Berlin: G. Reimer.

——— 1868. *Naturliche Schopfungsgeschichte.* Berlin: G. Reimer.

Haile-Selassie, Y. 2001. Late Miocene hominids from the Middle Awash, Ethiopa. *Nature* **412**: 178–81.

Haldane, J. B. S. 1932. *The Causes of Evolution.* London: Longmans.

Haldeman, D. C. 1994. The practice and ethics of sexual orientation. *Journal of Consulting and Clinical Psychology* **62**: 221–27.

Hall, B. K. 2000. Evo-devo or devo-evo – does it matter? *Evolution and Development* **2**, no. 4: 177–78.

——— 1998. *Evolutionary Developmental Biology.* London and New York: Chapman & Hall.

Hall, B. K., and W. M. Olson. 2003. *Keywords in Evolutionary Developmental Biology.* Cambridge, MA: Harvard University Press.

Halperin, D. M. 1990. *One Hundred Years of Homosexuality.* New York: Routledge.

Hamblin, M. T., E. E. Thompson, and A. DiRienzo. 2002. Complex signatures of natural selection at the duffy blood group locus. *American Journal of Human Genetics* **70**: 369–83.

Hamburger, V. 1988. *The Heritage of Experimental Embryology: Hans Spemann and the Organizer.* New York: Oxford University Press.

Hamer, D. H., S. Hu, V. L. Magnuson, N. Hu, and A. M. L. Pattatucci. 1993. A linkage between DNA markers on the X-chromosome and male sexual orientation. *Science* **261**: 321–37.

Hamilton, W. D. 1996 *Narrow Roads of Gene Land: The Collected Papers of W. D. Hamilton.* New York: W. H. Freeman Spektrum.

Hampe, M., and Morgan, S. R. 1988. Two consequences of Richard Dawkins' view of genes and organisms. *Studies in History and Philosophy of Science* **19**: 119–38.

Hardcastle, V. 1995. *How to Build a Theory in Cognitive Science.* Albany, NY: SUNY Press.

Hariharan, I. K., and D. A. Haber. 2003. Yeast, flies, worms, and fish in the study of human disease. *New England Journal of Medicine* **348**: 2457–63.

Harlin, M. 1999. The logical priority of the tree over characters and some of its consequences. *Biological Journal of the Linnaean Society* **68**: 497–503.

Harré, Rom. 1970. *The Principles of Scientific Thinking.* Chicago: University of Chicago Press.

Hartl, D., and A. Clark. 1989. *Theoretical Population Genetics*. Sunderland, MA: Sinauer

Hartwell, L.H., J.J. Hopfield, S. Leibler, and A.W. Murray. 1999. From molecular to modular cell biology. *Nature* **402**: 47–52.

Haught, J.F. 2000. *God After Darwin: A Theology of Evolution*. Boulder, CO: Westview Press.

Heisler, I.L., and J. Damuth. 1987. A method for analyzing selection in hierarchically structured populations. *American Naturalist* **130**: 582–602.

Hennig, W. 1950. *Grundzuge einer Theorie der Phylogenetischen Systematik*. Berlin: Deutscher Zentralverlag.

1966. *Phylogenetic Systematics*. Urbana: University of Illinois Press.

Henry, C.M. 2003. Systems biology. *Chemical and Engineering News* **81**, no. 20: 45–55. Available at http://pubs.acs.org/cen/coverstory/8120/8120biology.html.

Herdt, G. 1997. *Same Sex Different Cultures*. Boulder, CO: Westview Press.

Hesse, M. 1966. *Models and Analogies in Science*. Notre Dame, IN: University of Notre Dame Press.

Hill, M.O. 1973. Diversity and evenness: A unifying notation and its consequences. *Ecology* **54**: 427–32.

Hillis, D.M., J.P. Huelsenbeck, and D.L. Swofford. 1994. Hobgoblin of phylogenetics? *Nature* **369**: 363–64.

Hines, W.G.S. 1987. ESS theory: A basic review. *Theoretical Population Biology* **31**: 195–272.

Hodge, M.J.S. 1987. Natural selection as a causal, empirical, and probabilistic theory. *The Probabilistic Revolution*. Editor L. Kruger. Cambridge, MA: MIT Press.

Holmes, F.L. 2000. Symour Benzer and the definition of the Gene. *The Concept of the Gene in Development and Evolution*. Editors R. Falk, H.J. Rheinberger, and P. Beurton, 115–55. Cambridge: Cambridge University Press.

Holyoak, K.J., and P. Thagard. 1995. *Mental Leaps: Analogy in Creative Thought*. Cambridge, MA: MIT Press.

Hood, L., J.R. Heath, M.E. Phelps, and B. Lin. 2004. Systems biology and new technologies enable predictive and preventative medicine. *Science* **306**: 640–43.

Hopwood, N. 2002. *Embryos in Wax Models from the Ziegler Studio*. University of Cambridge and University of Bern: Whipple Museum of the History of Science and the Institute of the History of Medicine.

Hossfeld, U. 2005. *Geschichte der biologischen Anthropologie in Deutschland*. Stuttgart: Franz Steiner Verlag.

Hu, S., A. Pattatucci, C. Patterson, L. Li, D. Fulker, S. Cherny, L. Kruglyac, and D. Hamer. 1995. Linkage between sexual orientation and chromosome Xq28 in males but not females. *Nature Genetics* **11**: 248–56.

Hubbell, S. P. 2001. *The Unified Neutral Theory of Biodiversity and Biogeography*. Princeton, NJ: Princeton University Press.

Huelsenbeck, J. P., F. Ronquiest, R. Nielson, and J. P. Bollback. 2001. Bayesian inference of phylogeny and its impact on evolutionary biology. *Science* **294**: 2310–14.

Hull, D. 1976a. Are species really individuals? *Systematic Zoology* **25**: 174–91.

      1994. Contemporary systematic philosophies. *Conceptual Issues in Evolutionary Biology*. Editor E. Sober, 295–330. Cambridge, MA: MIT Press.

      1965. The effect of essentialism on taxonomy: Two thousand years of stasis. *British Journal for the Philosophy of Science* **15**: 314–26; **16**: 1–18.

      1980. Individuality and selection. *Annual Review of Ecology and Systematics* **11**: 311–32.

      1976b. *Informal aspects of theory reduction. PSA 1974*. Editor R. S. Cohen, 653–70. Dordrecht: Reidel.

      1988. *Science as a Process: An Evolutionary Account of the Social and Conceptual Development of Science*. Chicago: University of Chicago Press.

      1999. The use and abuse of Sir Karl Popper. *Biology and Philosophy* **14**: 481–504.

Hull, D., and M. V. H. Van Regenmortel, Editors. 2002. *Promises and Limits of Reductionism in the Biomedical Sciences*. Chichester, England: John Wiley.

Hume, D. 1990. *Dialogues Concerning Natural Religion*. Editor Martin Bell. London: Penguin.

Hurlbert, S. H. 1971. The nonconcepts of species diversity: A critique and alternative parameters. *Ecology* **52**: 577–86.

Hutcheson, K. 1970. "The Moments and Distribution for an Estimate of the Shannon Information Measure and Its Application to Ecology." Ph.D. dissertation. Virginia Polytechnic Institute.

Huxley, J. S. 1942. *Evolution: The Modern Synthesis*. London: Allen and Unwin.

Ideker, T., T. Galitski, and L. Hood. 2001. A new approach to decoding life: Systems biology. *Annual Review of Genomics and Human Genetics* **2**: 343–72.

Jablonka, E. 2002. Information: Its interpretaion, its inheritance and its sharing. *Philosophy of Science* **69**: 578–605.

Jacob, F. 1998 *Of Flies, Mice and Men: On the Revolution in Modern Biology, by One of the Scientists Who Helped Make It.* Cambridge, MA: Harvard University Press.

Janzen, D. H. 1986. The future of tropical ecology. *Annual Review of Ecology* **17**: 305–24.

Johansen, T. K. 2004. *Plato's Natural Philosophy: A Study of the Timaeus-Critias.* Cambridge: Cambridge University Press.

Johnson, P. E. 2004. Overestimating AIDS. *Touchstone Magazine,* October.
    1996. *Starting a conversation about evolution: A review of The Battle of the Beginnings: Why Neither Side Is Winning the Creation-Evolution Debate by Del Ratzsch* [Internet]. Access Research Network, 8/31/96. Available from www. arn.org/docs/johnson/ratzsch.htm.

Judson, H. F. 1979. *The Eight Day of Creation: Makers of the Revolution in Biology.* New York: Simon & Schuster.

Juengst, E. T. 1991. The Human Genome Project and bioethics. *Kennedy Institute of Ethics Journal* **1**: 71–74.

Justus, J., and S. Sarkar. 2002. The principle of complementarity in the design of reserve networks to conserve biodiversity: A preliminary history. *Journal of Biosciences* **27**: 421–35.

Karp, P. 1989. "Hypothesis Formation and Qualitative Reasoning in Molecular Biology." Ph.D. dissertation. Stanford University.

Kauffman, S. A. 1993. *The Origins of Order: Self-Organization and Selection in Evolution.* Oxford: Oxford University Press.

Kay, L. 2000. *Who Wrote the Book of Life? A History of the Genetic Code.* Palo Alto, CA: Stanford University Press.

Kearney, M., and O. Rieppel. 2006. Rejecting "the given" in systematics. *Cladistics* **22**: 369–77.

Keller, E. F. 2002. *Making Sense of Life: Explaining Biological Development with Models, Metaphors, and Machines.* Cambridge, MA: Harvard University Press.

Keller, R. A., R. N. Boyd, and Q. D. Wheeler. 2003. The illogical basis of phylogenetic nomenclature. *Botanical Review* **69**: 93–110.

Kendler, K. S., L. M. Thornton, S. E. Gilman, and R. C. Kessler. 2000. Sexual orientation in a U.S. national sample of twin and nontwin sibling pairs. *American Journal of Psychiatry* **157**: 1843–46.

Kettlewell, H. B. D. 1956. Further selection experiments on industrial melanism in the Lepidoptera. *Heredity* **10**: 287–301.
    1955. Selection experiments on industrial melanism in the Lepidoptera. *Heredity* **9**: 323–42.

Kilpatrick, A. M., and A. R. Ives. 2003. Species interactions can explain Taylor's power law for ecological time series. *Nature* **422**: 65–68.

Kincaid, H. 1996. *Philosophical Foundations of the Social Sciences: Analyzing Controversies in Social Research*. New York: Cambridge University Press.

Kingsland, S. 2002. Creating a science of nature reserve design: Perspectives from history. *Environmental Modeling and Assessment* 7: 61–69.

Kingslover, J. G., H. E. Hoekstra, J. M. Hoekstra, D. Berrigan, S. N. Vignieris, C. E. Hoang, A. Hill, P. Gibert, and P. Beerli. 2001. The strength of phenotypic selection in natural populations. *American Naturalist* **157**, no. 3: 245–61.

Kinsey, A., C. Pomeroy, B. Wardell, and C. E. Martin. 1948. *Sexual Behaviour in The Human Male*. Philadelphia: W. B. Saunders.

Kinzig, A. P., S. W. Pacala, and D. Tilman, Editors. 2002. *The Functional Consequences of Biodiversity: Empirical Progress and Theoretical Extensions*. Princeton, NJ: Princeton University Press.

Kirk, K. M., J. M. Dunne, M. P. Martin, and N. G. Bailey. 2000. Measurement models for sexual orientation in a community twin sample. *Behavior Genetics* **30**: 345–56.

Kirkpatrick, J. B., and C. E. Harwood. 1983. Conservation of Tasmanian macrophytic wetland vegetation. *Proceedings of the Royal Society of Tasmania* **117**: 5–20.

Kirschner, M., and J. Gerhart. 2005. *The Plausibility of Life: Great Leaps of Evolution*. New Haven, CT: Yale University Press.

Kirschner, M. W. 2005. The meaning of systems biology. *Cell* **121**: 503–04.

Kitano, H. 2002a. Looking beyond the details: A rise in system-oriented approaches in genetics and molecular biology. *Current Genetics* **41**: 1–10.

2002b. Systems biology: A brief overview. *Science* **295**: 1662–64.

Kitcher, P. 1984. 1953 and all that: A tale of two sciences. *Philosophical Review* **93**: 335–73.

2001. Battling the undead: How (and how not) to resist genetic determinism. *Thinking About Evolution: Historical, Philosophical and Political Perspectives*. Editors R. Singh, D. Paul, J. Beatty, and C. Krimbas. Cambridge: Cambridge University Press.

1988. Explanatory unification. *Theories of Explanation*. Editor J. Pitt, 167–87. Oxford: Oxford University Press.

1999. Games social animals play: Commentary on Brian Skyrms' *Evolution of the Social Contract*. *Philosophy and Phenomenological Research* **59**, no. 1: 221–28.

1999. The hegemony of molecular biology. *Biology and Philosophy* **14**, no. 2: 195–210.

1996. *The Lives to Come: The Genetic Revolution and Human Possibilities*. New York: Simon & Schuster.

2001. *Science Truth and Democracy*. New York: Oxford University Press.

Kluge, A. G. 2001. Philosophical conjectures and their refutation. *Systematic Biology* **50**: 322–30.

2003. The repugnant and the mature in phylogenetic inference: Atemporal similarity and historical identity. *Cladistics* **19**: 356–68.

1999. The science of phylogenetic systematics: Explanation, prediction, and test. *Cladistics* **15**: 429–36.

1997. Testability and the refutation and corroboration of cladistic hypotheses. *Cladistics* **13**: 81–96.

Kluge, A. G., and J. S. Farris. 1969. Quantitative phyletics and the evolution of anurans. *Systematic Zoology* **19**: 356–68.

Knight, R., S. Freeland, and L. Landweber. 1999. Selection, history and chemistry: The three faces of the genetic code. *Trends in Biochemical Sciences* **24**: 241–47.

Knoll, A. H. 2003. *Life on a Young Planet*. Princeton, NJ: Princeton University Press.

Knoll, A., and S. B. Carroll. 1999. Early animal evolution: Emerging views from comparative biology and geology. *Science* **284**: 2129–37.

Koleff, P., K. J. Gaston, and J. J. Lennon. 2003. Measuring beta diversity for presence-absence data. *Journal of Animal Ecology* **72**: 367–82.

Kowalevsky, A. 1867. Entwicklungsgeschichte des Amphioxus lanceolatus. *Memoirs of the Academy of Sciences, St. Petersburg* **11**: 1–17.

Kreitman, M., and R. R. Hudson. 1991. Inferring the evolutionary theories of the Adh and Adh-dup loci in *Drosophila melanogaster* from patterns of polymorphism and divergence. *Genetics* **127**: 565–82.

Kreps, D. 1990. *Game Theory and Economic Modeling*. Oxford: Clarendon Press.

Kühn, A. 1955. *Vorlesungen uber Entwicklungsphysiologie*. Berlin: Springer.

Laland, K., and G. R. Brown. 2002. *Sense and Nonsense: Evolutionary Perspectives on Human Behaviour*. Oxford: Oxford University Press.

Landecker, H. L. 2004. Building a new type of body in which to grow a cell: The origins of tissue culture. *Creating a Tradition of Biomedical Research: Contibutions to the History of the Rockefeller University*. Editor Darwin Stapleton, 151–74. New York: Rockefeller University Press.

Laubichler, M. D. 2003. Carl Gegenbaur (1826–1903): Integrating comparative anatomy and embryology. *Journal of Experimental Zoology; Part B: Molecular and Developmental Evolution* **300**, no. 1: 23–31.

2005. Evolutionare Entwicklungsbiologie. *Philosophie der Biologie.* Editors U. Krohs and G. Toepfer, 322–37. Frankfurt/Main: Suhrkamp.

2000. The organism is dead. Long live the organism! *Perspectives on Science* 8: 286–315.

Laubichler, M., and J. Maienschein. 2007. *From Embryology to Evo-Devo.* Cambridge, MA: MIT Press.

Laubichler, M. D., and J. Maienschein. 2003. Ontogeny, anatomy and the problem of homology: Carl Gegenbaur and the American tradition of cell lineage studies. *Theory in Biosciences* 112: 194–203.

Lauder, G. V. 1996. The argument from design. *Adaptation.* Editors M. R. Rose and G. V. Lauder, 55–91. San Diego: Academic Press.

Laumann, E. O., J. H. Michael, R. T. Michaels, and S. Gagnon. 1994. *The Social Organization of Sexuality: Sexual Practices in the United States.* Chicago: University of Chicago Press.

Lea, H. 1904. Ethical values in history. *American Historical Review* 9: 233–46.

Lee, M. 1999. Molecular clock calibrations and metazoan divergence dates. *Journal of Molecular Evolution* 49, no. 3: 385–91.

Lehman, C. L., and D. Tilman. 2000. Biodiversity, stability, and productivity in competitive communities. *American Naturalist* 156: 534–52.

Leibold, M. A. 1996. A graphical model of keystone predators in food webs: Trophic regulation of abundance, incidence, and diversity patterns in communities. *American Naturalist* 147: 784–812.

Leibold, M. A., J. M. Chase, J. B. Shurin, and A. L. Downing. 1997. Species turnover and the regulation of trophic structure. *Annual Review of Ecology and Systematics* 28: 467–94.

Lennox, J. G. 1993. Darwin *was* a teleologist. *Biology and Philosophy* 8: 409–21.

1985. Plato's unnatural teleology. *Platonic Investigations.* Editor D. O'Meara, 195–218. Pittsburgh: Mathesis Publications.

Lenoir, T. 1982. *The Strategy of Life: Teleology and Mechanics in Nineteenth-Century German Biology.* Chicago: University of Chicago Press.

Leroi, A. M. 2000. The scale independence of evolution. *Evolution & Development* 2, no. 2: 67–77.

Lewens, T. 2002. Adaptationism and engineering. *Biology and Philosophy* 17: 1–31.

2004. *Organisms and Artifacts: Design in Nature and Elsewhere.* Cambridge, MA: MIT Press.

Forthcoming. Seven types of adaptationism. *Twenty-Five Years of Spandrels.* Editor D. M. Walsh. Oxford: Oxford University Press.

Lewicki, M. S. 1998. A review of methods for spike sorting: The detection and classification of neural action potentials. *Network: Computational Neural Systems* **9**: R53–R78.

Lewis, D. 1973b. Causation. *Journal of Philosophy* **70**: 556–67.

——— 1969. *Convention: A Philosophical Study*. Cambridge, MA: Harvard University Press.

——— 1973a. *Counterfactuals*. Oxford: Oxford University Press.

Lewontin, R. C. 1978. Adaptation. *Scientific American* **239**, no. 3: 213–30.

——— 1984. Adaptation. *Conceptual Issues in Evolutionary Biology*. Editor E. Sober. Cambridge, MA: MIT Press.

——— 1985. Adaptation. *The Dialectical Biologist*. Editors R. Levins and R. Lewontin, 65–84. Cambridge, MA: Harvard University Press.

——— 1992. The dream of the human genome. *New York Review of Books* **39**, no. 10 (May 28): 31–40.

——— 1958. A general method for investigating the equilibrium of gene frequency in a population. *Genetics* **43**: 421–33.

——— 1969a. The meaning of stability. *Brookhaven Symposia in Biology* **22**: 13–24.

——— 1969b. The organism as subject and object of evolution. *The Dialectical Biologist*. Editors R. C. Lewontin and R. Levins. Cambridge, MA: Harvard University Press.

——— 1985. Population genetics. *Annual Review of Genetics* **19**: 81–102.

——— 2000a. *The Triple Helix: Gene, Organism, and Environment*. Cambridge, MA: Harvard University Press.

——— 1970. The units of selection. *Annual Review of Ecology and Systematics* **1**: 1–18.

——— 2000b. What do population geneticists know and how do they know it? *Biology and Epistemology*. Editors R. Creath and Jane Maienschein, 191–214. Cambridge: Cambridge University Press.

Lewontin, R., and R. Levins. 1985. *The Dialectical Biologist*. Cambridge: Cambridge University Press.

Likens, G. E., F. H. Bormann, N. M. Johnson, D. W. Fisher, and R. S. Pierce. 1970. Effects of forest cutting and herbicide treatment on nutrient budgets in the Hubbard Brook watershed-ecosystem. *Ecological Monographs* **40**: 23–47.

Lippman, A. 1992. Led (astray) by genetic maps. *Social Science and Medicine* **35**: 1469–76.

Lipton, P. 2004. *Inference to the Best Explanation*. 2nd ed. London: Routledge.

Livingstone, D. N. 1987. *Darwin's Forgotten Defenders: The Encounter Between Evangelical Theology and Evolutionary Thought*. Grand Rapids, MI: William B. Eerdmans.

Lloyd, E. 1987. Confirmation of ecological and evolutionary models. *Biology and Philosophy* **2**: 277–93.

    1988. *The Structure and Confirmation of Evolutionary Theory.* New York: Greenwood.

    1992. Unit of selection. *Keywords in Evolutionary Biology.* Editors E. F. Keller and E. A. Lloyd, 334–40. Cambridge, MA: Harvard University Press.

    2001. Units and levels of selection: An anatomy of the units of selection debates. *Thinking About Evolution: Historical, Philosophical and Political Perspectives.* Editors R. Singh, C. Krimbas, D. Paul, and J. Beatty, 267–91. New York: Cambridge University Press.

Lloyd, E. A., and S. J. Gould. 1993. Species selection on variability. *Proceedings of the National Academy of Sciences* **90**: 595–99.

Logan, G. A., et al. 1995. Terminal protoerozoic reorganization of biogeochemical cycles. *Nature* **376**: 53–56.

Love, A. C., and R. A. Raff. 2003. Knowing your ancestors: Themes in the history of evo-devo. *Evolution and Development* **5**, no. 4: 327–30.

Lovejoy, A. O. 1965. *The Great Chain of Being: A Study of the History of an Idea.* New York: Harper & Row.

MacArthur, R. H. 1957. On the relative abundance of bird species. *Proceedings of the National Academy of Sciences* **43**: 293–95.

MacArthur, R. H. 1965. Patterns of species diversity. *Biologica Review* **40**: 510–33.

MacArthur, R. H., and E. O. Wilson. 1967. *The Theory of Island Biogeography.* Princeton, NJ: Princeton University Press.

Machamer, P. 2004. Activities and causation: The metaphysics and epistemology of mechanisms. *International Studies in the Philosophy of Science* **18**: 27–39.

Machamer, P., C. Craver, and L. Darden. 2000. Thinking about mechanisms. *Philosophy of Science* **67**: 1–25.

Magurran, A. E. 1988. *Ecological Diversity and Its Measurements.* Princeton, NJ: Princeton University Press.

    2004. *Measuring Biological Diversity.* Oxford: Blackwell.

Maienschein, J. 1991. *Transforming Traditions in American Biology: 1880–1915.* Baltimore: Johns Hopkins University Press.

    2005. *Whose View of Life? Embryos, Cloning and Stem Cells.* Cambridge, MA: Harvard University Press.

Maienschein, J., M. Glitz, and G. E. Allen, Editors. 2004. *Centenniel History of the Carnegie Institution of Washington.* Vol. 5, *The Department of Embryology.* Cambridge: Cambridge University Press.

Marcus, G. 2004. *The Birth of the Mind: How a Tiny Number of Genes Creates the Complexities of Human Thought.* New York: Basic Books.

Marcus, S. J., Editor. 2004. *Neuroethics: Conference Proceedings.* New York: Dana Press.

Margalef, R. 1958. Information theory in ecology. *General Systems Yearbook* **3**: 36–71.

Margules, C. R., A. O. Nicholls, and R. L. Pressey. 1988. Selecting networks of reserves to maximize biological diversity. *Biological Conservation* **43**: 63–76.

Margules, C. R., and R. L. Pressey. 2000. Systematic conservation planning. *Nature* **405**: 242–53.

Marquet, P. A., R. A. Quinones, S. Abades, F. Labra, M. Tognelli, M. Arim, and M. Rivadeneira. 2005. Scaling and power-laws in ecological sytems. *Journal of Experimental Biology* **208**: 1749–69.

Matessi, C., and S. D. Jayakar. 1976. Conditions for the evolution of altruism under Darwinian selection. *Theoretical Population Biology* **9**: 360–87.

Matthen, M., and A. Ariew. 2002. Two ways of thinking about fitness and natural selection. *Journal of Philosophy* **99**, no. 2: 55–83.

May, R. M. 1973. *Stability and Complexity in Model Ecosystems.* Princeton, NJ: Princeton University Press.

Maynard Smith, J. 2000. The concept of information in biology. *Philosophy of Science* **67**: 177–94.

1982. *Evolution and the Theory of Games.* Cambridge: Cambridge University Press.

1976. Group selection. *Quarterly Review of Biology* **51**: 277–83.

1964. Group selection and kin selection. *Nature* **201**, no. 4924: 1145–47.

1987. How to model evolution. *The Latest on the Best: Essays on Evolution and Optimality.* Editor J. Dupre, 119–31. Cambridge, MA: MIT Press.

2001. Reconciling Marx and Darwin. *Evolution* **55**, no. 7: 149–98.

1969. The status of neo-Darwinism. *Towards a Theoretical Biology.* Editor C. H. Waddington. Edinburgh: Edinburgh University Press.

1974. The theory of games and the evolution of animal conflicts. *Journal of Theoretical Biology* **47**: 209–21.

Maynard Smith, J., R. Burian, S. Kauffman, P. Alberch, J. Campbell, B. Goodwin, R. Lande, D. Raup, and L. Wolpert. 1985. Developmental constraints and evolution. *Quarterly Review of Biology* **60**: 265–87.

Maynard Smith, J., and G. R. Price. 1973. The logic of animal conflict. *Nature* **246**: 15–18.

Maynard Smith, J., and E. Szathmary. 1995. *The Major Transitions in Evolution.* New York: W. H. Freeman.

1999. *The Origins of Life: From the Birth of Life to the Origin of Language.* Oxford: Oxford University Press.

Mayr, E. 1974. Behavior programs and evolutionary strategies. *American Scientist* **62**: 650–59.

1961. Cause and effect in biology. *Science* **134**: 1501–06.

1976. *Evolution and the Diversity of Life*. Cambridge, MA: Harvard University Press.

1982. *The Growth of Biological Thought: Diversity, Evolution, and Inheritance*. Cambridge, MA: Belknap Press.

1986. Natural selection: The philosopher and the biologist. *Paleobiology* **12**: 233–39.

1965. Numerical phenetics and taxonomic theory. *Systematic Zoology* **14**: 73–97.

1987. The ontological status of species: Scientific progress and philosophical terminology. *Biology and Philosophy* **2**: 145–66.

1969. *Principles of Systematic Zoology*. New York: McGraw-Hill.

1942. *Systematics and the Origin of Species*. New York: Columbia University Press.

1988. *Towards a New Philosophy of Biology: Observations of an Evolutionist*. Cambridge, MA: Belknap Press.

1959. Typological versus population thinking. *Evolution and Anthropology: A Centennial Appraisal*. Editor B. J. Meggers, 409–12. Washington, DC: Anthropological Society of Washington.

Mayr, E., and W. Provine, Editors. 1980. *The Evolutionary Synthesis: Perspectives on the Unification of Biology*. Cambridge, MA: Harvard University Press.

McCann, K. S. 2000. The diversity-stability debate. *Nature* **405**: 228–33.

McGue, M. 1999. The behavioral genetics of alcoholism. *Current Directions in Psychological Science* **8**: 109–15.

McKelvey, R. D., and T. R. Palfrey. 1992. "An Experimental Study of the Centipede Game." *Econometrica* **60**, no. 4: 803–36.

McKnight, J., and J. Malcom. 2000. Is male homosexuality maternally linked? *Psychology, Evolution & Gender* **2**: 229–39.

McMenamin, M., and D. McMenamin. 1990. *The Emergence of Animals: The Cambrian Breakthrough*. New York: Columbia University Press.

McMullin, E. 2000. Life and intelligence far from Earth: Formulating theological issues. *Many Worlds*. Editor Steven Dick, 151–75. Philadelphia: Templeton Press.

McShea, D. W. 1996. Metazoan complexity and evolution: Is there a trend? *Evolution* **50**: 477–92.

1998a. Possible largest-scale trends in organismal evolution: Eight "live hypotheses." *Annual Review of Ecology and Systematics* **29**: 293–318.

2000. Trends, Tools and Terminology. *Paleobiology* **26**, no. 3: 330–33.

Merzenich, M. M., J. H. Kaas, M. Sur, R. J. Nelson, and D. J. Fellemen. 1983. Progression of change following median nerve section in the cortical representation of the hand in areas 3b and 1 in adult owl and squirrel monkeys. *Neuroscience* **10**: 639–65.

Michod, R. W. 1999. *Darwinian Dynamics: Evolutionary Transitions in Fitness and Individuality*. Princeton, NJ: Princeton University Press.

Mikkelson, G. M. 2004. Biological diversity, ecological stability, and downward causation. *Philosophy and Biodiversity*. Editors M. Oksanen and J. Pietarinen, 119–29. New York: Cambridge University Press.

2003. Ecological kinds and ecological laws. *Philosophy of Science* **70**: 1390–1400.

In Press. Realism vs. instrumentalism in a new statistical framework. *Philosophy of Science*.

Miles, F. A., and S. G. Lisberger. 1981. Plasticity in the vestibulo-ocular reflex: A new hypothesis. *Annual Review of Neuroscience* **4**: 279–99.

Millikan, R. 1984. *Language, Thought and Other Biological Categories*. Cambridge, MA: MIT Press.

Mills, S., and J. Beatty. 1979. The propensity interpretation of fitness. *Philosophy of Science* **46**: 263–86.

Millstein, R. L. 2002. Are random drift and natural selection conceptually distinct? *Biology and Philosophy* **17**: 33–53.

2005. Selection vs. drift: A response to Brandon's reply. *Biology and Philosophy* **20**: 171–75.

Mishkin, M., L. G. Ungerleider, and K. A. Macko. 1983. Object vision and spatial vision: Two cortical pathways. *Trends in Neuroscience* **6**: 273–99.

Mishler, B., and M. Donoghue. 1982. Species concepts: A case for pluralism. *Systematic Zoology* **31**: 491–503.

Moore, J. H., E. M. Boczko, and M. L. Summar. 2005. Connecting the dots between genes, biochemistry, and disease susceptibility: Systems biology modeling in human genetics. *Molecular Genetics and Metabolism* **84**: 104–11.

Morange, M. 1998. *A History of Molecular Biology*. Cambridge, MA: Harvard University Press.

Morowitz, H. 1985. *Models for Biomedical Research: A New Perspective. Report of the Committee on Models for Biomedical Research*. Washington, DC: National Academy Press.

Moss, L. 1992. A kernel of truth? On the reality of the genetic program. *PSA 1992*. Vol. 1. Editors D. Hull, M. Forbes, and K. Okruhlik, 335–48. East Lansing, MI: Philosophy of Science Association.

2003. *What Genes Can't Do*. Cambridge, MA: MIT Press.

Moyers, B. 2004. "On receiving Harvard Medical School's Global Environment Citizen Award." Available at http://www.commondreams.org.

Mueller, L. D., and A. Joshi. 2000. *Stability in Model Populations.* Princeton, NJ: Princeton University Press.

Muller, G. 1989. Ancestral patterns in bird development. *Journal of Evolutionary Biology* 2: 31–47.

2005. Evolutionary developmental biology. *Handbook of Evolution.* Editors F. M. Wuketits, and F. J. Ayala, Vol. 2, 87–115. Weinheim: Wiley-VCH.

Muller, G., and S. Newman. 1999. Generation, integration, autonomy: The steps in the evolution of homology. *Homology.* Editors G. Cardew and G. R. Bock, 65–73. Chichester, England: John Wiley.

2003. *Origination of organismal form: Beyond the gene in developmental and evolutionary biology.* Cambridge, MA: MIT Press.

Muller, G., and G. P. Wagner. 1991. Novelty in evolution: Restructuring the concept. *Annual Review of Ecological Systems* 22: 229–56.

Mustanski, B. S., M. L. Bailey, and J. M. Chivers. 2002. A critical review of recent biological research on human sexual orientation. *Annual Review of Sexual Research* 12: 89–140.

Naeem, S. 2002. Biodiversity equals instability. *Nature* 416: 23–24.

2002. Ecosystem consequences of biodiversity loss: The evolution of a paradigm. *Ecology* 83: 1537–52.

Nagel, E. 1961. *The Structure of Science: Problems in the Logic of Scientific Explanation.* New York: Harcourt, Brace & World.

Narbonne, G. 2005. The Ediacara biota: Neoproterozoic origin of animals and their ecosystems. *Annual Review of Earth and Plant Science* 33: 421–42.

Nash, G., and C. Crabtree, Supervisors. 1996. *National Standards for History, Basic Edition.* Los Angeles: National Center for History in the Schools.

Nash, J. 1950. The bargaining problem. *Econometrica* 18: 155–62.

1951. Non-cooperative games. *Annals of Mathematics* 54, no. 2: 286–95.

Neander, K. 1995. Pruning the tree of life. *British Journal for the Philosophy of Science* 46: 59–80.

1991. The teleological notion of "function." *Ausralasian Journal of Philosophy* 69: 454–68.

Nelkin, D., and M. S. Lindee. 1995. *The DNA Mystique.* New York: W. H. Freeman.

Nelson, G., and N. Platnick. 1981. *Systematics and Biogeography: Cladistics and Vicariance.* New York: Columbia University Press.

Neumann-Held, E. M. 1998. The gene is dead – long live the gene: Conceptualising the gene the constructionist way. *Sociobiology and*

Bioeconomics: The Theory of Evolution in Biological and Economic Theory. Editor P. Koslowski, 105–37. Berlin: Springer-Verlag.

Newberg, A. B., et al. 2002. *Why God Won't Go Away: Brain Science and the Biology of Belief.* New York: Ballantine Books.

Newman, S. A. 2003. "The fall and rise of systems biology." Available at http://www. gene-watch.org/genewatch/articles/16-4newman.html.

Nicholson, J. K., and I. D. Wilson. 2003. Understanding "global" systems biology: Metabonomics and the continuum of metabolism. *Nature Reviews Drug Discovery* 2: 668–76.

Nixon, K. C., and J. M. Carpenter. 2000. On the other "phylogenetic systematics." *Cladistics* 16: 298–318.

Nixon, K. C., and Q. D. Wheeler. 1990. An amplification of the phylogenetic species concept. *Cladistics* 6: 211–23.

Norton, B. G. 1987. *Why Preserve Natural Variety?* Princeton, NJ: Princeton University Press.

Novick, P. 1988. *That Noble Dream: The "Objectivity Question" and the American Historical Profession.* Cambridge: Cambridge University Press.

Nursall, J. R. 1959. Oxygen as a prerequisite to the origin of the Metazoa. *Nature* 183: 1170–72.

Nyhart, L. K. 1995. *Biology Takes Form: Animal Morphology and the German Universities, 1800–1900.* Chicago: University of Chicago Press.

——— 2002. Learning from history: Morphology's challenges in Germany ca. 1900. *Journal of Morphology* 252, no. 1: 2–14.

Odenbaugh, J. 2001. Ecological stability, model building, and environmental policy: A reply to some of the pessimism. *Philosophy of Science* 68: S493–S505.

Odling-Smee, J., et al. 2003. *Niche construction: The Neglected Process in Evolution.* Princeton, NJ: Princeton University Press.

Okasha, S. 2004a. The "averaging fallacy" and the levels of selection. *Biology and Philosophy* 19: 167–84.

——— 2003. The concept of group heritability. *Biology and Philosophy* 18, no. 3: 445–61.

——— 2004b. Multilevel selection, covariance and contextual analysis. *British Journal for the Philosophy of Science* 55: 481–504.

——— 2004c. Multilevel selection and the partitioning of covariance: A comparison of three approaches. *Evolution* 58, no. 3: 486–94.

Olby, R. 1985. *The Origins of Mendelism.* 2nd ed. Chicago: University of Chicago Press.

——— 1974. *The Path to the Double Helix: The Discovery of DNA.* Seattle: University of Washington Press.

Oppenheim, P., and H. Putnam. 1958. Unity of science as a working hypothesis. *Concepts, Theories, and the Mind-Body Problem*. Editors H. Feigl, M. Scriven, and G. Maxwell. Minnesota Studies in the Philosophy of Science. Vol. 2, 3–36. Minneapolis: University of Minnesota Press.

Orr, H. 2005. Master planned: Why intelligent design isn't. *New Yorker*, May 30.

Ortona, G. 1991. The ultimatum game. *Economic Notes* **20**, no. 2: 324–34.

Orzack, S. H., and E. Sober. 1994. Optimality models and the test of adaptationism. *American Naturalist* **143**: 361–80.

Oyama, S. 2001. *Cycles of Contingency: Developmental Systems and Evolution*. Editors S. Oyama, P. E. Griffiths, and R. D. Gray. Cambridge, MA: MIT Press.

——— 1985. *The Ontogeny of Information*. Cambridge: Cambridge University Press.

Paine, R. T. Food web complexity and species diversity. *The American Naturalist* **100**: 65–75.

Paley, W. 1828. *Natural Theology*. 2nd ed. Oxford: J. Vincent.

Parker, A. 2003. *In the Blink of an Eye*. Cambridge: Perseus.

Patil, G. P., and C. Taillie. 1976. Biological diversity: Concepts, indices, and applications. *Proceedings of the 9th International Biometric Conference*, Raleigh, NC, 383–411.

——— 1982. Diversity as a concept and its measurement. *Journal of the American Statistical Association* **77**: 548–61.

——— 1979. An overview of diversity. *Ecological Diversity in Theory and Practice*. Editors J. F. Grassel, G. Patil, P. Smith, and C. Taillie, 3–27. Fairland, MD: International Cooperative Publishing House.

Patil, G. P., and C. Taillie, Editors. 1994. *Handbook of Statistics 12: Environmental Statistics*. Amsterdam: Elsevier Science.

Pattatucci, A. M. L., and D. H. Hamer. 1995. Development and familiarity of sexual orientation in females. *Behavior Genetics* **25**: 407–20.

Patten, B. C. 1962. Species diversity in net phytoplankton of Raritan Bay. *Journal of Marine Research* **20**: 57–75.

Patterson, C. 1982. Morphological characters and homology. *Problems of Phylogenetic Reconstruction*. Editors K. A. Joysey and A. E. Friday, 21–74. London: Academic Press.

Peet, R. K. 1974. The measurement of species diversity. *Annual Review of Ecology and Systematics* **5**: 285–307.

Pennisi, E. 2003. Tracing life's circuitry. *Science* **302**: 1646–49.

Pennock, R. T. 1999. *Tower of Babel: The Evidence Against the New Creationism*. Cambridge, MA: MIT Press.

Peters, T. 2003. *Playing God?* New York: Routledge.

Peterson, K., and N. Butterfield. 2005. Origin of the Eumetazoa: Testing ecological predictions of molecular clocks against the Proterozoic fossil record. *Proceedings of the National Academy of Sciences* **102**, no. 27: 9547–52.

Peterson K., and E. Davidson. 2000. Regulatory evolution and the origin of Bilaterians. *Proceedings of the National Academy of Science* **97**, no. 9: 4430–33.

Peterson, K., et al. 2000. Bilateran origins: Significance of new experimental observations. *Developmental Biology* **219**: 1–17.

2004. Estimating metazoan divergence times with a molecular clock. *Proceedings of the National Academy of Science* **101**, no. 17: 6536–41.

Pielou, E.C. 1975. *Ecological Diversity*. London: John Wiley.

1969. *An Introduction to Mathematical Ecology*. London: Wiley Interscience.

Pietilainen, K.H., J. Rissanen, A. Winter, T. Rimpela, A. Viken, R.J. Rose, and R.J. Kaprio. 1999. Distribution and heritability of BMI in Finnish adolescents aged 16y and 17y: A study of 4884 twins and 2509 singletons. *International Journal of Obesity* **23**: 107–15.

Pimental, D. 1961. Species diversity and insect population outbreaks. *Annals of the Entomological Society of America* **54**: 76–86.

Pimm, S.L. 1991. *The Balance of Nature? Ecological Issues in the Conservation of Species and Communities*. Chicago: University of Chicago Press.

Pinker, S., and P. Bloom. 1990. Natural language and natural selection. *Behavioral and Brain Sciences* **13**.

Pinto-Correia, C. 1997. *The Ovary of Eve: Egg and Sperm and Preformation*. Chicago: University of Chicago Press.

Platnick, N.I. 1979. Philosophy and the transformation of cladistics. *Systematic Zoology* **28**: 537–46.

Plato. 2000. *Timeaus*. Translator D.Zeyl. Indianapolis: Hackett.

Poole, R.W. 1974. *An Introduction to Quantitative Ecology*. New York: McGraw-Hill.

Popper, K.R. 1962. *Conjectures and Refutations*. New York: Basic Books.

1959. *The Logic of Scientific Discovery*. London: Hutchinson.

Portin, P. 1993. The concept of the gene: Short history and present status. *The Quarterly Review of Biology* **68**, no. 2: 173–223.

Preston, F.E. 1962. The canonical distribution of commonness and rarity. *Ecology* **43**: 185–215, 410–32.

1948. The commonness, and rarity, of species. *Ecology* **29**, no. 3: 254–83.

Przibram, H. 1907. *Experimental-Zoologie*. Teil 1. Embryogenese. Leipzig and Wien: Franz Deuticke.

Pusey, A., and C. Packer. 1986. Dispersal and philopatry. *Primate Societies.* Editors B. Smuts, D. Cheney, M. Seyfarth, R. Wrangham, and T. Struhsaker, 250–66. Chicago: University of Chicago Press.

Putnam, H. 2002. *The Collapse of the Fact-Value Dichotemy.* Cambridge, MA: Harvard University Press.

Rabin, M. 1993. Incorporating fairness into game theory. *American Economic Review* **83**: 1281–1302.

Ramsey, G. 2006. Block fitness. *Studies in History and Philosophy of Science* **C37**: 484–98.

Rao, C. R. 1982. Diversity and dissimilarity indices: A unified approach. *Theoretical Population Biology* **21**: 24–43.

Reeve, H. K., and P. W. Sherman. 1993. Adaptation and the goals of evolutionary research. *Quarterly Review of Biology* **68**: 1–32.

Reynolds, V., and R. Tanner. 1983. *The Biology of Religion.* London: Longman.

1995. *The Social Ecology of Religion.* New York: Oxford University Press.

Rheinberger, H.-J. 1997. *Towards a History of Epistemic Things: Synthesizing Proteins in the Test Tube.* Stanford, CA: Stanford University Press.

Rice, G., C. Risch, N. Ebers, and G. Anderson. 1999. Male homosexuality: Absence of linkage to microsatellite markers at Xq28. *Science* **284**: 665–67.

Rice, W. R., and G. W. Salt. 1990. The evolution of reproductive isolation as a correlated character under sympatric conditions: Experimental evidence. *Evolution* **44**: 1140–52.

Richards, R. 2005. The aesthetic and morphological foundations of Ernst Haeckel's evolutionary project. *The Many Faces of Evolution in Europe, 1860–1914.* Editors M. Kemperink and P. Dassen, 1–16. Amsterdam: Peeters.

2003. Character individuation in phylogenetic inference. *Philosophy of Science* **70**: 264–79.

2004. If this be heresy: Haeckel's conversion to Darwinism. *Darwinian Heresies.* Editors A. Lusting, R. Richards, and M. Ruse, 101–30. Cambridge: Cambridge University Press.

2002. Kuhnian, values and cladistic parsimony. *Perspectives on Science* **10**: 1–27.

1992. The structure of narrative explanation in history and biology. *History and Evolution.* Editors M. Nitecki and D. Nitecki, 19–53. Albany, NY: SUNY Press.

Richerson, P. G., and R. Boyd. 2005. *Not by Genes Alone: How Culture Transformed Human Evolution.* Chicago: University of Chicago Press.

Ricotta, C. 2005. Through the jungle of biological diversity. *Acta Biotheoretica* **53**: 29–38.

Ridley, M. 2000. *Mendel's Demon: Gene Justice and the Complexity of Life*. London: Weidenfeld and Nicolson.

Riedl, R. 1975. *Die Ordnung des Lebendigen: Systembedingungen d. Evolution*. Hamburg: Parey.

Rieppel, O. 2006. The Phylocode: A critical discussion of its theoretical foundation. *Cladistics* **22**: 186–97.

⎯⎯⎯. 2003. Popper and systematics. *Systematic Biology* **52**: 259–71.

Rieppel, O., and M. Kearney. 2002. Similarity. *Biological Journal of the Linnaean Society* **75**: 59–82.

Robert, J. S. 2004. *Embryology, Epigenesis, and Evolution: Taking Development Seriously*. New York: Cambridge University Press.

Robert, J. S., and F. Baylis. 2003. Crossing species boundaries. *The American Journal of Bioethics* **3**, no. 3: 1–13.

Robert, J. S., J. Maienschein, and M. Laubichler. 2006. Systems bioethics and stem cell biology. *Journal of Bioethical Inquiry* **3**: 19–31.

Robert, J. S., and A. Smith. 2004. Toxic ethics: Environmental genomics and the health of populations. *Bioethics* **18**: 493–514.

Roe, S. 1981. *Matter, Life, and Generation*. Cambridge: Cambridge University Press.

Rohwer, Y. Forthcoming. Evolutionary origins of altruistic punishment. *Philosophy of Science* (Supplemental).

Rolston, H. 1999. *Genes, Genesis and God: Values and Their Origins in Natural and Human History*. Cambridge: Cambridge University Press.

Rosenberg, A. 2000. Reductionism in a historical science. *Philosophy of Science* **68**: 135–63.

⎯⎯⎯. 1997. Reductionism redux: Computing the embryo. *Biology and Philosophy* **12**, no. 4: 445–70.

⎯⎯⎯. 1985. *The Structure of Biological Science*. Cambridge: Cambridge University Press.

⎯⎯⎯. 1978. The supervenience of biological concepts. *Philosophy of Science* **45**: 368–86.

Rosenzweig, M. L. 2003. Reconciliation ecology and the future of species diversity. *Oryx* **37**: 194–205.

⎯⎯⎯. 1995. *Species Diversity in Space and Time*. New York: Cambridge University Press.

Rousseau, R., P. van Hecke, D. Nijssen, and J. Bogaer. 1999. The relationship between diversity profiles, evenness, and species richness based on partial ordering. *Environmental and Ecological Statistics* **6**: 211–23.

Rudge, D.W. 1998. A Bayesian analysis of strategies in evolutionary biology. *Perspectives on Science* 6: 341–60.

——— 1999. Taking the peppered moth with a grain of salt. *Biology and Philosophy* 14: 9–37.

Rudwick, M.J.S. 1964. The inference of function from structure in fossils. *British Journal for the Philosophy of Science* 15: 27–40.

Ruse, M. 1981. Are there gay genes? Sociobiology and homosexuality. *Journal of Homosexuality* 6, no. 4: 5–34.

——— 1970. Are there laws in biology? *Australasian Journal of Philosophy* 48: 234–46.

——— 1987. Biological species: Natural kinds, individuals, or what? *The British Journal for the Philosophy of Science* 38: 225–42.

——— 2003. *Darwin and Design: Does Evolution Have a Purpose?* Cambridge, MA: Harvard University Press.

——— 1979. *The Darwinian Revolution: Science Red in Tooth and Claw.* Chicago: University of Chicago Press.

——— 1990. *Homosexuality: A Philosophical Inquiry.* Oxford: Blackwell.

——— 1977. Is biology different from physics? *Logic, Laws and Life.* Editor R.G. Colodny, 89–127. Pittsburgh: University of Pittsburgh Press.

——— 1996. *Monad to Man: The Concept of Progress in Evolutionary Biology.* Cambridge, MA: Harvard University Press.

——— 1988. *Philosophy of Biology Today.* Albany, NY: SUNY Press.

——— 1976. Reduction in genetics. *PSA 1974.* Editor R.S. Cohen, 633–51. Dordrect: Reidel.

Russell, R.J., W.R. Stoeger, and F.J. Ayala, Editors. 1999. *Evolutionary and Molecular Biology: Scientific Perspectives on Divine Action.* Rome: Vatican Observatory.

Rutimeyer, L. 1868. Review of Ernst Haeckel, "Ueber die Enstehung und den Stammbaum des Menschengeschlechts" und Naturliche Schopfungsgeschichte. *Archiv für Anthropologie* 3: 301–02.

Salais, D., and R.B. Fischer. 1995. Sexual preference and altruism. *Journal of Homosexuality* 40: 51–77.

Salmon, W. 1984. *Scientific Explanation and the Causal Structure of the World.* Princeton, NJ: Princeton University Press.

Samuelson, L. 1997. *Evolutionary Games and Equilibrium Selection.* Cambridge, MA: MIT Press.

Sanders, A.R. 1998. Poster of presentation 149, annual meeting of the American Psychiatric Association, Toronto, Ontario Canada. Cited in Hamer, D. 1999. Genetics and male sexual orientation. *Science* 285: 803.

Sarkar, S. 2005. *Biodiversity and Environmental Philosophy: An Introduction to the Issues.* New York: Cambridge University Press.

2004. "Conservation Biology." Available at http://plato.stanford.edu/archives/win2004/entrie/conservation-biology.

1996. Decoding "coding" – information and DNA. *BioScience* **46**: 857–64.

2002. Defining "biodiversity": Assessing biodiversity. *Monist* **85**: 131–55.

1998. *Genetics and Reductionism*. Cambridge: Cambridge University Press.

2005. *Molecular Models of Life: Philosophical Papers on Molecular Biology*. Cambridge, MA: MIT Press.

1994. The selection of alleles and the additivity of variance. *PSA: Proceedings of the Biennial Meeting of the Philosophy of Science Association* **1**: 3–12.

Sarkar, S., C. Pappas, J. Garson, A. Aggarwal, and S. Cameron. 2004. Place prioritization for biodiversity conservation using probabilistic surrogate distribution data. *Diversity and Distribution* **10**: 125–33.

Schaffner, K. 1993. *Discovery and Explanation in Biology and Medicine*. Chicago: University of Chicago Press.

2000. Behavior at the organismal and molecular levels: The case of C-elegans. *Philosophy of Science* **67**: 273–88.

2001. Extrapolation from animal models: Social life, sex, and super models. *Theory and Method in the Neurosciences*. Editors Peter Machamer, P. McLaughlin, and R. Grush, 200–30. Pittsburgh: University of Pittsburgh Press.

1988. Model organisms and behavioral genetics: A rejoinder. *Philosophy of Science* **65**: 276–88.

1976. Reductionism in biology: Prospects and problems. *PSA 1974*. Editor R. S. Cohen, 613–32. Dordrecht: Reidel.

Schlosser, G., and G. P. Wagner, Editors. 2004. *Modularity in Development and Evolution*. Chicago: University of Chicago Press.

Schmalz, J. 1993. Poll finds an even split on homosexuality's cause. *New York Times*, A1, March 5.

Schoener, T. W. 1986. Mechanistic approaches to ecology: A new reductionism? *American Zoologist* **26**: 81–106.

Schrodinger, E. 1944–1992. *What Is Life?* Cambridge, MA: MIT Press.

Schweber, S. 1977. The origin of the *Origin* revisited. *Journal of the History of Biology* **10**: 229–316.

Scott, J. M., F. Davis, B. Csuti, R. Noss, B. Butterfield, C. Groves, H. Anderson, S. Caicco, F. D'Erchia, T. Edwards, J. Ulliman, and G. Wright. 1993. Gap analysis: A geographic approach to protection of biological diversity. *Journal of Wildlife Management* **57**, no. 1: 123.

Selten, R. 1975. Reexamination of the perfectness concept for equilibrium points in extensive games. *International Journal of Game Theory* **4**: 25–55.

Senut, B., M. Pickford, D. Gommery, P. Mein, K. Cheboi, and Y. Coppens. 2001. First hominid from the Miocene (Lukeino Formation, Kenya). *Comptes Rendus de l'Academie Jes Sciences* **332**: 137–44.

Shaefer, K. P., and D. L. Meyer. 1974. Compensation of vestibular lesions. *Handbook of Sensory Physiology*. Editor H. H. Kornhuber, Vol. 6. 462–90. New York: Plenum.

Shannon, C. 1948. A mathematical theory of communication. *Bell Systems Technical Journal* **27**: 279–423, 623–56.

Shea, N. Forthcoming. Representation in the Genome.

Simberloff, D. S., and E. O. Wilson. 1969. Experimental zoogeography of islands: The colonization of empty islands. *Ecology* **50**: 278–96.

Simpson, E. H. 1949. Measurement of diversity. *Nature* **163**: 688.

Simpson, G. G. 1953. *The Major Features of Evolution*. New York: Columbia University Press.

    1961. *Principles of Animal Taxonomy*. New York: Columbia University Press.

Sirkin, D. W., W. Precht, and J. H. Courjon. 1984. Intitial, rapid phase of recovery from unilateral vestibular lesion in rat not dependent on survival of central portion of vestibular nerve. *Brain Research* **302**: 245–56.

Skipper, R. A. Jr. 2004. Calibration of laboratory models in population genetics. *Perspectives on Science* **12**: 369–93.

    2002. The persistence of the R. A. Fisher-Sewall Wright controversy. *Biology and Philosophy* **17**: 341–67.

Skipper, R. A., and R. L. Millstein. 2005. Thinking about evolutionary mechanisms: Natural selection. *Studies in the History and Philosophy of Biological and Biomedical Sciences. Special Issue: Mechanisms in Biology* **36**: 327–47.

Skipper, R. A. Jr., C. Allen, R. Ankeny, C. F. Craver, L. Darden, G. M. Mikkelson, and R. C. Richardson, Editors. In press. *Philosophy Across the Life Sciences*. Cambridge, MA: MIT Press.

Sklar, L. 1999. The reduction(?) of thermodynamics to statistical mechanics. *Philosophical Studies* **95**: 187–202.

Skyrms, B. 1994. Darwin meets the logic of decision: Correlation in evolutionary game theory. *Philosophy of Science* **61**: 503–28.

    1996. *Evolution of the Social Contract*. Cambridge: Cambridge University Press.

    2004. *The Stag Hunt and the Evolution of the Social Contract*. Cambridge: Cambridge University Press.

Skyrms, B., and J. Alexander. 1999. Bargaining with neighbors: Is justice contagious? *Journal of Philosophy* **96**: 588–98.

Slatkin, M., and M. J. Wade. 1978. Group selection on a quantitative character. *Proceedings of the National Academy of Sciences of the United States of America* **75**: 3531–34.

Slobodkin, L. B., and A. Rapoport. 1974. An optimal strategy of evolution. *Quarterly Review of Biology* **49**: 181–200.

Smith, B. 2002. The foundations of computing. *Computationalism: New Directions*. Editor M. Scheutz. Cambridge, MA: MIT Press.

Smith, B., and J. B. Wilson. 1996. A consumer's guide to evenness indices. *Oikas* **76**: 70–82.

Smith, E. A. 2000. Three styles in the evolutionary analysis of human behavior. *Adaptation and Human Behavior: An Anthropological Perspective*. Editors L. Cronk, N. Chagnon, and W. Irons, 27–46. New York: Aldine de Gruyter.

Smith, E. A., Monique Bergerhoff Mulder, and Kim Hill. 2001. Controversies in the evolutionary social sciences: A guide for the perplexed. *Trends in Ecology and Evolution* **16**: 128–35.

Sneath, P. H. A., and R. R. Sokal. 1973. *Numerical Taxonomy*. San Fransisco: W. H. Freeman.

Snyder, M., and M. Gerstein. 2003. Defining genes in the genomics era. *Science* **300**, no. 5617: 258–60.

Sober, E. 2001. *Core Questions in Philosophy*. 3rd ed. Englewood-Cliffs, NJ: Prentice-Hall.

1999. Instrumentalism revisited. *Critica* **31**: 3–39.

1999. The multiple realizability argument against reductionism. *Philosophy of Science* **66**: 542–64.

1995. Natural selection and distributive explanation: A reply to Neander. *British Journal for the Philosophy of Science* **46**: 384–87.

1984. *The Nature of Selection*. Cambridge, MA: MIT. Press.

2000. *Philosophy of Biology*. 2nd ed. Boulder, CO: Westview Press.

1988. *Reconstructing the Past: Parsimony, Evolution, and Inference*. Cambridge, MA: MIT Press.

1998. Six sayings about adaptationism. *The Philosophy of Biology*. Editors D. Hull and M. Ruse. Oxford: Oxford University Press.

2001. The two faces of fitness. *Thinking About Evolution: Historical, Philosophical and Political Perspectives*. Editors R. S. Singh, C. B. Krimbas, D. B. Paus, and J. Beatty, 309–21. New York: Cambridge University Press.

Sober, E., and D. S. Wilson. 1997. *Unto Others: The Evolution of Altruism*. Cambridge, MA: Harvard University Press.

Sokal, R. R., and P. H. A. Sneath. 1963. *Principles of Numerical Taxonomy*. San Francisco: W. H. Freeman.

Sopher, B. 1993. A laboratory analysis of bargaining power in a random ultimatum game. *Journal of Economic Behavior and Organization* **21**, no. 1: 324–34.

Sorger, P. K. 2005. A reductionist's systems biology. *Current Opinion in Cell Biology* **17**: 9–11.

Soule, M. E. 1985. What is conservation biology? *Bioscience* **35**: 727–34.

Stegman, U. 2004. The arbitrariness of the genetic code. *Biology and Philosophy* **19**: 205–22.

Stein, E. 1999. *The Mismeasure of Desire: The Science, Theory, and Ethics of Sexual Orientation.* Oxford: Oxford University Press.

Stent, G. 1968. That was the molecular biology that was. *Science* **160**: 390–95.

Stephens, C. 2004. Selection, drift, and the "forces" of evolution. *Philosophy of Science* **71**, no. 4: 550–70.

Sterelny, K. 1999a. Bacteria at the high table. *Biology and Philosophy* **14**, no. 3: 459–70.

———. 2003. Last will and testament: Steven J. Gould's *The Structure of Evolutionary Theory. Philosophy of Science* **70**: 255–63.

———. 1999b. Species as ecological mosaics. *Species: New Interdisciplinary Essays.* Editor R. A. Wilson, 119–38. Cambridge, MA: MIT Press.

———. 2004. Symbiosis, evolvability and modularity. *Modularity in Development and Evolution.* Editors G. Schlosser and G. Wagner, 490–516. Chicago: University of Chicago Press.

Sterelny, K., and P. E. Griffiths. 1999. *Sex and Death: An Introduction to Philosophy of Biology.* Chicago: University of Chicago Press.

Sterelny, K., and P. Kitcher. 1988. The return of the gene. *Journal of Philosophy* **85**, no. 7: 339–62.

Sterelny, K., K. Smith, and M. Dickison. 1996. The extended replicator. *Biology and Philosophy* **11**: 377–403.

Stotz, K. 2006. With genes like that who needs an environment: Postgenomics argument for the "ontogeny of information." *Philosophy of Science* **73**, no. 5.

Stotz, K., A. Bostanci, and P. E. Griffiths. 2006. Tracking the shift to "postgenomics." *Community Genetics* **9**, no. 3: 190–96.

Stotz, K., and P. E. Griffiths. 2004. Genes: Philosophical analyses put to the test. *History and Philosophy of the Life Sciences* **26**, no. 1: 5–28.

Stotz, K., P. E. Griffiths, and R. Knight. 2004. How scientists conceptualise genes: An empirical study. *Studies in History & Philosophy of Biological and Biomedical Sciences* **35**, no. 4: 647–73.

Strange, K. 2005. The end of "naive reductionism": Rise of systems biology or renaissance physiology? *American Journal of Cell Physiology* **288**: 968–74.

Suppe, F. 1989. *The Semantic Conception of Theories and Scientific Realism*. Chicago: University of Illinois Press.

Swenson, W., D. S. Wilson, and R. Elias. 2000. Artificial ecosystem selection. *Proceedings of the National Academy of Sciences* **97**: 9110–14.

Symons, D. 1992. On the use and misuse of Darwinism in the study of human behavior. *The Adapted Mind: Evolutionary Psychology and the Generation of Culture*. Editors J. H. Barkow, L. Cosmides, and J. Tooby, 137–59. New York: Oxford University Press.

Tabery, J. G. 2004. Synthesizing activities and interactions in the concept of a mechanism. *Philosophy of Science* **71**: 1–15.

Takacs, D. 1996. *The Idea of Biodiversity: Philosophies of Paradise*. Baltimore: Johns Hopkins University Press.

Tauber, A., and S. Sarkar. 1992. The Human Genome Project: Has blind reductionism gone too far? *Perspectives in Biology and Medicine* **35**: 220–35.

Taylor, P., and L. Jonker. 1978. Evolutionary stable strategies and game dynamics. *Mathematical Biosciences* **40**: 145–56.

Templeton, A. R. 2002. Out of Africa again and again. *Nature* **416**: 45–51.

Thagard, P. 1999. *How Scientists Explain Disease*. Princeton, NJ: Princeton University Press.

2003. Pathways to biomedical discovery. *Philosophy of Science* **70**: 235–54.

Thaler, R. 1988. Anomolies: The ultimatum game. *Journal of Economic Perspectives* **2**: 195–206.

Thoday, J. M. 1953. Components of fitness. *Symposia for the Society for Experimental Biology*, 96–114. Cambridge: Cambridge University Press.

Tilman, D. 1999. Diversity and production in European grasslands. *Science* **286**: 1099–1100.

Tilman, D., P. B. Reich, J. Knops, D. Wedin, T. Mielke, and C. Lehman. 2001. Diversity and productivity in a long-term grassland experiment. *Science* **294**: 843–45.

Tooby, J., and L. Cosmides. 1992. The psychological foundations of culture. *The Adapted Mind: Evolutionary Psychology and the Generation of Culture*. Editors J. H. Barkow, L. Cosmides, and J. Tooby, 19–136. New York: Oxford University Press.

Turner, D. 2000. The functions of fossils: Inference and explanation in functional morphology. *Studies in History and Philosophy of Biological and Biomedical Sciences* **31**: 193–212.

Tygart, C. E. 2000. Genetic causation attribution and public support of gay rights. *International Journal of Public Opinion Research* **12**: 259–75.

Uyenoyama, M. K. 1979. Evolution of altruism under group selection in large and small populations in fluctuating environments. *Theoretical Population Biology* **15**: 58–85.

Valentine, J. 2004. *On the Origin of Phyla*. Chicago: University of Chicago Press.

van der Weele, C. 1999. *Images of Development: Environmental Causes in Ontogeny*. Albany, NY: SUNY Press.

Vane-Wright, R. I., C. J. Humphries, and P. H. Williams. 1991. What to protect? Systematics and the agony of choice. *Biological Conservation* **55**: 235–54.

van Ommen, G. J. B., E. Bakker, and J. T. Dunnen. 1999. The Human Genome Project and the future of diagnostics, treatment, and prevention. *Lancet (Supplement)* **354**: S5–S10.

Vermeig, G. J. 1995. Economics, volcanoes, and Phanerozoic revolutions. *Paleobiology* **21**, no. 2: 125–52.

——— 1987. *Evolution and Escalation*. Princeton, NJ: Princeton University Press.

——— 1999. Inequality and the directionality of history. *American Naturalist* **153**, no. 3: 243–53.

Vignaud, P., et al. 2002. Geology and palaeontology of the Upper Miocene Toros-Menalla hominid locality, Chad. *Nature* **418**: 152–55.

Vrba, E. 1989. Levels of selection and sorting with special reference to the species problem. *Oxford Surveys in Evolutionary Biology* **6**: 111–68.

——— 1983. Macroevolutionary trends: New perspectives on the roles of adaptation and incidental effect. *Science* **221**: 387–89.

——— 1984. What is species selection? *Systematic Zoology* **33**: 318–28.

Vrba, E. S., and S. J. Gould. 1986. The hierarchical expansion of sorting and selection: Sorting and selection cannot be equated. *Paleobiology* **12**: 217–28.

Wachtel, S. S. 1983. *H-Y Antigen and the Biology of Sex Determination*. New York: Grune & Stratton.

Waddington, C. H. 1940. *Organisers and Genes*. Cambridge: Cambridge University Press.

Wade, M. J. 1978. A critical review of the models of group selection. *Quarterly Review of Biology* **53**: 101–14.

——— 1977. An experimental study of group selection. *Evolution* **31**: 134–53.

——— 1980. Kin selection: Its components. *Science* **210**: 665–67.

——— 1985. Soft selection, hard selection, kin selection, and group selection. *American Naturalist* **125**: 61–73.

Wade, M. J., and D. E. McCauley. 1980. Group selection: The phenotypic and genotypic differentiation of small populations. *Evolution* **34**: 799–812.

Wagner, G. P. 2000. What is the promise of developmental evolution? Part I. Why is developmental biology necessary to explain evolutionary innovations? *Journal of Experimental Zoology* **288**, no. 2: 95–98.

———. 2001. What is the promise of devolopmental evolution? Part II. A causal explanation of evolutionary innovations may be impossible. *Journal of Experimental Zoology* **291**, no. 4: 305–09.

Wagner, G. P., et al. 2000. Developmental evolution as a mechanistic science: The inference from developmental mechanisms to evolutionary processes. *American Zoologist* **40**: 819–31.

Wagner, G. P., and L. Altenberg. 1996. Complex adaptations and the evolution of evolvability. *Evolution* **50**: 967–76.

Wagner, G. P., and H. C. Larsson. 2003. What is the promise of developmental evolution? Part III. The crucible of developmental evolution. *Journal of Experimental Zoology; Part B: Molecular and Developmental Evolution* **300**, no. 1: 1–4.

Wagner, G. P., and M. D. Laubichler. 2004. Rupert Riedl and the re-synthesis of evolutionary developmental biology: Body plans and evolvability. *Journal of Experimental Zoology; Part B: Molecular and Developmental Evolution* **302**, no. 1: 92–102.

Walsh, D. M. 2000. Chasing shadows: natural selection and adaptation. *Studies in History and Philosophy of Biological and Biomedical Sciences* **31**: 135–53.

Walsh, D. M., T. Lewens, and A. Ariew. 2002. The trials of life: Natural selection and random drift. *Philosophy of Science* **69**, no. 3: 452–73.

Waters, C. K. 2003. The arguments in the *Origin of Species*. *Cambridge Companion to Darwin*. Editors J. Hodge and G. Radick, 116–42. Cambridge: Cambridge University Press.

———. 1998. Causal regularities in the biological world of contingent distributions. *Biology and Philosophy* **13**: 5–36.

———. 1994. Genes made molecular. *Philosophy of Science* **61**: 163–85.

———. 2000. Molecules made biological. *Revue Internationale de Philosophie* **4**, no. 214: 539–64.

———. 2004. What was classical genetics? *Studies in History and Philosophy of Science* **35**, no. 4: 783–909.

———. 1990. Why the anti-reductionist consensus won't survive: The case of classical Mendelian genetics. *PSA 1990*. Vol. 1, 125–39. East Lansing, MI: Philosophy of Science Association.

Waters, M. D., and J. M. Fostel. 2004. Toxicogenomics and systems toxicology: Aims and prospects. *Nature Reviews Genetics* **5**: 936–48.

Watson, J. D., and F. H. C. Crick 1953. A structure for deoxyribose nucleic acid. *Nature* **171**: 737–38.

Weber, M. 2005. *Philosophy of Experimental Biology*. Cambridge: Cambridge University Press.

Weibull, J. 1995. *Evolutionary Game Theory*. Cambridge, MA: MIT Press.

Weikart, R. 2004. *From Darwin to Hitler: Evolutionary Ethics, Eugenics, and Racism*. New York: Palgrave Macmillan.

Weinrich, J.D. 1987. A new sociobiological theory of homosexuality applicable to societies with universal marriage. *Ethology and Sociobiology* **8**: 37–47.

Weisberg, M. Forthcoming. Who is a modeler? *British Journal for the Philosophy of Science*.

Weitzman, M.L. 1992. On diversity. *Quarterly Journal of Economics* **107**: 363–405.

West-Eberhard, M.J. 2003. *Developmental Plasticity and Evolution*. Oxford: Oxford University Press.

Weston, A.D., and L. Hood. 2004. Systems biology, proteomics, and the future of health care: Toward predictive, preventative, and personalized medicine. *Journal of Proteome Research* **3**: 179–96.

Whewell, W., and M. Ruse. 2001. *Of the Plurality of Worlds: A Facsimile of the First Edition of 1853; Plus Previously Unpublished Material Excised by the Author Just Before the Book Went to Press; and Whewell's Dialogue Rebutting His Critics, Reprinted from the Second Edition*. Editor M. Ruse. With introductory material by Michael Ruse. Chicago: University of Chicago Press.

White, L. Jr. 1967. The historical roots of our ecologic crisis. *Science* **155**, no. 3767: 1203–07.

Whiten, A. 2005. The second inheritance system of chimpanzees and humans. *Nature* **437**: 52–55.

Whiten, A., J. Goodall, W.C. McGrew, T. Nishida, V. Reynolds, Y. Sugiyama, G.E.G. Tutin, R.W. Wrangham, and C. Boesch. 1999. Cultures in chimpanzees. *Nature* **399**: 682–85.

Whittaker, R.H. 1975. *Communities and Ecosystems*. New York: Macmillan.

1960. Vegetation of the Siskiyou Mountains, Oregon and California. *Ecological Monographs* **30**: 279–338.

Wilkins, A.S. 2002. *The Evolution of Developmental Pathways*. Sunderland, MA: Sinauer.

Williams, G.C. 1966. *Adaptation and Natural Selection*. Princeton, NJ: Princeton University Press.

1985. A defense of reductionism in evolutionary biology. *Oxford Surveys in Evolutionary Biology* **2**: 1–27.

1992. *Natural Selection: Domains, Levels, and Challenges*. Oxford: Oxford University Press.

Williams, P. A. 2001. *Doing Without Adam and Eve: Sociobiology and Original Sin*. Minneapolis: Augsburg Fortress Publishers.

Wilmut, I., K. Campbell, and C. Tudge. 2000. *The Second Creation: Dolly and the Age of Biological Control*. New York: Farrar, Straus, Giroux.

Wilson, D. S., and R. K. Colwell. 1981. Evolution of sex ratio in structured demes. *Evolution* 35: 882–97.

Wilson, E. O. 1997. Introduction. *Biodiversity II*. Editors M. L. Reaka-Kudla, D. E. Wilson, and E. O. Wilson, 1–3. Washington, DC: Joseph Henry Press.

1978. *On Human Nature*. Cambridge, MA: Harvard University Press.

1975. *Sociobiology: The New Synthesis*. Cambridge, MA: Harvard University Press.

Wilson, E. O., Editor. 1988. *BioDiversity*. Washington, DC: National Academy Press.

Wilson, E. O., and F. M. Peter, Editors. 1986. *Biodiversity*. Washington, DC: National Academy Press.

Wimsatt, W. 1987. False models as means to truer theories. *Neutral Models in Biology*. Editors M. Nitecki and A. Hoffman, 23–55. Oxford: Oxford University Press.

1980a. Reductionistic research strategies and their biases in the units of selection controversy. *Scientific Discovery: Case Studies*. Editor T. Nickles, 218–59. Dordrecht: Reidel.

1981. Robustness, reliability, and overdetermination. *Scientific Inquiry and the Social Sciences*. Editors M. Brewer and B. Collins, 124–63. San Francisco: Jossey-Bass.

1980b. Units of selection and the structure of the multi-level genome. *Proceedings of the Philosophy of Science Association* 2: 122–83.

Wolpoff, M. H., J. Hawks, D. W. Frayer, and K. Hunley. 2001. Modern human ancestry at the peripheries: A test of the replacement theory. *Science* 291: 293–97.

Woodward, J. 1989. Data and phenomena. *Synthese* 79: 393–472.

Wright, L. 1973. Functions. *Philosophical Review* 82: 139–68.

1976. *Teleological Explanations: An Etiological Analysis of Goals and Functions*. Berkeley: University of California Press.

Wright, R. T. 1989. *Biology Through the Eyes of Faith*. San Francisco: Harper Publishers.

Wright, S. 1968. *Evolution and the Genetics of Populations: A Treatise*. 4 Vols. Chicago: University of Chicago Press.

1929. Evolution in a Mendelian Population. *Anatomical Record* 44: 287.

1931. Evolution in Mendelian Populations. *Genetics* 16: 97–159.

1980. Genic and organismic evolution. *Evolution* 34: 825–43.

1943. Isolation by distance. *Genetics* 28: 114–38.

1948. On the roles of directed and random changes in gene frequency in the genetics of populations. *Evolution* **2**: 279–94.

1945. Tempo and mode in evolution: A critical review. *Ecology* **26**: 415–19.

Wulff, P., and W. Wisden. 2005. Dissecting neural circuitry by combining genetics and pharmacology. *Trends in Neurosciences* **28**: 44–50.

Yang, Z., and J.P. Bielawski. 2000. Statistical methods for detecting molecular adaptation. *Trends in Ecology and Adaptation* **15**: 496–503.

Zeyl, D. 2006. "Plato's *Timeaus.*" Available at http://plato.stanford.edu/entries/plato-timaeus/.

INDEX

SPINOZA *Edited by* DON GARRETT
THE STOICS *Edited by* BRAD INWOOD
WITTGENSTEIN *Edited by* HANS SLUGA *and*
DAVID STERN